Complexity

MIT-Pappalardo Series in Mechanical Engineering
Series editors: Rohan C. Abeyaratne and Nam P. Suh

James A. Fay and Dan S. Golomb: Energy and the Environment
Nam P. Suh: Axiomatic Design, Advances and Applications
Nam P. Suh: Complexity: Theory and Applications
Gang Chen: Nanoscale Heat Transfer

Complexity
Theory and Applications

Nam P. Suh

2005

OXFORD
UNIVERSITY PRESS

Oxford   New York

Auckland   Bangkok   Buenos Aires   Cape Town   Chennai
Dar es Salaam   Delhi   Hong Kong   Istanbul   Karachi   Kolkata
Kuala Lumpur   Madrid   Melbourne   Mexico City   Mumbai   Nairobi
São Paulo   Shanghai   Taipei   Tokyo   Toronto

Library of Congress Cataloging-in-Publication Data

Suh, Nam P., 1936–
    Complexity: theory and application / Nam P. Suh.
        p.   cm.
    Includes bibliographical references and index.
    ISBN-13 978-0-19-517876-0
    ISBN 0-19-517876-9
    1. Design, Industrial—Methodology.   2. Axiomatic set theory.
    3. Computational complexity.   I. Title

    TS171.4.S86 2005
    620′.0042—dc22   2004057597

9 8 7 6 5 4 3 2 1

Printed in the United States of America
on acid-free paper

To

Nathan H. Cook and Milton C. Shaw

whose mentorship made this book possible

▇ Preface

The complexity theory presented in this book is a result of my attempt to apply axiomatic design theory to a variety of problems. Axiomatic design theory provides a systematic means of designing complex systems, but it cannot deal with such questions as: "Why is there so much wasted effort in developing new products?," "How can we predict and guarantee the long-term behavior of engineered systems?," "Why do certain things appear to be so complex but actually are not complex at all once we understand them?," "Why do people think that a product with many parts is complex?," "Is the complexity in engineering any different from the complexity in natural science or social science?," "How do we reduce complexity?," and "What is complexity?"

Everyone—engineers, natural scientists, social scientists, business leaders, artists, and even politicians—deals with "complexity" all the time. Yet, to these basic questions on complexity, we often receive many different answers. "Complexity" appears to have different meanings in each specific field. Sometimes, even colleagues in the same discipline use the word "complexity" to mean different things. This certainly is the case in engineering and science.

"Complexity" has been an intriguing topic to engineers, natural scientists, and social scientists. They have known intuitively that complexity is an important topic. Yet we have not had a general theoretical framework that can provide engineers and scientists with a unified tool to deal with complexity. Many of them have a general idea of what they mean by "complexity" when referring to their own fields, but their understanding is not precise enough to be useful in solving scientific, technological, and social problems.

One of the reasons for the difficulties encountered in the complexity field has been that the word "complex" has a variety of different meanings. In Webster's dictionary (2001 edition), "complex" is defined as: (1) composed of many interconnected parts; compound; composite; (2) characterized by a very complicated or involved arrangement of parts or units; (3) so complicated or intricate as to be hard to understand; (4) *psychology,* a system of interrelated or emotion-charged ideas or feelings or memories and impulses that is usually repressed and that gives rise to abnormal or pathological behavior, etc. These definitions do not enlighten us when we deal with the complexity associated with designing an object such as an orbital space plane or in understanding the causality between the physiological functions of our body and the molecular behavior of proteins and genes. They are also not very helpful in solving organizational or socioeconomic problems.

The complexity theory presented in this book provides a general theoretical framework that may be used to solve complexity problems in engineering, science, and even in certain nontechnical areas. To achieve this goal, the word "complexity" has been defined narrowly as a measure of uncertainty in achieving a set of design goals (or what we want to know) that a system must satisfy. This definition makes "complexity" a relative quantity rather than an absolute quantity, a major departure from other approaches to complexity.

The complexity theory presented in this book differs from other theories on complexity. According to this theory, complexity must be measured in the functional domain, not in the physical domain as many scientists have tried to do in the past. It also shows how the complexity of a system can be reduced through the introduction of "functional periodicity," which is a new and unique concept that is a direct consequence of the theory.

Chapters 1 through 3 are a general introduction to axiomatic design and complexity theory. Chapter 4 deals with the reduction of time-independent real complexity and time-dependent imaginary complexity. Chapter 5 presents a method of transforming a time-dependent combinatorial complexity to a time-dependent periodic complexity through the creation of "functional periodicity." Chapters 6 through 8 are applications of the materials presented in chapter 5 to engineering problems that arise in several diverse fields. Chapter 9, which is quite speculative, attempts to apply this theory of complexity to biological systems. Chapter 10 deals with the complexity of economic planning, government administration, and academic departments. This chapter draws heavily on the author's personal experience, which is, by definition, quite subjective.

This book was prepared to present the work done on complexity in a coherent manner, rather than through a series of papers. This edition of the book was quickly written to be used in the system analysis and design course at MIT and at a major aerospace company as a supplement to the main textbook, *Axiomatic Design: Advances and Applications.*

For a new theory to survive, it must be tested over a long period of time by many different means. The predictions made based on the theory must be consistent with observations and test results; the theory should provide a general framework in formulating and solving new problems; and ultimately, it should become a basis for creating other new theories and practices in many fields of human endeavor. One can only hope that the complexity theory presented in this book can survive the test of time and provide an intellectual platform for future scientific inquiries and technological advances.

Nam Pyo Suh
Sudbury, Massachusetts

Acknowledgments

It has been an enjoyable project to write this book on complexity, which is a challenging and fascinating subject. Because the topic is relevant in all areas of intellectual endeavor, it has forced me to explore some new subjects, such as biology and modern physics. Working with colleagues in other fields has been a rich learning experience, and I was fortunate to have been able to interact with many outstanding people. There were so many that I cannot thank them all. A few people will be listed here to acknowledge their special contributions.

The materials presented on the IC engine, electrical connectors, and the pin-joints were done with the support of Professor Alex d'Arbeloff and Dr. B.J. Park. I am most appreciative of their trust and support of the work. The actual technical work on the pin-joints and electrical connectors was done by the superb technologists and entrepreneurs, Dr. Matthew Sweetland and Mr. Don Bowers, of Tribotek, Inc. The detailed work on the IC engine was done by Dr. Nam-Hyo Cho of the Institute for Advanced Engineering of Korea. They took rough design concepts and made them to be interesting engineering artifacts.

Many of my former and current students and colleagues read many of the chapters of this book, making many important suggestions. I had the privilege of working with many extremely capable people: Dr. Taesik Lee, Hrish V. Deo, Dr. Jason W. Melvin, Dr. Jeffrey D. Thomas, Professor George Barbastathis, Professor Jung-Hoon Chun, Professor Seth Lloyd, and Denise Y. McCort. Professor Stephen Lu of the University of Southern California also provided valuable inputs. Their specific contributions are noted in various parts of the book.

Part of this book was written while I was working intensively with aerospace engineers and managers at Lockheed Martin Astronautics (LMA) and the National Astronautics and Space Agency (NASA). I am grateful for the support of Robert Ford and Raymond Demaso, who made this relationship possible. It was a stimulating experience for me to work with Joanne Beckham, Richard Freeman, Randall K. Munkres, Chip Woods, and Gregory J. Kehrl. Richard Freeman did a wonderful job of directing the effort of many of his colleagues in Axiomatic Design. To all those engineers at LMA and NASA who permitted me to intrude in their lives by keeping them in classrooms, I would like to say "thank you."

The impetus for getting into complexity in the field of biology was my interaction with Professor Ravi Iyengar of Mount Sinai Medical Center. His article in *Science* provided an opportunity for us to work together, which was later reinforced by the participation of Professor Michael Sheetz of Columbia University. This project also enabled me to work with Jeffrey D. Thomas, M.D., Ph.D., who kept me in line whenever my imagination overtook my real knowledge of biology. I also enjoyed being a student in a freshman biology course at MIT. I learned a great deal from Professors Eric Lander and Robert Weinberg, two great teachers. Our work has been partly funded by the National Institutes of Health. I gratefully acknowledge the support of NIH.

The work reported in this book was partly stimulated by the research conducted in the KIMM-MIT Research Alliance. Professor Jung-Hoon Chun and Dr. Sang-Ro Lee established this cooperative program between MIT and the Korea Institute of Metals and Machinery (KIMM). They provided the leadership to make this collaboration productive. They promoted interactions among the MIT faculty members as well as between the researchers at MIT and KIMM. Dr. Byung Sun Kim of KIMM has done a great job of strengthening this Alliance.

I am most grateful to Dr. Carol Vale, who has edited a rough manuscript into a readable book. She made many substantive modifications in addition to making many editorial changes. She has a deep and unique insight into science and scientific writing. I am also indebted to Peter Gordon of Cambridge University Press (CUP), who made many editorial comments and suggestions. I am also grateful to CUP for its willingness to publish this book.

I decided to publish this book as a part of the MIT-Pappalardo Series in Mechanical Engineering of Oxford University Press (OUP) as a token of our appreciation to Neil and Jane Pappalardo for their unquestioning support of the MIT Department of Mechanical Engineering. Danielle Christensen, Engineering Editor of OUP, worked hard to make the MIT-Pappalardo Series successful.

I am not sure exactly when I began to think about complexity. It might have begun when I was on the board of directors of Silicon Valley Group, Inc. (SVG), a manufacturer of semiconductor manufacturing equipment. SVG had many challenging tasks in optics, machine design, scheduling, and business strategy. As the chairman of the Technical Advisory Committee of the board of directors, I had extensive interactions with engineers and managers, learning first-hand about the difficulties of satisfying the functional requirements of these complicated machines. My friendship with Papken Der Torossian, CEO of SVG, gave me a wonderful opportunity to work on many challenging problems. William Hightower, President of SVG, was a strong advocate for Axiomatic Design within SVG. I am deeply indebted to them.

This book was written mostly at home late at night, disrupting the schedule of everyone else in the house. The love, understanding, and tolerance of my wife, Young, were essential for the completion of this book-writing task. Our lifelong partnership would have been less than stellar without her love and patience. No one could have asked for a better environment for thinking and writing. I owe much to my wife.

My book-writing effort always took a back seat whenever I had an opportunity to spend time with our grandchildren. I hope that someday they will read this book and wonder how Poppop had time to simplify such a complex topic and still had time for them.

N.P.S.

▉▉ Contents

Complexity

1

▨▨▨ Introduction

▨▨▨ The Why and What of Complexity

The *New York Times* on May 6, 1997, carried an article entitled "Researchers on Complexity Ponder What It's All About." The appearance of such an article in a daily newspaper indicates that the issue of complexity is moving toward center stage in science and technology. The article stated that

> Some of the grandest phenomena, like the coursing of comets around the sun, are marvelously predictable. But some of the most mundane, like weather, are so convoluted that they continue to elude the most diligent forecasters. They are what scientists call complex systems. Though made up of relatively simple units—like the molecules in the atmosphere—the pieces interact to yield behavior that is full of surprise[s].

We must ask: "Is the weather really convoluted and complex?" If so, why?

The topic of "complexity" is a challenging issue for intellectuals, scientists, and technologists. Scientists encounter complexity issues as they probe the limits of our understanding of nature. For instance, despite major advances, there is still a gap in our understanding of the relationship between the way living beings function and the

3

fundamental molecular behavior of proteins, DNA molecules, and the like. Engineers confront complexity issues as they create engineered artifacts in such diverse fields as space exploration and nanotechnology. Vast human and financial resources are wasted because of our inability to understand and address engineering complexities. Humanists struggle to understand the complexities that arise in human, political, and socioeconomic development. The miseries and conflicts between people and nations may be due to our inability to deal with complexities.

Advances in dealing with complexity will have a profound effect on humanity in all aspects of human endeavor. The challenge is to create the new discipline of complexity with unifying concepts for these diverse and seemingly disparate issues. The new discipline of complexity requires fundamental understanding, logical rationale, consistent methodologies, and intellectual linkage with other learned disciplines of natural sciences, engineering, and social sciences. Once a unifying concept for complexity is established, complexity encountered in all scientific, technological, and non-technical fields may be treated using the same basic principles. A well-established discipline of complexity will yield rich dividends of diverse benefits to humanity.

Confusion about complexity

Most people seem to know intuitively what "complexity" is, but when we probe their deeper perception and understanding, we find many different views on the subject matter. This situation exists even among students and scholars in complexity. They use the word "complexity" to describe different ideas and perceptions.

The meaning of complexity is further obscured by lack of clarity in the difference between the words "complexity" and "complicatedness." Sometimes they are used to describe the same thing, but in other cases they imply different things. Searching through dictionaries does not enlighten one's understanding. One of the reasons for confusion stems from the diverse use of these English words. Therefore, we need to define the term "complexity" in specific terms if we are to advance the complexity field.

1.2   Definition of Complexity as a Relative Quantity

A major departure of the complexity theory described in this book from various other notions of complexity stems from the observation that complexity must be defined in the "functional domain," not in the "physical domain." In the past, "physical things" (e.g., machines, lines of codes, computation time, biological cells) were examined to understand their complexity, which has resulted in many different definitions of complexity. When we try to achieve a certain function within a desired accuracy (or equivalently, if we want to know certain behavior of natural systems within a desired accuracy), our ability to achieve the desired function determines the complexity. This is illustrated in figure 1.1. When our goal is always achieved using selected physical implements, the task would not be regarded as being complex. When we cannot achieve the function (i.e., what we want to achieve or what we want to know), the task would appear to be very complex.

Complexity is defined as:

A measure of uncertainty in understanding what it is we want to know or in achieving a functional requirement (FR).

Probability density

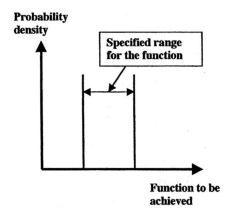

Figure 1.1 We want to achieve the function within the specified range. When it is always achieved, we would say the task is not complex. When it cannot be achieved, we would say that the task is very complex.

Functional requirements (FRs) are defined, as in axiomatic design, as a minimum set of independent requirements that completely characterize the functional needs of the product (or software, organization, system, etc.) in the functional domain (Suh, 1990, 2001).

In designing engineered systems, an FR is a function to be achieved by the designed system (e.g., FR = Measure the time). In natural science, an FR might be the function performed by an organ, tissue, cell, and so on (e.g., FR = Diffuse CO_2 into the lung from the blood stream). An FR is specified in terms of its nominal value with allowable variations or desired accuracy (e.g., measure the time within a billionth of a second). The allowable variation of FR in the functional domain is called the *design range*.

We satisfy the FRs by means of design parameters (DPs) or physical parameters. FR is "what it is we want to achieve" and DP is "how we are going to satisfy the FR." The potential DPs that can satisfy the FR (measure the time within a billionth of a second) are many (e.g., "frequency/period of cesium laser" or "the period of a pendulum") and we have to choose the DP that is the best.

When we try to fulfill the FR, there is an uncertainty, thus complexity, of satisfying it within the specified accuracy or tolerance. When a given DP is chosen to satisfy the FR, the uncertainty is characterized by the system's ability to satisfy the FR within its design range (Suh, 1999). The system performance is shown by its probability density function (pdf), which defines the *system range*. The FR is satisfied only when the system range is within the design range, a region called the *common range*. This is shown in figure 1.2.

EXAMPLE 1.1: COMPLEXITY OF CUTTING A ROD TO A SPECIFIED LENGTH

Consider the task of cutting a rod to 2 m $\pm$ 10 μm. The task (i.e., the goal) we want to achieve is defined as the FR. Although we have not yet determined how we can achieve the FR, the design range of the FR is shown in figure 1.2. To achieve the goal, the FR must be satisfied within the design range, which is the nominal value of FR $\pm$ 10 μm.

When we figure out how to cut the rod within the specified tolerance using a selected tool (i.e., system), we find that the task of cutting the rod within the specification may or may not appear to be complex depending on the capability of the

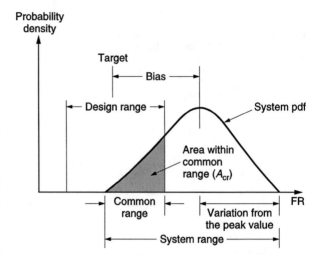

Figure 1.2 Probability density function (pdf) of an FR. The design range is the specification of the desired value of FR and its tolerance. The system range is defined by the pdf of the actual system chosen to satisfy the FR. The FR is satisfied only when the system range is in the design range.

system chosen. If the tool can always cut the rod to within the specified length, the task appears to be simple and, thus, the complexity is zero. That is, the capability of the chosen implement (e.g., a hacksaw versus a precision lathe) determines how easily we can satisfy the FR. If the implement is such that the resulting length of the rod does not lie completely inside the specified design range, the task appears to be complex. In this case, the FR is satisfied only when the system range overlaps the design range indicated by the common range. When the system range is not completely in the design range, there is a finite uncertainty that the FR may not be satisfied. Therefore, the system has a finite complexity.

EXAMPLE 1.2: COMPLEXITY IN BIOLOGY

Consider the complexity associated with controlling the glucose level of a Type II diabetic patient. It is important that the sugar level in human beings be maintained between 90 and 120 mg/dL at all times. That is, the design range for the FR (control sugar level) is between 90 and 120. However, the patient's body has lost the ability to control its glucose level and therefore it must be controlled by taking medication (three 25 mg Glyburide pills) to control calorie intake and exercising regularly. Despite the effort, he or she finds that only for 3 days out of 10 does the blood-glucose testing indicate that the glucose level is within the design range. What is the complexity associated with the task of controlling glucose level?

Complexity is defined as the measure of uncertainty in achieving the FR. The uncertainty, and thus complexity, associated with controlling the glucose level of the patient is large because we can be certain of achieving the FR only three out of ten times.[1]

Unlike the example of cutting a rod, in the case of this FR related to the control of sugar level, the body continues to deteriorate with time and the medication can no longer control the sugar level. We will refine these basic ideas further in chapters 2 and 3.

What examples 1.1 and 1.2 demonstrated is that *complexity can only be defined in the functional domain rather than the physical domain.* This is a major departure from the effort made to understand complexity in the past. Many researchers have attempted to understand complexity by analyzing physical interactions of natural and engineered systems rather than their functions.

When we deal with a large system, we may be required to satisfy many FRs at the same time. We have to choose appropriate DPs to minimize complexity. The complexity is determined by our ability to satisfy all of the FRs within their specified design ranges. As will be shown in chapters 2 and 3, the complexity of the system is a function of the relationship between FRs, that is, if FRs are independent from one another, the complexity is less than when FRs are dependent on each other.

EXAMPLE 1.3: COMPLEXITY OF ECONOMIC GROWTH OF A NATION

Managing the economic growth of a nation appears to be a complex task. There are many FRs that must be satisfied at the same time. The gross domestic product (GDP) must increase at least at the rate of population increase to maintain the status quo. A maximum employment rate must be maintained for socioeconomic–political stability. The inflation rate must be controlled within reasonable lower and upper bounds so as to avoid deflation and an unstable economy. Savings and investment must be controlled for increase in productivity and new economic activity. The social and defense costs must be carefully managed within reasonable bounds. There may be other FRs that are particularly important in certain countries.

To satisfy these FRs, various branches of the government develop policy instruments (DPs). Depending on which DPs are chosen, the FRs may or may not be satisfied. In many cases, the attempt to improve one of the FRs may adversely affect other FRs, thus increasing the complexity of managing the national economy. To decrease the complexity, the policy must be designed correctly and implemented in a systematic way. The complexity of these socioeconomic–political issues is further discussed in chapter 10.

Some of these ideas will be discussed further in chapters 2 and 3, which present axiomatic design theory and complexity theory, respectively.

1.3 The Foundation of Relative Complexities

Complexity is a function of the relationship between the design range and the system range, which is affected by the relationships among the FRs. Thus, there are different types of complexity.

Once we define complexity as *a measure of uncertainty in achieving the FRs*, it follows that *complexity is a function of the relationship between the design range and the system range.* Based on this observation, the following important conclusions can be drawn:

1. The overlap of the design range and the system range provides the quantitative measure of complexity. For instance, complexity is zero if the system range is completely inside the design range. Complexity is infinite if the system range is completely outside the design range. When the system range partially overlaps

the design range, complexity is finite. This is real complexity because it is not possible to satisfy the FR at all times with the system chosen.

2. The relationship between the design range and the system range may be static or dynamic, depending on whether or not the system range drifts as a function of time relative to the design range. Therefore, there are two kinds of complexity: *time-dependent complexity* and *time-independent complexity*.

3. When there are many FRs to be satisfied at the same time, we must know the relationships among the FRs, which are dictated by the DPs chosen to control the FRs. When the DPs couple the FRs in such a manner that the FRs must be satisfied in a specified order, time-independent complexity can further be divided into time-independent *real* complexity and time-independent *imaginary* complexity. Real complexity exists when the system range is not completely within the design range. Imaginary complexity arises when we do not know the specific order of the FRs that must be followed to satisfy the FRs. It is called imaginary, because this complexity is not real and arises only because we do not know the relationship between the FRs and the DPs that control them.

4. When a system consists of many FRs that must be satisfied at the same time and when the system range of FRs can change with respect to the design range as a function of time, time-dependent complexity can be divided into two kinds: time-dependent *combinatorial* complexity and time-dependent *periodic* complexity. Time-dependent combinatorial complexity is caused by the following two different reasons:

   (a) When the system range of an FR continues to move away from the design range because of a physical phenomenon that affects the FR, making it increasingly difficult to satisfy the FR, the complexity increases as a function of time. For example, if the FR is to have low friction at a sliding interface under normal load, we may find that it is increasingly difficult to satisfy it with further sliding. This would be the case if wear particles, which are generated at a sliding interface by plowing, agglomerate and increase in size. Then the friction coefficient and the wear rate increase, generating more wear particles. Eventually the system may always fail, that is, the system range may move completely outside of the design range.

   (b) When the number of possible sequences of FRs increases combinatorially as a function of time, the system range may continue to move away from the design range because it becomes increasingly difficult to find the best combination. This results in time-dependent combinatorial complexity. For example, consider the functional requirement of a manufacturing system that must process identical parts by carrying out various manufacturing processes. If the FR is to maximize the productivity by letting a robot transport the finished parts from machine to machine as fast as possible, the FR will be a function of the scheduling of the robot motion. Depending on which part the robot moves first, the subsequent operations will be affected. In this case, the probability of having the best combination decreases continuously, eventually bringing about system failure. Similarly, Type II diabetes may be an example of time-dependent combinatorial complexity, since it tends to worsen with aging of the patient unless it is controlled.

5. When a set of FRs that a system must satisfy undergoes cyclic changes, for each cycle we may be able to make the system perform in a predictable manner. Then if the system ranges of the set of FRs may be made to oscillate cyclically with respect to the design ranges, time-dependent periodic complexity[2] results. If a system with time-dependent combinatorial complexity can be changed to one with time-dependent periodic complexity, the complexity of the system can be reduced.

Four different types of complexity

As described in the preceding paragraphs, there are four types of complexity:

1. Time-independent real complexity.
2. Time-independent imaginary complexity.
3. Time-dependent combinatorial complexity.
4. Time-dependent periodic complexity.

It should be noted here again that these complexities refer to functional complexity, where the function may be related to geometry, materials, and manufacturing processes as well as time. For example, time-dependent periodic complexity refers to time-dependent functional periods, not necessarily temporal periods. To emphasize the functional nature of these complexities, these four complexities will sometimes be simply referred to as

1. Real complexity.
2. Imaginary complexity.
3. Combinatorial complexity.
4. Periodic complexity.

As the subsequent chapters illustrate, these four complexities provide an important framework for understanding the complexity and behavior of natural systems and engineered systems.

(a) Time-independent real complexity (real complexity)
In relation to the rod-cutting example (figure 1.2), it was stated that when the pdf of the system range is not completely within the design range, there is a finite uncertainty and the task is complex because the FR cannot always be achieved with the available implement. This finite uncertainty results in a type of complexity called *time-independent real complexity* (or simply *real complexity*). Time-independent real complexity is a measure of the uncertainty involved in achieving a task. Real complexity can change only if the pdf of the system range changes with respect to the design range.

It will be shown later that the real complexity of the system, C_R, is equal to the informational content I_{sys} as

$$C_R = I_{sys} \tag{1.1}$$

where I_{sys} is the information content defined in axiomatic design theory as the negative logarithm of the probability of successfully satisfying the FRs. To decrease the real complexity, we must reduce the information content by eliminating the bias and the variance of the system range shown in figure 1.2.

To eliminate the time-independent real complexity of engineered systems with many FRs, it will be shown in chapter 4 that the system must be designed correctly by creating *uncoupled* or *decoupled* designs so that the FRs can be satisfied within their specified accuracy. When there are two or more FRs, the independence of FRs must be preserved to minimize the uncertainty, especially in the presence of random variations of physical systems. When the highest-level FRs must be decomposed for several layers, the real complexity may become unacceptably large if the design is coupled at the highest level of the design hierarchy.

(b) Time-independent imaginary complexity (imaginary complexity)

Imaginary complexity is a complexity that is not real, but exists because of our lack of understanding about the system design, system architecture, and/or system behavior. The imaginary complexity can exist even though the real complexity is zero—that is, the system range is completely inside the design range. For example, many of us have the experience of assembling toys using the parts supplied by a store with an instruction manual. It takes a long time to assemble the first one, because we do not know the assembly sequence well enough. Once we assemble one, the next one will be very simple. The first one was difficult because of imaginary complexity; in other words, our lack of knowledge made it complex.

This type of complexity exists only when there are many FRs that must be satisfied in a particular sequence (i.e., decoupled design) or when the FRs are coupled so that changing one FR requires all other FRs to be readjusted. We will see in chapter 2 that when the FRs of a system are completely independent from each other as in an uncoupled design, imaginary complexity is zero. It is highly probable that many industrial firms waste effort and money because of imaginary complexity. In industry, cost overruns and missed schedules in developing new products are partly caused by time-independent imaginary complexity.

(c) Time-dependent combinatorial complexity (combinatorial complexity)

A system can have a time-dependent combinatorial complexity when the system range changes with time. If the system range drifts away from the design range with time, the system performance cannot be predicted. An example is the airline-scheduling problem. When there is a major storm in Chicago, the airplanes that were to arrive and depart from Chicago cannot operate according to their schedules. Some flights may be canceled. Eventually, Chicago's weather affects the flight schedules at all the airports that are connected to Chicago either directly or indirectly. As a result, there will be increasing uncertainty in dispatching aircraft as a function of time in many airports. This is an example of time-dependent combinatorial complexity. When a system is affected by time-dependent combinatorial complexity, it eventually breaks down because it can go into a chaotic state.

As was briefly discussed, sometimes a certain physical phenomenon that affects an FR continues to evolve in such a manner that the system range continues to move away from the design range. In this case, the combinatorial complexity increases as a function of time. Such a phenomenon must be stopped by transforming the system to time-dependent periodic complexity.

(d) Time-dependent periodic complexity (periodic complexity)

Periodic complexity can be understood by analogy if we go back to the airline-scheduling problem caused by a major snowstorm in Chicago. Although the airline-scheduling problem continues to worsen throughout the day because of the cascading effect of airlines having to make decisions on the deployment of available airplanes, the airline typically recovers from the fiasco the next morning, since most airplanes do not fly all night. Therefore, they can "reinitialize" the schedule the next morning and resume the regular schedule. As discussed in later chapters, we can use any functions (e.g., functions related to geometry, chemistry, etc.) to create periodicity and to initialize the system.

One important conclusion of complexity theory is that complexity is greatly reduced when a system with time-dependent combinatorial complexity is transformed to a system with periodic complexity. The period is the *functional period*—not necessarily a temporal period—where the same set of functions repeats, that is, the periodicity may or may not have a constant time period.

1.4 Functional Periodicity

One of the important results of the complexity theory presented in this book is the concept of *functional periodicity*. For a system to operate stably for a long time, functional periodicity must exist in the system or must be built into the system. There are many examples of functional periodicity in nature and in engineered systems created by humans over the centuries. In nature, the things that have survived billions of years are stable and therefore tend to have functional periodicity. Conversely, we may state that the long-term stability and, thus, fundamental periodicity is a fundamental requirement for some natural systems.

When a set of FRs of a system repeats itself cyclically (not necessarily in a temporal cycle), we can introduce the functional periodicity to transform a time-dependent combinatorial complexity into a periodic complexity and thus reduce the complexity.

As discussed in later chapters, there are many different kinds of functional periodicity such as those shown in table 1.1.

1.5 Reduction of Complexity

The complexity of some systems can be reduced based on complexity theory. We can reduce the time-independent real complexity and eliminate imaginary complexity. We can also transform the time-dependent combinatorial complexity into a periodic complexity to reduce complexity. This means of reducing time-dependent complexity is a powerful way of increasing the reliability and robustness of engineered systems. Biological systems have survived through effective use of time-dependent periodic complexity or, conversely, natural systems with time-dependent combinatorial complexity have been eliminated through the evolutionary process.

1.6 Source and Nature of Functional Periodicity

Time-dependent periodic complexity exists in a system with a *functional* period. A functional period can be a temporal period, but it can also be of many other types that are

Table 1.1 Functional periodicity in natural systems and engineered systems

Examples of functional periodicity	Examples in nature	Examples in engineered systems
Temporal periodicity	Planetary system, solar calendar	Airline/train schedules, computers, pendulum
Geometric periodicity	Crystalline solids, surfaces of certain leaves	Undulated surface for low friction, "woven" electric connectors
Biological periodicity	Cell cycle, life–death cycle, plants, grains	Fermentation processes such as wine making
Manufacturing/processing periodicity	Biological systems	Scheduling a clustered manufacturing system
Chemical periodicity	Periodic table of chemical elements/atoms	Polymers
Thermal periodicity	Temperature of Earth	Heat cycles (e.g., Carnot cycle)
Information processing periodicity	Language	Reinitialization of software systems, music
Electrical periodicity	Thunderstorm	LCD, alternating current
Circadian periodicity	Living beings	Light-sensitive sensors
Material periodicity	Wavy nature of matter, atomic structure, crystallinity	Fabric, wire drawing, microcellular plastics

driven by a set of events, which may or may not be temporally periodic. Time-dependent periodic complexity makes a system robust and reliable. By introducing "functional periodicity" and eliminating combinatorial complexity through "reinitialization," the system returns to its initial state on a periodic basis, which prevents an undesirable phenomenon from continuing to create damage to the system. Intentional introduction of functional periodicity should prolong the life of hardware, software, and biological systems, as well as increase their reliability and safety.

Functional periodicity can be introduced in various spaces: time, geometry, temperature, chemistry, biology, electric charges, and processes. For example, geometric periodicity can be used to lower friction and increase the life of sliding wear surfaces (Suh, 1986). Electric charge reversal can provide high-quality performance of liquid crystal displays. The survival of biological systems may depend on functional periodicity induced by electric charge and protein synthesis. Manufacturing systems can be made to have a higher throughput rate by introducing periodicity based on completion of critical processes (Suh and Lee, 2002, 2004).

Typically, the failure of a system follows the solid curve in figure 1.3, which shows that the system functions normally until a critical time is reached and then suddenly the failure occurs. Such a system may contain time-dependent combinatorial complexity. The failure of such a system may be delayed by introducing functional periodicity, as shown by the dotted line in figure 1.3. Throughout this book, examples will show how introducing functional periodicity improves the performance and reliability of many systems.

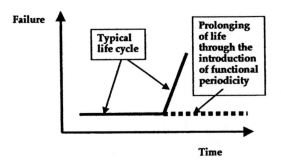

Figure 1.3 Life cycle of a typical system. Failure of a system with a time-dependent combinatorial complexity (solid line) may be prevented by introducing time-dependent periodic complexity (dotted line).

1.7 Past and Present Views on Complexity

Until complexity theory based on axiomatic design was proposed (Suh, 1999), the approaches developed to deal with complexity in science and engineering may be roughly divided into the following three categories: *probabilistic, algorithmic,* and *computational.* These different approaches to complexity were confined to the immediate needs of each specific subfield (i.e., information technology, computer science, and computation). All these approaches of the past to complexity are different from the one presented in this book in that the complexity was assumed to reside in the "physical" interactions and that the time-dependent nature of complexity was not recognized. One common thread in these earlier approaches has been the search for an "absolute measure" of complexity. However, this view may have been misguided.

Computational approaches to complexity include a quantitative measure of process, for example, how much computing resources—time, memory, and others—it would take to solve a problem (Cover, 1991). Computational complexity measures the difficulty of executing a task in terms of computational resources required.

The basic idea of algorithmic complexity[3] (Chaitin, 1987) is that simple tasks can be done by short computer programs while complex tasks require longer programs. According to this view, one should be able to measure the complexity of a task by the length of its most compact description. Similarly, Bennet (1985) has developed a measure of complexity called "logical depth." Bennet's idea is to gauge how long it would plausibly take for a computer to go from a simple blueprint to the final product. The problem with this idea of "minimal length of description" is that the length of even the shortest computer program depends on the *design* of the software as well as the coding. This approach begs the question of how we can identify the shortest computer program on an absolute basis, given diverse program styles and languages.

The probabilistic approach to complexity includes the idea of "schema" to identify the system's regularities as a means of defining complexity (Gell-Mann and Lloyd, 1996). According to this view, the length of the schema measures "effective complexity," which is roughly the length of a compact description of the identified regularities of an entity. In the case of language, the schema is its grammar. Gell-Mann and Lloyd (1996) also proposed the concept of total information, which is effective complexity plus an entropy term that measures the information required to describe the random aspects of the entity. Other probabilistic approaches may include the work of Huberman and Hogg (1994), who equate complexity with "a phase transition" between order and randomness.

The origin of these ideas may be traced to the information theory introduced by Shannon (1948), which dealt with uncertainty in transmitting information—how many bits of information it takes to describe an object or a message (Shannon and Weaver, 1949).

There are other ideas as well (Flood and Carson, 1993). For example, Lloyd and Pagels (1988) equated complexity to free energy. There are also other definitions of complexity that are specific to other fields, such as social sciences, architecture, and biology. (For a survey of complexity measures, see Feldman and Crutchfield, 1998.)

The journal *Science*[4] in 1999 devoted a special issue to the topic of complex systems, which dealt with complex systems in many fields of science, including the life sciences, chemistry, mathematics, biology, physiology, geology, meteorology, and economics. Diverse views were presented with no attempt to present a unified definition of complexity. The notions of complexity used by these authors include the following:

- In the introductory article by R. Gallagher and T. Appenzeller, a "complex system" is taken to be one whose properties are not fully explained by an understanding of its component parts.
- In their article entitled "Simple Lessons from Complexity," N. Goldenfield and L.P. Kadanoff state that "complexity means that we have structure with variations. Thus, a living organism is complex because it has many different working parts, each formed by variation in the working out of the same genetic coding."
- In their article on "Complexity in Chemistry," G.M. Whiteside and R.F. Ismagilov state: "[A] complex system is one whose evolution is very sensitive to initial conditions or to small perturbations, one in which the number of independent interacting components is large, or one in which there are multiple pathways by which the system can evolve. Analytical descriptions of such systems typically require nonlinear differential equations." Whiteside and Ismagilov's second characterization is more informal: "the system is 'complicated' by some subjective judgement and is not amenable to exact description, analytical or otherwise."
- In the abstract of their article entitled "Complexity in Biological Signaling Systems," G. Weng, U.S. Bhalla, and R. Iyengar state: "Complexity arises from the large number of components, many with isoforms that have partially overlapping functions, from the connections among components, and from the spatial relationship between components."

These views on complexity assume that there is an absolute measure of complexity. There are many difficulties with this view. For example, human beings perform a limited number of functions, which are supported by interactions among millions of molecules—proteins, DNA, and low-molecular-weight molecules—to perform various tasks. Most people regard the way human beings function as less complex than the way molecules inside human beings interact, although molecules are merely subsets of human physiology. This seemingly paradoxical disparity occurs when an absolute measure of complexity is sought. When complexity is viewed as a relative concept, the function of the human being is less complex than the molecular behavior if we can answer the way human beings function better than we can answer how DNA and protein molecules interact within the required scientific precision.

All of these efforts are attempts to discover the basic absolute measure for complexity. This book offers a different view of complexity, which treats complexity in the functional domain (not in the physical domain) as a relative quantity based on the concept of information used in axiomatic design (Suh, 1990, 2001).

1.8   Summary

In this chapter, we defined complexity as the measure of uncertainty in satisfying functional requirements (FRs), caused by poor design or by lack of knowledge (or understanding) about the system under consideration. The system may be an engineered system or a natural system. In both systems, complexity is defined in the functional domain rather than the physical domain, which may be a function of time—an important distinction from other theories of complexity.

Complexity is a *relative* quantity, which is determined by the overlap between the system range (actual performance) of FRs based on the system chosen and the design range (desired performance) of FRs.

There are four different types of complexity: time-independent real complexity, time-independent imaginary complexity, time-dependent combinatorial complexity, and time-dependent periodic complexity.

To create an engineered system with high reliability, the complexity of the system must be reduced. This can be done by eliminating time-independent real and imaginary complexity and by transforming time-dependent combinatorial complexity to time-dependent periodic complexity through the introduction of a functional periodicity. In addition, real complexity (expressed as information content as discussed in chapter 2) may be reduced by eliminating the coupling of FRs in an engineered system.

The reliability and safety of systems are governed by these complexities, once the FRs and constraints of the system are correctly established. When time-independent real complexity and time-independent imaginary complexity are zero, safety and reliability of a system may still be adversely affected by time-dependent combinatorial complexity. To reduce complexities of engineered systems, design must be done based on axiomatic design principles.

In chapter 2 of this book, axiomatic design theory is introduced since the complexity theory presented in this book is an extension of axiomatic design theory. Chapter 3 outlines the basic theory of complexity, which is the basic foundation on which the contents of the subsequent chapters of the book are developed. Chapter 4 shows how time-independent real complexity can be reduced either through the elimination of bias and the reduction of variance of uncoupled or decoupled designs or through redesign to eliminate coupling. The design of a knob, injection-molded parts, and an internal-combustion engine are used to illustrate the process of creating a design that satisfies the independence Axiom. Chapter 5 presents the basic principles involved in transforming a system with time-dependent combinatorial complexity into a system with time-dependent periodic complexity.

Chapters 6 through 9 treat the application of this complexity theory to various systems. In particular, these chapters show how the concept of functional periodicity can be introduced to various systems. Chapter 6 deals with the use of functional periodicity in manufacturing systems. Chapter 7 shows how functional periodicity based on geometry can be used to control the friction and wear behavior of materials. It also discloses new electrical connectors that have been invented, which have geometric functional periodicity. Chapter 8 covers the functional periodicity in materials to improve material properties. Chapter 9 shows the functional periodicity in biological systems, which cannot survive without functional periodicity. Chapter 10 presents the complexity of sociopolitical–economic systems. It shows how these systems can be revitalized through redesign or through reinitialization—an important aspect of functional periodicity.

Notes

1. Type I diabetes is a juvenile diabetes in which victims' immune systems attack the insulin-making cells in the pancreas early in life. In 2004, researchers led by Dr. Denise Faustman at Massachusetts General Hospital discovered that when diabetic mice are injected with the spleen cells, the cells migrated to their pancreases, prompting the damaged organs to regenerate into healthy, insulin-making organs and ending their diabetes. When this process is fully understood and routinely implemented, the complexity associated with curing Type I diabetes will be reduced to zero.

2. In time-dependent periodic complexity, the "period" is a "functional period," not necessarily a "temporal period." Functional periodicity may include geometric functional period, thermal functional period, and so on.

3. Sometimes referred to as Kolmogorov complexity or Kolmogorov–Solomonoff–Chaitin complexity.

4. *Science*, Vol. 284, No. 5411, April 2, 1999.

References

Bennet, C.H. 1985. "Dissipation, information, computational complexity and definition of organization," in Pines, D. (ed.), *Emerging Syntheses in Science* (Proceedings of the Founding Workshops of Santa Fe Institute, Santa Fe, New Mexico), Addison-Wesley, Redwood City, CA, pp. 215–233.

Chaitin, G.J. 1987. *Algorithmic Information Theory*, Cambridge University Press, Cambridge.

Cover, T.M. 1991. *Elements of Information Theory*, Wiley, New York.

Feldman, D.P. and Crutchfield, J. 1998. "A survey of 'complexity measures,'" SFI 1998 Complex Systems, Summer School, June 11.

Flood, R.L. and Carson, E.R. 1993. *Dealing with Complexity: An Introduction to the Theory and Application of Systems Science*, 2nd ed., Plenum Press, New York.

Gell-Mann, M. and Lloyd, S. 1996. "Information measures, effective complexity, and total information," *Complexity*, Vol. 2, pp. 44–52.

Huberman, B.A. and Hogg, T. 1994. In Nørretranders, T., *Märk världen. En bok om vetenskap och intuition*, Bonnier Alba, Stockholm.

Lloyd, S. and Pagels, H. 1988. "Complexity as thermodynamic depth," *Annals of Physics,* Vol. 188, pp. 186–213.

Shannon, C.E. 1948. "A mathematical theory of communication," *The Bell System Technical Journal*, Vol. 27, pp. 379–623.

Shannon, C.E. and Weaver, W. 1949. *The Mathematical Theory of Communication*, University of Illinois Press, Urbana.

Suh, N.P. 1986. *Tribophysics*, Prentice-Hall, Englewood Cliffs, NJ.

Suh, N.P. 1990. *The Principles of Design*, Oxford University Press, New York.

Suh, N.P. 1999. "A theory of complexity, periodicity, and design axioms," *Research in Engineering Design*, Vol. 11, pp. 116–131.

Suh, N.P. 2001. *Axiomatic Design: Advances and Applications*, Oxford University Press, New York.

Suh, N.P. and Lee, T. 2002. "Reduction of complexity in system integration through re-initialization", International Manufacturing Systems Seminar, CIRP, Seoul, Korea, May 14, 2002 (Keynote address).

Suh, N.P. and Lee, T. 2004. "System integration based on time-dependent periodic complexity," U.S. Patent 6,701,205 B2, March 2.

2

Introduction to Axiomatic Design Principles

2.1 Complexity and Axiomatic Design

To understand the complexity theory presented in this book, one has to understand axiomatic design theory, on which this complexity theory is partly based. Axiomatic design theory provides a conceptual basis for the complexity theory: the importance of functions, the need to define complexity in the functional domain rather than in the physical domain, the significance of the Independence Axiom and the Information Axiom in understanding complexity, and the measurement of uncertainty based on the Information

Axiom. The purpose of this chapter is to introduce axiomatic design for engineered systems as a prelude to the complexity theory presented in chapter 3. For a more complete treatise on axiomatic design, readers are encouraged to refer to Suh (1990, 2001). The materials presented in this chapter are similar to the materials presented in chapter 1 of Suh (2001).

Complexity theory provides a broad theoretical framework for understanding and designing complicated systems. Complexity theory is applicable to the design of engineered systems and to understanding of the behavior of natural systems such as biological systems. The complexity theory presented in this book augments axiomatic design theory in the design of engineered systems. It is the 40,000-foot view that the theory provides to the designers of engineered systems and to natural scientists. The theory gives guidelines for what is possible and desirable in these systems. It also provides a stability criterion for all engineered and natural systems.

Axiomatic design theory was advanced to provide a scientific basis for the design of engineered systems. It has been used in developing software, hardware, machines and other products, manufacturing systems, materials and materials processing, organizations, and large systems such as space ships.[1] It has provided designers with logical and rational thought processes and design tools. Axiomatic design theory has been used for the following specific purposes:

1. To provide a systematic way of designing products and large systems.
2. To make human designers more creative.
3. To reduce the random search process.
4. To minimize the iterative trial-and-error process.
5. To determine the best designs among those proposed.
6. To create systems architecture that completely captures the construction of the system functions and provides ready documentation.
7. To endow the computer with creative power.

Although it has been applied in designing many different kinds of engineered systems, mechanical examples will be used to explain the basic concept, since it is easier to illustrate axiomatic design with visual help.

When large systems are designed and developed by traditional means—the repetition of the "design/build/test/redesign/build/test" cycle—the cost of development is high and the development time long. Moreover, the reliability of such a system is often less than acceptable and the cost of ownership is high. These problems—high development cost, high cost of ownership, low reliability and safety, and high life-cycle cost—are closely tied to the complexity of these traditionally engineered systems. They probably have large time-independent real and imaginary complexities and also have large time-dependent combinatorial complexity. These complexities should be reduced to make the system more reliable at an affordable cost.

2.2   Elements of Axiomatic Design

Axioms have played a key role in the development of modern science. Many fields of science and technology owe their advances to the development and existence of axioms. Fields as diverse as mathematics, physical sciences, and engineering have gone through the transition from experience-based practices to the use of scientific theories and methodologies that are based on axioms.

It is only recently that axioms were developed for the field of synthesis in the form of design axioms (see Suh, 1990, for a history of axiomatic design). Nevertheless, axiomatic design follows a historical trend in science and technology.

2.2.1 A brief historical perspective

Axioms, which are truths that cannot be derived but for which there are no counter-examples or exceptions, have played a major role in developing mathematics and natural science, which includes fields such as physics, chemistry, and biology. These fields of natural science deal with energy, matter, living organisms, and their transformations and interrelations. Axioms were more easily accepted in these fields because the predictions that were made based on the axioms could be objectively measured and witnessed as natural phenomena.

The scientific field of thermodynamics was born as a result of attempts to generalize how "good steam engines" work. Before the field of thermodynamics emerged, many people might have said that the steam engine was too complicated to explain and that it could be designed only by experienced, ingenious designers and through trial-and-error processes. The first law of thermodynamics, which states that energy is conserved, is believed to be true because no observations or measurements contradict either the law or predictions based on the law. It defines the universal concept of energy for all sorts of diverse situations and matter. Similarly, the second law of thermodynamics was not derived. It is a generalization of the commonly observed fact that no net mechanical work can be done by a heat engine unless it exchanges heat with two other bodies. It is also an axiom in that it is believed to be a universal truth for which there are no counterexamples or exceptions. Based on the second law of thermodynamics, the concept of entropy could be derived. Thermodynamics has become a fundamental pillar of science and engineering.

Sir Isaac Newton (1642–1727) formulated three laws or axioms of mechanics. The first law states that if there is no force acting on a body, it will remain at rest or move with constant velocity in a straight line. The second law states that the product of mass and acceleration is equal to the force acting on the body. The third law states that the force that one body exerts on another must always be equal in magnitude and opposite in direction to the force that the second body exerts on the first. These were axioms. Newton's laws established the concept of force. To prove the validity of these laws or axioms, Newton applied all three of his laws to the motion of planets around the sun. Newton predicted Kepler's three laws of planetary motion based on his own three laws. From this work, he could determine the gravitational force acting between two masses. Newton's three laws are universally accepted because they predict observed natural phenomena and physical measurements.

These examples show that the development of natural science has been possible because of the advent of important axioms or laws that could generalize the behavior of nature. The validity of these axioms is tested by comparing the theoretical predictions of given phenomena with experimental measurements, by testing hypotheses based on these axioms, and by analysis of observed phenomena using the axioms.

Design axioms are presumed to be valid if they lead to better designs that satisfy the functional requirements and are more reliable and robust at low cost. Also if we can take a design that violates the design axioms and, using the axioms, create better-designed products and systems, the validity of the axioms can be claimed. Furthermore, theories, such as complexity theory, that are derived from the axioms and can predict the behavior of unknown systems provide further support for the verification of the design axioms.

2.2.2 Axiomatic approach versus algorithmic approach

There are two ways to deal with design and complexity: *axiomatic* and *algorithmic*. In an ideal world, the development of knowledge should proceed from axioms to algorithms to tools.

In a purely algorithmic process, we try to identify or prescribe the process, so that in the end it will lead to a solution. Generally, the algorithmic approach is founded on the notion that the best way of advancing a given field is to understand the process by following the best practice. The algorithmic approach is ad hoc for specific situations. It is difficult to come up with algorithms for all situations. Algorithms are generally useful at the detail level, because they are manageable.

The axiomatic approach to any subject begins with the premise that there are general principles that govern the underlying behavior of the system being investigated. Axioms generate new abstract concepts, such as force, energy, and entropy. The axiomatic approach to design is based on the *abstraction* of good design decisions and processes. The design axioms were created by identifying common elements that are present in all good designs. Once the common elements could be stated, they were reduced to two axioms through a logical reasoning process (Suh, 1990). They are general principles. These principles in turn can be used to create innovative products and systems. The complexity theory presented in chapter 3 is a consequence of these design principles, which has generated such an abstract concept as functional periodicity.

2.3   Axiomatic Design Framework

There are several key concepts that are fundamental to axiomatic design. They are the existence of domains, mapping, axioms, decomposition by zigzagging between the domains, theorems, and corollaries. Axiomatic design requires that one think "functionally" first before considering physical attributes or parameters.

2.3.1 The concept of domains

The design world consists of four domains.
Design involves an interplay between "what we want to achieve" and "how we choose to satisfy the need (i.e., the what)." To systematize the thought process involved in this interplay, the concept of *domains* that create demarcation lines between four different kinds of design activities provides an important foundation of axiomatic design.

The world of design is made up of four domains: the *customer domain,* the *functional domain*, the *physical domain*, and the *process domain*. The domain structure is

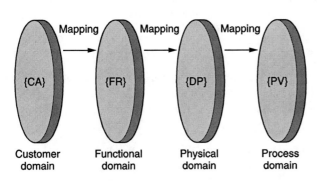

Figure 2.1 Four domains of the design world. The $\{x\}$ are the characteristic vectors of each domain. During the design process we map from a left domain (i.e., what we want to know or achieve) to a domain on its right (i.e., how we hope to satisfy "what"). The process is iterative in the sense that the designer can go back to the domain on the left based on the ideas generated in the right domain.

illustrated schematically in figure 2.1. The domain on the left represents "what we want to achieve," relative to the domain on the right, which represents the design solution, "how we propose to satisfy the requirements specified in the left domain."

The *customer domain* is characterized by the attributes (CAs) that the customer is looking for in a product or process or system or materials or organizations. In the *functional domain*, the customer needs are specified in terms of *functional requirements* (FRs) and *constraints* (Cs). In order to satisfy the specified FRs, we conceive *design parameters* (DPs) in the *physical domain*. Finally, to produce the product specified in terms of DPs, we develop a process that is characterized by *process variables* (PVs) in the *process domain*.

Many different fields—software, hardware, systems, materials, organizations, and manufacturing systems—can be described in terms of the four design domains. In the case of *product* design, the customer domain consists of the needs or attributes that the customer is looking for in a product. The functional domain consists of FRs, often defined as engineering specifications in the case of product design, and constraints. The physical domain is the domain in which the key DPs are chosen to satisfy the FRs. Finally, the process domain specifies the manufacturing PVs that can produce the DPs in the case of products.[2] Depending on the specific design tasks (e.g., materials, organizations, software, machines), FRs, DPs, and PVs take different characters (see Suh, 2001).

All design activities can be generalized in terms of the same principles. Because of this logical structure of the design world, the generalized design principles can be applied to all design applications and we can consider all design issues that arise in the four domains systematically. Similarly, the complexity theory presented in this book should be applicable to all fields.

2.3.2 Definitions

Axioms are valid only within the bounds established by the definitions of the key terms:

Axiom: Self-evident truth or fundamental truth for which there are no counterexamples or exceptions. An axiom cannot be derived from other laws or principles of nature.

Theorem: A proposition that is not self-evident but that can be proven from accepted premises or axioms and so is established as a law or principle.

Corollary: Inference derived from axioms or from propositions (theorems) that follow from axioms or from other propositions that have been proven.

Functional requirement: Functional requirements (FRs) are a minimum set of independent requirements that completely characterize the functional needs of the product (or software, organization, system, etc.) in the functional domain. By definition, each FR is independent of every other FR at the time the FRs are established.

Constraint: Constraints (Cs) are bounds on acceptable solutions. There are two kinds of constraints: *input* constraints and *system* constraints. Input constraints are imposed as part of the design specifications. System constraints are constraints imposed by the system in which the design solution must function.

Design parameter: Design parameters (DPs) are the key physical variables (or other equivalent terms in the case of software design, etc.) in the physical domain that characterize the design that satisfies the specified FRs.

Process variable: Process variables (PVs) are the key variables (or other equivalent terms in the case of software design, etc.) in the process domain that characterize the process that can generate the specified DPs.

2.3.3 Mapping from domain to domain

Once we identify and define the perceived customer needs (or the attributes the customer is looking for in a product), these needs must be translated into FRs. This must be done within a "solution-neutral environment." This means that the FRs must be defined without ever thinking about something that has already been designed or what the design solution should be. This is very difficult to do, especially if the designer has many years of experience in the specific field. If FRs are chosen thinking about an existing product, the new design will be a slight variation of the existing product.

After the FRs are chosen, we map them into the physical domain to conceive a design with specific DPs that can satisfy the FRs. The mapping process is typically a one-to-many process, that is, for a given FR there can be many possible DPs. We must choose the right DP by making sure that other FRs are not affected by the chosen DP as per the Independence Axiom and that the FR can be satisfied within its design range as per the Information Axiom.

2.3.4 Axioms

The basic postulate of the axiomatic approach to design is that there are fundamental axioms that govern the design process. Two axioms were identified by examining the common elements that are always present in good designs.

The first axiom is called the *Independence Axiom*. It states that the independence of FRs must always be maintained. The second axiom is called the *Information Axiom*, and it states that among those designs that satisfy the Independence Axiom, the design that has the smallest information content is the best design. Because the information content is defined in terms of probability of achieving the FRs, the second axiom also states that the design with the highest probability of success is the best design. In an ideal design, the information content should be zero to satisfy the FR every time and all the time. The complexity theory presented in chapter 3 depends on both the Independence Axiom and the Information Axiom.

The axioms are formally stated as:

AXIOM 1: THE INDEPENDENCE AXIOM

Maintain the independence of the functional requirements (FRs).

AXIOM 2: THE INFORMATION AXIOM

Minimize the information content of the design.

During the mapping process, we must make the right design decisions using the Independence Axiom. When several designs that satisfy the Independence Axiom are available, the Information Axiom can be used to select the best design. When only one FR is to be satisfied by having an acceptable DP, the Independence Axiom is always satisfied and the Information Axiom is the only axiom the one-FR design must satisfy. When there are many FRs, the Independence Axiom must always be satisfied by choosing the right set of DPs.

The case studies presented in Suh (1990, 2001) showed that the performance, robustness, reliability, and functionality of products, processes, software, systems, and organizations are all significantly improved when these axioms are satisfied.

2.4 The First Axiom: The Independence Axiom

A set of FRs is a description of the design goals. The Independence Axiom states that when there are two or more FRs, the design solution must be such that each of the FRs can be satisfied without affecting any of the other FRs. This means that we have to choose a correct set of DPs to be able to satisfy the FRs and maintain their independence.

After the FRs are established, the next step in the design process is the conceptualization process, which occurs during the mapping process going from the functional domain to the physical domain.

To design, we have to go from "what" in the functional domain to "how" in the physical domain, which requires *mapping*. After the overall design concept is generated by mapping, we must identify the DPs and complete the mapping process. During this process, we must think of the different ways of fulfilling each of the FRs by identifying plausible DPs. Sometimes it is convenient to think about a specific DP to satisfy a specific FR, repeating the process until the design is completed. Identifying a DP for a given FR is somewhat straightforward, but when there are many FRs that we must satisfy, the design task becomes more difficult since the Independence Axiom must be satisfied by the chosen set of DPs.

The mapping process between the domains can be expressed mathematically in terms of the characteristic vectors that define the design goals and design solutions. At a given level of the design hierarchy, the set of functional requirements that define the specific design goals constitutes the {FR} vector in the functional domain. Similarly, the set of design parameters in the physical domain that has been chosen to satisfy the FRs constitutes the {DP} vector. The relationship between these two vectors can be written as

$$\{FR\} = [A] \{DP\} \tag{2.1}$$

where $[A]$ is called the *design matrix* that relates FRs to DPs and characterizes the product design. Equation (2.1) is a design equation for the design of a product. The design matrix is of the following form for a design that has three FRs and three DPs:

$$[A] = \begin{bmatrix} A_{11} & A_{12} & A_{13} \\ A_{21} & A_{22} & A_{23} \\ A_{31} & A_{32} & A_{33} \end{bmatrix} \tag{2.2}$$

When equation (2.1) is written in a differential form as

$$\{dFR\} = [A] \{dDP\}$$

the elements of the design matrix are given by

$$A_{ij} = \frac{\partial FR_i}{\partial DP_j}$$

With three FRs and three DPs, equation (2.1) may be written in terms of its elements as

$$\begin{aligned} FR_1 &= A_{11} DP_1 + A_{12} DP_2 + A_{13} DP_3 \\ FR_2 &= A_{21} DP_1 + A_{22} DP_2 + A_{23} DP_3 \\ FR_3 &= A_{31} DP_1 + A_{32} DP_2 + A_{33} DP_3 \end{aligned} \tag{2.3}$$

In general,

$$FR_i = \sum_{i=1}^{n} A_{ij} DP_j$$

where n = the number of DPs.

For a linear design, A_{ij} are constants; for a nonlinear design, A_{ij} are functions of the DPs. There are two special cases of the design matrix:

1. The diagonal matrix, where all $A_{ij} = 0$ except those where $i = j$:

$$[A] = \begin{bmatrix} A_{11} & 0 & 0 \\ 0 & A_{22} & 0 \\ 0 & 0 & A_{33} \end{bmatrix} \tag{2.4}$$

2. The triangular matrix; in a lower triangular (LT) matrix, all upper triangular elements are equal to zero, as shown below:

$$[A] = \begin{bmatrix} A_{11} & 0 & 0 \\ A_{21} & A_{22} & 0 \\ A_{31} & A_{32} & A_{33} \end{bmatrix} \tag{2.5}$$

In an upper triangular (UT) matrix, all lower triangular elements are equal to zero. A UT matrix can always be changed to a LT matrix.

For the design of processes involving mapping from the {DP} vector in the physical domain to the {PV} vector in the process domain, the design equation may be written as

$${DP} = [B] {PV} \qquad (2.6)$$

where [B] is the design matrix that defines the characteristics of the process design and is similar in form to [A].

To satisfy the Independence Axiom, the design matrix must be either diagonal or triangular. When the design matrix [A] is diagonal, each of the FRs can be satisfied independently by means of its respective DP. Such a design is called an *uncoupled* design. When the matrix is triangular, the independence of FRs can be guaranteed if and only if the DPs are determined in a proper sequence. Such a design is called a *decoupled* design. Any other form of the design matrix is called a full matrix and results in a *coupled* design.

When the matrix is a full matrix producing a coupled design, we may get a unique solution that gives the right values for FRs, but such a design has many problems. Coupled designs are not robust and cannot survive random variations of DPs and the environment surrounding the design. For example, when one of the FRs is changed, all DPs must be changed to create a new system that satisfies the new set of FRs. Also whenever the DPs are not exact and deviate from the desired (or set) values so as to satisfy the FRs within their specified ranges, the FRs may not be satisfied. Therefore, when several FRs must be satisfied, we must develop designs that will enable us to create either a diagonal or a triangular design matrix.

What are constraints?
The design goals are often subject to constraints (Cs). Constraints provide bounds on acceptable design solutions and differ from FRs in that they do not have to be independent.

There are two kinds of constraints: *input* constraints and *system* constraints. Input constraints are specific to the overall design goals (i.e., all designs that are proposed must satisfy these). System constraints are specific to a given design; they are the result of design decisions made.

The designer often has to specify input constraints at the beginning of the design process because the designed product (or process or system or software or organization) must satisfy external boundary conditions, such as the voltage and the maximum current of the power supply. The environment within which the design must function may also impose many constraints. All of these constraints must be satisfied by all proposed design embodiments regardless of the specific details of the design.

Some constraints are generated because of design decisions made as the design proceeds. All higher-level decisions act as constraints at lower levels. For example, if we have chosen to use a diesel engine in a car, all subsequent decisions related to the vehicle must be compatible with this decision. These are system constraints.

2.4.1 Ideal design, redundant design, and coupled design—a matter of relative numbers of DPs and FRs

Depending on the relative numbers of DPs and FRs, the design can be classified as coupled, redundant, or ideal.

Case 1. Number of DPs < Number of FRs: Coupled Design
When the number of DPs is less than the number of FRs, we always have a coupled design. This is stated as Theorem 1, which is given below:

> **THEOREM 1 (Coupling Due to Insufficient Number of DPs)**
>
> When the number of DPs is less than the number of FRs, either a coupled design results or the FRs cannot be satisfied.

Case 2. Number of DPs > Number of FRs: Redundant Design
When there are more DPs than there are FRs, the design is called a redundant design. A redundant design may or may not violate the Independence Axiom as illustrated below.
    Consider the following two-dimensional case:

$$\begin{Bmatrix} FR_1 \\ FR_2 \end{Bmatrix} = \begin{bmatrix} A_{11} & 0 & A_{13} & A_{14} & A_{15} \\ A_{21} & A_{22} & 0 & A_{24} & 0 \end{bmatrix} \begin{Bmatrix} DP_1 \\ DP_2 \\ DP_3 \\ DP_4 \\ DP_5 \end{Bmatrix}$$

This design takes on various characteristics, depending on which design parameters are varied and which ones are fixed. If DP_1 and DP_4 are varied after DP_2, DP_3, and DP_5 are fixed to control the values of FRs, the design is a coupled design. On the other hand, if we fix the values of DP_1, DP_4, and DP_5, the design is an uncoupled design. If DP_3, DP_4, and DP_5 are fixed, then the design is a decoupled design. If DP_1 and DP_4 are set first, then the design behaves as an uncoupled redundant design. Theorem 3 states this fact:

> **THEOREM 3 (Redundant Design)**
>
> When there are more DPs than FRs, the design is a redundant design, which can be reduced to an uncoupled design or a decoupled design, or a coupled design.

Case 3. Number of DPs = Number of FRs: Ideal Design
When the number of FRs is equal to the number of DPs, the design is an ideal design, provided that the Independence Axiom is satisfied. This is stated as Theorem 4.

> **THEOREM 4 (Ideal Design)**
>
> In an ideal design, the number of DPs is equal to the number of FRs, and the FRs are always kept independent from each other.

Many other theorems and corollaries are presented in the Appendix (page 44 below) They may be used as design rules for specific cases.

2.4.2 Decomposition, zigzagging, and hierarchy

When the design details are missing at the highest level of design, the design equation represents the design *intent*. We must decompose the highest-level design to develop design details that can be implemented. As we decompose the highest-level design, the lower-level design decisions must be consistent with the highest-level design intent.

When the Independence Axiom is violated by design decisions made, we should go back and redesign rather than proceed with a flawed design.

How do we decompose FRs and DPs?

To decompose FR and DP characteristic vectors, we must zigzag between the domains. That is, we start out in the "what" domain and go to the "how" domain. This is illustrated in figure 2.2. From an FR in the functional domain, we go to the physical domain to conceptualize a design and determine its corresponding DP. Then we come back to the functional domain to create FR_1 and FR_2 at the next level that collectively satisfy the highest-level FR. FR_1 and FR_2 are the FRs for the highest-level DP. Then we go to the physical domain to find DP_1 and DP_2, which satisfy FR_1 and FR_2, respectively. This process of decomposition is continued until the highest-level FR can be satisfied without further decomposition, that is, when all of the branches reach the final state. The final state is indicated by the thick boxes in figure 2.2, which are called "leaves."

To be sure that we have made the right design decision, we must write down the design equation—{FR} = [A]{DP}—at each level of decomposition. For example, in the case shown in figure 2.2, after FR and DP are decomposed into FR_1, FR_2 and DP_1, DP_2, we must write down the design equation to indicate our design *intent* at this level. At this high level of the design process, we can only state our design intent, since we have not yet developed the lower-level detailed designs. We know that the design must be either uncoupled or decoupled, and therefore the intended design must have either a diagonal or a triangular matrix.

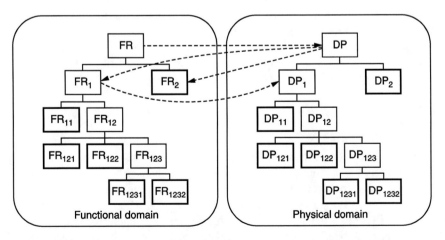

Figure 2.2 Zigzagging to decompose FRs and DPs in the functional and the physical domains to create the FR and DP hierarchies. Boxes with thick lines represent "leaves" that do not require further decomposition.

Suppose that the designer wanted to have a decoupled design represented by the design equation

$$\begin{Bmatrix} FR_1 \\ FR_2 \end{Bmatrix} = \begin{bmatrix} X & 0 \\ X & X \end{bmatrix} \begin{Bmatrix} DP_1 \\ DP_2 \end{Bmatrix}$$

Since design details are unknown at this stage of the design process, the triangular matrix represents the design *intent*. All subsequent lower-level design decisions must be consistent with this high-level design decision. The consistency of all lower-level design decisions can be checked by constructing the master design matrix.

Through the design decomposition process, the designer is transforming design intent into realizable design details.

At the highest level of the design process, the designer develops the design concept based on the available data; that is, the designer develops design *intent*. To complete the detailed design, the FR and DP vectors must be decomposed to the lowest level of FRs and DPs, that is, to leaf-level FRs and DPs. Throughout the decomposition process, the designer is transforming the design *intent* expressed by the higher-level design matrices into realizable detailed designs given by the lowest-level design matrices.

At each level of decomposition, the design decisions made must be consistent with all higher-level design decisions that were previously made. That is, if the highest-level design matrix is a diagonal matrix, all lower-level decisions must not make the off-diagonal elements of the highest-level design matrix nonzeroes—either intentionally or inadvertently. To check this fidelity and consistency of design decisions, the master design matrix must be constructed by combining all lower-level design matrices into a single master matrix. This process can be facilitated using Acclaro.[3]

How does this design process affect inventions and innovations?

As a designer tries to develop detailed designs that do not violate the original design intent, the designer may find that existing technologies cannot be used. Then the designer may develop a new technology that can achieve the original design goals. This process of recognizing the shortcomings of existing technologies and/or designs often leads to inventions and innovations. When a coupled design is replaced by an uncoupled or a decoupled design, major improvements can be made. These novel solutions often constitute inventions or innovations.

What is the current state of design practice as far as the decomposition process is concerned?

In some large organizations, there exists a "division for engineering specification" that is charged with creating FRs at all levels. The major task of the division is to develop functional requirements or specifications for their products. These divisions are typically organized so that they have to create FRs at all levels without zigzagging, that is, by remaining only in the functional or physical domain. As should be quite obvious by now, these divisions cannot do their job right, since FRs cannot be decomposed by remaining in one domain, in other words, without zigzagging. Thus, when designers/engineers are forced to work in such an organization, they often develop FRs or specifications by thinking of an already existing design, which results in respecifying that which already exists.

To decompose FRs and DPs, the designer must zigzag. For example, suppose you want to design a vehicle that satisfies the following four FRs: go forward, go backward, stop, and turn. We cannot decompose these FRs unless we first conceptualize DPs that can satisfy these highest-level FRs. If we decide to use an electric motor as a DP to satisfy the FR of moving forward, the decomposed FRs at the next level would be quite different from those that would have resulted had we chosen gas turbines as the DP. Therefore, when we define the FRs in a solution-neutral environment, we have to "zig" to the physical domain, and after proper DPs are chosen, we have to "zag" to the functional domain for further decomposition. Organizations that have created a division for the specific task of specifying FRs at all levels without zigzagging between the domains will not get the results they are looking for and will miss important opportunities for innovation.

When does analysis come into the picture during the design process?
To refine the design, we must model and analyze the proposed design whenever possible. In the preceding examples, the design matrix was formulated in terms of X and 0. In some cases it may be sufficient to complete the design using simply X and 0. In many cases we may take further steps to determine the precise values of design parameters. After the conceptual design is done in terms of X and 0, we need to model the design more precisely to replace the Xs with equations or numbers. Through modeling, we can replace each X with either a constant or a function that involves the DP. We then have a set of equations that relates the FRs to the DPs. This set of equations can be solved separately for uncoupled designs or by following the sequence given by the design matrix for decoupled designs.

Summary of a case study
In designing a space plane that can take astronauts into space, the survivability of the crew is the most important issue, since there can be many unexpected accidents such as the malfunctioning of the rocket during the initial ascent, structural failures, loss of vehicle control, and so on. These events involve numerous components and subsystems that interact in complicated ways. A small group of engineers (3 full-time and 4 part-time workers) used axiomatic design to design a system that can satisfy the requirements for crew survivability in a short time (5 months). They developed five designs at the system, subsystem and component levels to ensure that the crew survives the launch ascent into orbit. The system they designed possessed the least number of FRs that are collectively exhaustive to ensure that no FR is missed or frivolous, provided design solutions to satisfy the FRs that are mutually exclusive for the best tunability of the design, and gave a concise representation of the system for clarity in the connectivity between FRs and DPs through the design matrix (DM).

At the system level, they developed a comprehensive crew survivability system that ensures the crew survives launch ascent into orbit. A total of 663 FRs across nine levels of design hierarchy were defined in a solution-neutral environment through the use of the zigzagging process outlined in this chapter. Design solutions were provided for the FRs. The connectivity in the form of the design matrix (DM) of the FRs to DPs was examined for coupling to reduce complexity and enhance tunability of the design. The logic and the thought process that accompanied the design were documented in a commercial software system, Acclaro.

At the subsystem level, they developed the design for the thermal protection subsystem (TPS) and the design for the airbag landing subsystem. Because the TPS has the inherent conflict between the need for structural constraints to withstand external loads and the need to accommodate thermal expansion, their design had to make sure that there were no coupled designs through proper design. These DPs must be integrated to reduce information content to come up with a design that is functionally independent but physically integrated.

At the system and subsystem levels, designs were developed based on qualitative reasoning that ensures independence of FRs and the minimization of information content. As design decompositions reached the leaf level, the qualitative decisions were replaced with quantitative models. Once quantitative models were developed for the leaf-level modules, they were combined to create a comprehensive quantitative model for the entire system. The system model can then be used to simulate the system behavior and evaluate the system effectiveness. This was done for the design of sensing systems and others.

This case study illustrated that a small number of engineers can develop a complicated and reliable system for very demanding FRs using axiomatic design in a short period of time.

2.5 The Second Axiom: The Information Axiom

In the preceding section, the Independence Axiom was discussed and its implications were presented. The design effort may produce several designs, all of which may be acceptable in terms of the Independence Axiom. Even for the same task defined by a given set of FRs, it is likely that different designers will come up with different designs because there can be many designs that satisfy a given set of FRs. However, one of these designs is likely to be superior to the others. The Information Axiom provides a quantitative measure of the merits of a given design, and thus is useful in selecting the best among those designs that are acceptable. In addition, the Information Axiom provides the theoretical basis for design optimization and robust design.

Among the designs that are equally acceptable from the functional point of view, one may be superior to others in terms of the probability of achieving the design goals as expressed by the functional requirements. The Information Axiom states that the design with the highest probability of success is the best design.

Information content I_i for a given FR_i is defined in terms of the probability P_i of satisfying FR_i:

$$I_i = \log_2 \frac{1}{P_i} = -\log_2 P_i \qquad (2.7)$$

The information is given in units of bits.[4] The logarithmic function is chosen so that the information content will be additive when there are many FRs that must be satisfied simultaneously. Either the logarithm based on 2 (with the unit of bits) or the natural logarithm (with the unit of nats) may be used.

In the general case of m FRs, the information content for the entire system I_{sys} is

$$I_{sys} = -\log_2 P_{\{m\}} \qquad (2.8)$$

where $P_{\{m\}}$ is the joint probability that all m FRs are satisfied.

When all FRs are statistically independent, as is the case for an uncoupled design,

$$P_{\{m\}} = \prod_{i=1}^{m} P_i$$

then I_{sys} may be expressed as

$$I_{sys} = \sum_{i=1}^{m} I_i = -\sum_{i=1}^{m} \log_2 P_i \tag{2.9}$$

When all FRs are not statistically independent, as is the case for a decoupled design,

$$P_{\{m\}} = \prod_{i=1}^{m} P_{i|\{j\}} \qquad \text{for } \{j\} = \{1, \ldots, i-1\}$$

where $P_{i|\{j\}}$ is the conditional probability of satisfying FR$_i$ given that all other relevant (correlated) $\{FR_j\}_{j=1,\ldots,i-1}$ are also satisfied. In this case, I_{sys} may be expressed as

$$I_{sys} = -\sum_{i=1}^{m} \log_2 P_{i|\{j\}} \qquad \text{for } \{j\} = \{1, \ldots, i-1\} \tag{2.10}$$

The Information Axiom states that the design with the smallest I is the best design, since it requires the least amount of information to achieve the design goals. When all probabilities are equal to 1.0, the information content is zero, and conversely, the information required is infinite when one or more probabilities are equal to zero. That is, if the probability is small, we must supply more information to satisfy the FRs.

A design is called *complex* when its probability of success is low, that is, when the information content required to satisfy the FRs is high. This occurs when the tolerances of FRs for a product (or DPs for a process) are small, requiring high accuracy. This situation also arises when there are many parts because as the number of parts increases, the likelihood that some of the components do not meet the specified requirements also increases, such as when the interface between the parts introduces additional errors. In this sense, the quantitative measure for complexity is the information content because complex systems may require more information to make the system function. A physically large system is not necessarily complex if the information content is low. Conversely, even a small system can be complex if the information content is high. Therefore, the notion of complexity is tied to the design range for the FRs—the tighter the design range, the more difficult it becomes to satisfy the FRs.

The probability of success is governed by the intersection of the design range defined by the designer to satisfy the FRs and the ability of the system to produce the part within the specified range. For example, if the design specification for cutting a rod is 1 meter plus or minus 1 micron and the available tool (i.e., system) for cutting the rod consists of only a hacksaw, the probability of success will be extremely low. In this case, the information required to achieve the goal would approach infinity. Therefore, this may be called a complex design. On the other hand, if the rod needs to be cut within an accuracy of 10 cm, the hacksaw may be more than adequate,

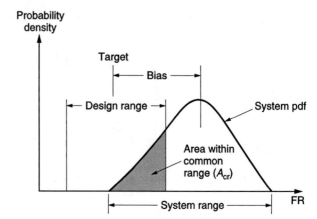

Figure 2.3 Design range, system range, common range, and system pdf for an FR.

and therefore the information required is close to zero. In this case, the design is simple.

The probability of success can be computed by specifying the *design range* (*dr*) for the FR and by determining the *system range* (*sr*) that the proposed design can provide to satisfy the FR. Figure 2.3 illustrates these two ranges graphically. The vertical axis (the ordinate) represents the probability density and the horizontal axis (the abscissa) represents either the FR or DP, depending on the mapping domains involved. When the mapping is between the functional domain and the physical domain as in product design, the abscissa is for the FR. When the mapping is between the physical domain and the process domain as in process design, the abscissa is for the DP.

In figure 2.3, the system probability density function (pdf) is plotted over the system range for the specified FR. The overlap between the design range and system range is called the *common range* (*cr*), and this is the only region where the FR is satisfied. Consequently, the area under the system pdf within the common range, A_{cr}, is the design's probability of achieving the specified goal. Then the information content may be expressed as (Suh, 1990):

$$I = \log_2 \frac{1}{A_{cr}} \qquad (2.11)$$

In terms of the system pdf $p_s(\mathrm{FR}_i)$, the probability p_i of satisfying FR_i is the integral of $p_s(\mathrm{FR}_i)$ over the FR_i design range, which may be expressed as

$$p_i = \int_{\text{design range}} p_s(\mathrm{FR}_i) d\mathrm{FR}_i \qquad (2.12)$$

When there are many FRs, the probability $p_{1,2,\dots,n}$ of satisfying all the FRs is given by integrating the joint density function $p_s(\mathrm{FR}_1, \mathrm{FR}_2, \dots, \mathrm{FR}_n)$ over the design space, which may be expressed as

$$p_{1,2,\dots,n} = \int_{\text{design space}} p_s(\mathrm{FR}_1, \mathrm{FR}_2, \dots, \mathrm{FR}_n) d\mathrm{FR}_1, d\mathrm{FR}_2, \dots, d\mathrm{FR}_n \qquad (2.13)$$

When the outcome for FR is binary, that is, either 0 or 1, such as in software (see Suh, 2001), the probability p_i is estimated by the ratio of the number of positive outcomes divided by the total number of trials. The larger the number of trials, the more accurate will be the probability estimate.

The Information Axiom is a powerful tool for selecting the best set of DPs when there are many FRs to be satisfied, but should we also use weighting factors?
Often design decisions must be made when there are many FRs that must be satisfied at the same time. The Information Axiom provides a powerful criterion for making such decisions without the arbitrary weighting factors used in other decision-making theories. In equation (2.9), the information content for each FR is simply summed with all other information terms without a weighting factor for two reasons. First, if we sum the information terms, each of which has been modified by multiplying it by a weighting factor, the total information content no longer represents the total probability (exercise 1.1). Second, the intention of the designer and the importance assigned to each FR by the designer are represented by the design range. If the design ranges for all of the FRs are precisely specified and if every specified FR is satisfied within its design range, the goal of the design is fully satisfied. Then there is no need for rank ordering or giving weighting factors to FRs, since the design range specifies their relative importance.

When there is only one FR, the Independence Axiom is always satisfied if there is an appropriate DP that satisfies the FR. In the one-FR case, the only task left is the selection of the right values for the design matrix element and the DP to come up with a robust design based on the Information Axiom. In the case of one-FR nonlinear design, various optimization techniques have been developed to deal with the task of finding a maximum or minimum of an objective function. However, when there are more than two FRs that are coupled, some of these optimization techniques do not work.

To develop a design with more than one FR, we must first develop a design that is either uncoupled or decoupled. If the design is uncoupled, each FR can be satisfied and the optimum points for all FRs can be found because each FR is controlled only by its corresponding DP. If the design is decoupled, the FRs must be satisfied following a set sequence, which is further discussed in chapter 3. The Information Axiom provides a measurement that enables us to measure the information content and thus be able to judge a superior design.

2.5.1 Reduction of the information content—robust design

The ultimate goal of design is to reduce the additional information required to make the system function as designed, that is, to minimize the information content, as stated by the Information Axiom. To achieve this goal, the design must be able to accommodate large variations in design parameters and process variables and yet still satisfy the functional requirements. Such a design is called a *robust* design.

To achieve a robust design, the variance of the system must be small and the bias must be eliminated to make the system range lie inside the design range, thus reducing the information content to zero (see figure 2.3). The bias can be eliminated if the design satisfies the Independence Axiom. There are four different ways of reducing the variance of a design if the design satisfies the Independence Axiom.

2.5.1.1 Elimination of bias

In figure 2.3, the target value of the FR is shown at the middle of the design range. The distance between the target value and the mean of the system pdf is called the *bias*. In order to have an acceptable design, the bias associated with each FR should be very small or zero. That is, the mean of the system pdf should be equal to the target value inside the design range.

How can we eliminate bias? What are the prerequisites for eliminating bias?
In a one-FR design, the bias can be reduced or eliminated by changing the appropriate DP, because the DP controls only this FR so we do not have to worry about its effect on other FRs. Therefore, it is easy to eliminate the bias when there is only one FR.

When there is more than one FR to be satisfied, we may not be able to eliminate the bias unless the design satisfies the Independence Axiom. If the design is coupled, each time a DP is changed to eliminate the bias for a given FR, the bias for the other FRs changes also, making the design uncontrollable. If the design is uncoupled, the design matrix is diagonal and the bias associated with each FR can be changed independently as if the design were a one-FR design. When the design is decoupled, the bias for all FRs can be eliminated by following the sequence dictated by the triangular matrix. In a decoupled design, it is important to make the magnitude of the diagonal element larger than those of the off-diagonal elements in order to make the decoupled design robust. Otherwise, other DPs chosen to satisfy other FRs will affect the intended FR more than the chosen DP.

2.5.1.2 Reduction of variance

What is variance? What causes variance? How do we control it? How is it related to redundant design?
Variance is a statistical measure of the variability of a pdf. Variability is caused by a number of factors, such as noise, coupling, environment, and random variations in design parameters. In a multi-FR design, the prerequisite for variance reduction is the satisfaction of the Independence Axiom. In all situations, the variance must be minimized. The variation can be reduced in a few specific situations discussed below.

(a) *Reduction of the information content through reduction of stiffness.* Suppose there is only one FR that is related to its DP as

$$FR_1 = (A_{11}) DP_1 \tag{2.14}$$

In a linear design, the allowable tolerance for DP_1, given the specified design range for FR_1, depends on the magnitude of A_{11}, that is, the stiffness. As shown in figure 2.4, the smaller the stiffness, A_{11}, the larger is the allowable tolerance of DP_1.

(b) *Reduction of the information content through the design of a system that is immune to variation.* When the stiffness, as shown in figure 2.4, is zero, the system will be completely insensitive to variation in DP. However, if the goal is to vary the FR by changing the DP, the stiffness must be large enough to allow control of the FR, although from the robustness point of view, low stiffness is desired. When there are

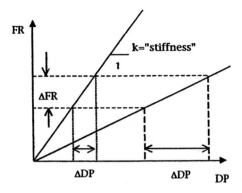

Figure 2.4 Allowable variation of DP as a function of stiffness. For a specified FR, the allowable variation of DP increases with a decrease in the stiffness, A_{11}.

many DPs that affect a given FR, design should be done so that the FR will be "immune" to variation of all these other DPs except the one specific DP chosen to control the FR. In the case of nonlinear design, we should search for such a design window where this condition is satisfied.

The variance is the statistical measure of the spread of the distribution of the output. If a number of DPs are affecting an FR, the total variance of the FR is equal to the sum of the separate variances of the DPs when these DPs are statistically independent.

Often the variation in the system range may be due to many factors that affect the FR. Consider the one-FR design problem. The designer might have created a redundant design as follows:

$$FR = f(DP_a, DP_b, DP_c)$$

or

$$FR = A_a \cdot DP_a + A_b \cdot DP_b + A_c \cdot DP_c \qquad (2.15)$$

where A_a, A_b, and A_c are coefficients and the DPs are design parameters that affect the FR. In this case, variation in FR can be introduced by any uncontrolled variation in all coefficients and DPs. The variance can be reduced by making the design so that the FR is not sensitive to (is immune to) changes in DP_b and DP_c, which can be done either if A_b and A_c are small or if DP_b and DP_c are fixed so that they remain constant. In this case, since the FR would be a function of only DP_a, the FR can be controlled by changing DP_a. In this case, the only source of variation is the random variation of A_a.

Now consider the case of the multi-FR design given by

$$\begin{Bmatrix} FR_1 \\ FR_2 \\ FR_3 \end{Bmatrix} = \begin{bmatrix} A_{11} & 0 & 0 \\ 0 & A_{22} & 0 \\ 0 & 0 & A_{33} \end{bmatrix} \begin{Bmatrix} DP_1 \\ DP_2 \\ DP_3 \end{Bmatrix} \qquad (2.16)$$

In this ideal design with a diagonal design matrix, the variance will be minimized if the random variation in A_{11}, A_{22}, and A_{33} can be eliminated. Therefore, the coefficients A_{11}, A_{22}, and A_{33} should be small, but large enough to exceed the required signal-to-noise ratio. It should be noted that any variation in DPs would also contribute to the variance and the bias.

(c) *Reduction of the information content by fixing the values of extra DPs.*   When the design is a redundant design, the variance of the FRs can be reduced by identifying the key DPs and preventing the extra DPs from varying by fixing their values.

Consider a multi-FR design given by

$$
\begin{Bmatrix} FR_1 \\ FR_2 \\ FR_3 \end{Bmatrix} = \begin{bmatrix} A_{11} & 0 & 0 & A_{14} & A_{15} & 0 \\ 0 & A_{22} & 0 & 0 & A_{25} & A_{26} \\ 0 & 0 & A_{33} & A_{34} & 0 & A_{36} \end{bmatrix} \begin{Bmatrix} DP_1 \\ DP_2 \\ DP_3 \\ DP_4 \\ DP_5 \\ DP_6 \end{Bmatrix} \tag{2.17}
$$

Equation (2.17) represents a redundant design because there are more DPs than FRs. The task now is to reduce the information content of this redundant design. The first thing we have to do is to find a way to make the design represented by equation (2.17) an ideal, uncoupled design as shown by equation (2.16). This can be done either by fixing DP_4, DP_5, and DP_6 so that they do not act as design parameters or by making the coefficients associated with these DPs equal to zero. Fixing DP_4, DP_5, and DP_6 also will minimize the variance in the FRs due to any variation of these three DPs. The variation can also be reduced by setting A_{14}, A_{15}, A_{25}, A_{26}, A_{34}, and A_{36} to zero so that the FRs will be immune to changes in DP_4, DP_5, and DP_6. If the design matrix were different from the one shown above, other appropriate design elements should be made zero or other appropriate DPs could be fixed to reduce the variance of the FRs.

(d) *Reduction of information content by minimizing the random variation of DPs and PVs.*   One way of reducing the variance of the FRs is to reduce the random variation of input parameters since they contribute to the total random variation of the FRs. The variance of FR may be expressed as

$$
\sigma_{FR_i}^2 = \sum_{j=1}^{n} A_{ij}^2 \sigma_{DP_j}^2 + 2 \sum_{j=1}^{n} \sum_{k=1}^{j-1} A_{ij} A_{ik} \, \text{Cov}(DP_j, DP_k) \tag{2.18}
$$

By reducing the variance of any of the DP_j, we can reduce the contributions to the variance of FR_i. Moreover, if some of the DPs are independent of one another, the relevant covariance terms disappear from equation (2.18), further reducing the contributions to the variance of FR_i. The second term in variance of equation (2.18) is always positive. If the covariance term is negative, then the design matrix elements will change their signs in such a way that the second term is always positive. So, variance of decoupled design is always smaller than that of a coupled design.

It is clear from equation (2.18) why it is easier to reduce the information content for uncoupled designs because only one DP contributes to the variance of FR*i* and there are no covariance terms.

(e) *Reduction of the information content by compensation.*   The one-FR design given by equation (2.15) was a redundant design, having three DPs rather than one DP. For the design given by equation (2.15), we could satisfy the FR with only one DP (Theorem 4 (Ideal Design)). Therefore, the best solution for dealing with random variation (noise) for one-FR design is to eliminate the unnecessary DPs and lower the stiffness of the one DP that has been selected to satisfy the FR. However, there may be situations where a given redundant design must be made to work.

Suppose that we have to work with the less than ideal design represented by equation (2.15), and that the design cannot be made to be "immune" to random variations by having low stiffness, because the coefficients associated with the redundant DPs cannot be made sufficiently small. In this case, the effect of random variation of the extra DPs on FR can be eliminated by "compensating" for the effects through the adjustment of the selected DP.

In equation (2.15), suppose that the following is true:

$$A_a \gg A_b \quad \text{and} \quad A_a \gg A_c$$

Then we should choose DP_a as the chosen DP and try to minimize the effect of the random variation of DP_b and DP_c on FR. The random variation will be represented as δDP_b and δDP_c. If we want to change FR from one state to another state, which is represented by ΩFR, it can be done by changing DP_a by ΩDP_a. For this change of state of FR, equation (2.15) may be written as

$$\Omega FR = A_a\,\Omega DP_a + \sum_{i=\text{noise terms}} A_i\,\delta DP_i \qquad (2.19)$$

If the allowable random variation of FR, that is, the design range of FR, is represented as ΔFR, the random noise term represented by the second term of the RHS of equation (2.19) can be compensated by adjusting DP_a. The necessary adjustment ΔDP_a to compensate for the random variation is given by

$$\Delta DP_a = \frac{\Delta FR - \sum\limits_{i=\text{noise terms}} A_i\,\delta DP_i}{A_a} \qquad (2.20)$$

In equation (2.20), if the noise term is larger than the allowable tolerance of FR, we have to look for a new design by choosing new DPs.

This means of compensating for the random error can be done with multi-FR designs as well as with one-FR designs if the Independence Axiom is satisfied by the multi-FR design. This kind of compensation scheme can be used to eliminate the effect of the random variation introduced during manufacturing.

(*f*) *Reduction of the information content by increasing the design range.*   In some special cases, the design range can be increased without jeopardizing the design goals. The system range may then be inside the design range.

2.5.2 Reduction of the information content through integration of DPs

The preceding section presented a means of reducing the information content of a design by making the system range fit inside the design range. This technique is normally called "robust design." Another equally significant means of reducing the information content is through integration of DPs in a single physical part without compromising the independence of FRs. In this way, the information content can be

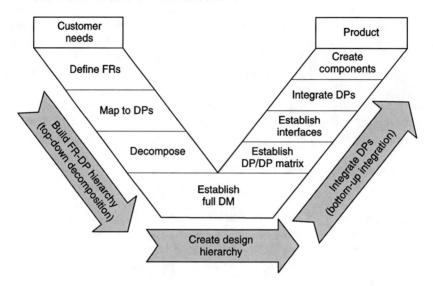

Figure 2.5 V-model for designing products.

made small by reducing the likelihood of introducing errors when many physical parts are assembled or by making the manufacturing operation simple.

A good example of DP integration is the beverage can, which has twelve FRs and DPs, but only three physical pieces. Another example is a can and bottle opener that must open bottles and cans, but not at the same time. In this case, the DP that opens the bottle and the DP that opens the can (by punching a triangular opening in the lid of the can) may be integrated in the same steel sheet stock—the can opener at one end and the bottle opener at the other end (figure 3.3 in *The Principles of Design* (Suh, 1990)).

When the design has been achieved by decomposing FRs and DPs to many levels, the integration of DPs can be done in the physical domain. In this case, only the leaf-level DPs of each branch need to be integrated, since higher-level DPs are made up of the leaf-level DPs. To create a physically integrated system, all physical parts that contain the leaf-level DPs must be integrated into a physically integrated system that satisfies the FRs and constraints (such as geometry and weight) and minimizes the information content. To achieve this goal, the relationship between DPs must be determined by means of a DP/DP matrix to place those DPs that ought to be in close geometric proximity near each other or in an integrated physical part (see Carrascosa et al., 1998). This integration process must reversibly follow the original decomposition process, since the hierarchical decomposition process provides a framework of the design intention at different levels of geometric and functional aggregation. This process of designing through decomposition and integration of the physical part is referred to as the V-model for product design, and is shown in figure 2.5.

2.5.3 Designing with incomplete information

During design, we encounter situations where the necessary knowledge about the proposed design is insufficient and thus design must be executed in the absence of complete information. The basic questions are:

- Under what circumstances can design decisions be made in the absence of sufficient information?
- What are the most essential kinds of information for making design decisions?

These questions will be explored in this section.

Throughout the design process, the designer collects, manipulates, creates, classifies, transforms, and transmits information. Information in design assumes a variety of forms—knowledge, databases, causality, paradigms, etc. The information necessary to design must be distinguished from the *information content* we need to minimize as required by the Information Axiom. Information is not as specific as the information content, which was specifically defined as a function of the probability of satisfying the FR in terms of design range and system range (see equations (2.7) and (2.8)).

For example, in mapping from the CAs of the customer domain to the FRs of the functional domain, the information needed is in the form of customer preference, potential FRs, and the relationship between the CAs and the FRs. Similarly, information is needed when FRs are mapped into the physical domain and when the DPs are mapped into the process domain.

The information we need is indicated by the design equations. First, we need information on the characteristic vectors (i.e., what they are, etc.). Given an FR, the most appropriate DP must be chosen, the likelihood of which increases with the size of the library of DPs that satisfy the FR. Similarly, given a DP, the more PVs we have, the more options we will have. Once DPs and PVs are chosen, information must be available on the elements of the design matrix, which define the relationship between "what we want to achieve" and "how we want to achieve it."

One of the central issues in the design process is: "What is the minimum information that is necessary and sufficient for making design decisions given a set of DPs for a given set of FRs." The necessary information depends on whether or not the proposed design satisfies the Independence Axiom. In the case of a coupled design, which violates the Independence Axiom, all of the information associated with all elements of the design matrix is required. That is, in the case of coupled designs, design cannot be done rationally without complete information. Repeated "design/build/test" cycles are needed for these coupled designs to discover all the flaws of the coupled design.

(a) Information required for an uncoupled design
Consider an ideal uncoupled design that satisfies the Independence Axiom and consists of three FRs. For this uncoupled design, which is the simplest case, the design equation may be written as

$$\begin{Bmatrix} FR_1 \\ FR_2 \\ FR_3 \end{Bmatrix} = \begin{bmatrix} A_{11} & 0 & 0 \\ 0 & A_{22} & 0 \\ 0 & 0 & A_{33} \end{bmatrix} \begin{Bmatrix} DP_1 \\ DP_2 \\ DP_3 \end{Bmatrix} \tag{2.21}$$

A_{11}, A_{22}, and A_{33} relate FRs to DPs. They are constants in the case of linear design, whereas in the case of nonlinear design, A_{11} is a function of DP_1, and so on. To proceed with this design, we must know the diagonal elements. Therefore, the minimum information required is the information associated with the diagonal elements. The information required for the uncoupled case is less than that for the coupled case because the off-diagonal elements are zeros.

(b) Information required for a decoupled design

Again consider the three-FR case, but this time the design is a decoupled design given by the following design equation:

$$\begin{Bmatrix} FR_1 \\ FR_2 \\ FR_3 \end{Bmatrix} = \begin{bmatrix} A_{11} & 0 & 0 \\ A_{21} & A_{22} & 0 \\ A_{31} & A_{32} & A_{33} \end{bmatrix} \begin{Bmatrix} DP_1 \\ DP_2 \\ DP_3 \end{Bmatrix} \tag{2.22}$$

As in the case of the uncoupled design given by equation (2.21), we need to know the diagonal elements A_{ii}. It is also desirable to know the off-diagonal elements A_{ij}. However, information on the off-diagonal elements may not be required to satisfy the given set of FRs with a given set of DPs. We can proceed with the design if the diagonal elements are known and if the magnitudes of the off-diagonal elements are smaller than those of the diagonal elements, that is, if $A_{ii} > A_{ij}$. This can be done because the value of FR_1 can be set first, and then the value of FR_2 can be set by varying the value of DP_2, regardless of the value of A_{21}. When DP_2 is chosen, we must be certain that it does not affect FR_1, but it is not necessary that any information for A_{21} be available if DP_2 has the dominant effect on FR_2, that is, if $A_{22} > A_{21}$. Similarly, as long as DP_3 does not affect FR_1 or FR_2, the design can be completed even if we do not have any information on A_{31} and A_{32}. This is the only case when design can proceed in the absence of complete information. This is stated as Theorem 17.

Suppose that the upper triangular elements are not quite equal to zero but have very small values, as shown in equation (2.23):

$$\begin{Bmatrix} FR_1 \\ FR_2 \\ FR_3 \end{Bmatrix} = \begin{bmatrix} A_{11} & a_{12} & a_{13} \\ A_{21} & A_{22} & a_{23} \\ A_{31} & A_{32} & A_{33} \end{bmatrix} \begin{Bmatrix} DP_1 \\ DP_2 \\ DP_3 \end{Bmatrix} \tag{2.23}$$

The absolute magnitudes of the elements a_{ij} are much smaller than those of A_{ij}, that is, $|a_{ij}| \ll |A_{ij}|$. In this case, FR_1 will still be affected by large state changes of DP_2 and DP_3 and this effect may not be negligible, since

$$\Omega FR_1 = A_{11}\, \Omega DP_1 + a_{12}\, \Omega DP_2 + a_{13}\Omega DP_3 \tag{2.24}$$

where Ω signifies a large change in the value of FR_i due to large state changes in the DPs. In this case, we must compensate for the effect of the DP state changes if the design range of FR_i is smaller than the variability caused by these state changes.

2.6   Allowable Tolerances of Uncoupled, Decoupled, and Coupled Designs

As stated in the preceding sections, an ideal design is an uncoupled design with a diagonal design matrix with zero information content. In this case, a multi-FR design is almost identical to a one-FR design problem. For each FR, we can write a design equation relating the FR to a single DP. If there are m FRs, there are m design equations, each of which can be solved independently. Modeling of the design also becomes simple because the modeling can be limited to relating one FR to one DP. The element of the design matrix can be expressed quantitatively or analytically.

Furthermore, the design can be made robust using the techniques discussed in section 2.5.

Decoupled designs can also be modeled similarly, although this involves additional consideration of the off-diagonal elements and the sequence of the operation. However, there is a substantial difference between the uncoupled design and the decoupled design in the allowable DP and PV tolerances.

How does the tolerance propagate from domain to domain in the case of an uncoupled design?
Tolerance specification is simple in the case of an uncoupled design. If the specified design range for FR_i is ΔFR_i, then the allowable tolerance for DP_i is simply

$$\Delta DP_i = \frac{\Delta FR_i}{A_{ii}} \tag{2.25}$$

Because the goal of a robust design is to make ΔDP_i as large as possible, A_{ii} should be made small. Similarly, the tolerance for PV_i is

$$\Delta PV_i = \frac{\Delta DP_i}{B_{ii}} \tag{2.26}$$

The *design range* is defined by ΔFR. The actual variation of FR, which is determined by the variation of DPs and PVs as well as by the magnitude of the design matrix elements, defines the system range. If the system range determined by the random variation of FR is completely contained within the specified design range ΔFR_i, then the information content is equal to zero.

How does the tolerance propagate from domain to domain in the case of a decoupled design?
Is the propagation of tolerance different for a decoupled design from the case of an uncoupled design discussed so far? Can a decoupled design be as robust as an uncoupled design? Why? How are they different?

Consider the decoupled design shown below:

$$\begin{Bmatrix} FR_1 \\ FR_2 \\ FR_3 \end{Bmatrix} = \begin{bmatrix} A_{11} & 0 & 0 \\ A_{21} & A_{22} & 0 \\ A_{31} & A_{32} & A_{33} \end{bmatrix} \begin{Bmatrix} DP_1 \\ DP_2 \\ DP_3 \end{Bmatrix} \tag{2.27}$$

The Independence Axiom can be satisfied if we change the DPs in the order shown. However, to have a robust design, we must be sure that the off-diagonal elements are much smaller than the diagonal elements, that is, $A_{ii} \gg A_{ij}$.

If the specified design ranges for the FRs are ΔFR_1, ΔFR_2, and ΔFR_3, the maximum allowable tolerances for the DPs may be expressed as

$$\Delta DP_1 = \frac{\Delta FR_1}{A_{11}}$$

$$\Delta DP_2 = \frac{\Delta FR_2 - |A_{21}\, \Delta DP_1|}{A_{22}} \tag{2.28}$$

$$\Delta DP_3 = \frac{\Delta FR_3 - |A_{31}\, \Delta DP_1| - |A_{32}\, \Delta DP_2|}{A_{33}}$$

The fluctuation of ΔDP_2 due to the term $A_{21}\,\Delta DP_1$ can make ΔDP_2 larger or smaller depending on its sign. However, the maximum allowable ΔDP_2 corresponds to the worst possible case, that is, when ΔDP_2 is made smaller by the term $A_{21}\,\Delta DP_1$. A similar argument holds for ΔDP_3. Therefore, the absolute value represented by $|x|$ is used to represent the worst possible case.

According to equation (2.28), the maximum tolerances for DPs of a decoupled design are less than the corresponding tolerances for DPs of an uncoupled design. This means that the decoupled design is inherently less robust than the uncoupled design. This may be stated as Theorem 22.[5]

THEOREM 22 (Comparative Robustness of a Decoupled Design)

Given the maximum tolerances for a given set of FRs, decoupled designs cannot be as robust as uncoupled designs in that the allowable tolerances for DPs of a decoupled design are less than those of an uncoupled design.

Equation (2.28) was for a decoupled design with three FRs and three DPs. Extending the argument given above to the case of m FRs and m DPs, it becomes obvious that as m increases, the allowable tolerance for the last DP of the triangular matrix becomes increasingly smaller. This means that the robustness of a decoupled design diminishes as the number of FRs increases.

THEOREM 23 (Decreasing Robustness of a Decoupled Design)

The allowable tolerance and thus the robustness of a decoupled design with a full triangular matrix diminish with an increase in the number of functional requirements.

How does the tolerance propagate in the case of coupled designs? Can a coupled design be robust?

In the case of a coupled design, the maximum allowable tolerance is even smaller than was the case for a decoupled design. Consider the following coupled design:

$$\begin{Bmatrix} FR_1 \\ FR_2 \\ FR_3 \end{Bmatrix} = \begin{bmatrix} A_{11} & A_{12} & A_{13} \\ A_{21} & A_{22} & A_{23} \\ A_{31} & A_{32} & A_{33} \end{bmatrix} \begin{Bmatrix} DP_1 \\ DP_2 \\ DP_3 \end{Bmatrix} \qquad (2.29)$$

The above equation may be solved for DPs if the determinant of the design matrix $|A|$ is not equal to zero, which is likely to be the case. The solution for DP_1 is

$$DP_1 = \frac{1}{|A|}\{\alpha\,FR_1 - \beta\,FR_2 - \gamma\,FR_3\} \qquad (2.30)$$

where

$$\alpha = A_{22}\,A_{33} - A_{23}\,A_{32}$$
$$\beta = A_{12}\,A_{33} - A_{32}\,A_{13}$$
$$\gamma = A_{22}\,A_{13} - A_{12}\,A_{23}$$

The expressions for DP_2 and DP_3 are of a similar form.

For a given set of design ranges of FRs, the maximum allowable tolerances for DPs may be expressed as

$$\Delta DP_1 = \frac{1}{|A|}\{\alpha\ \Delta FR_1 - |\beta\ \Delta FR_2| - |\gamma\ \Delta FR_3|\} \tag{2.31}$$

As argued before, although the magnitudes of $\beta\ \Delta FR_2$ and $\gamma\ \Delta FR_3$ can be either positive or negative, the maximum allowable ΔDP_1 is given by equation (2.31). Therefore, the allowable tolerances of DPs for coupled designs are smaller than those for uncoupled or decoupled designs.

The main point of this section is to show that whenever coupling terms are present in the design matrix, the time-independent real complexity is likely to increase since the allowable tolerances of DPs and PVs decrease.

2.7   Summary

Since the complexity theory presented in this book is based on axiomatic design theory, the basic elements of axiomatic design theory are reviewed in this chapter. The basic concepts and methodologies of axiomatic design include domains, mapping, the two design axioms (the Independence Axiom and the Information Axiom), decomposition, hierarchy, and zigzagging.

Several key terms, such as functional requirement (FR), design parameter (DP), and process variable (PV), are carefully defined, because strict adherence to definitions is required in an axiomatic treatise of the subject matter for internal consistency, logical deduction, and mathematical derivation of the resulting relationships. The acceptance of these definitions is a prerequisite in applying the axiomatic principles for design.

Mapping between the domains generates design equations and design matrices. The design equation models the relationship between the design objectives (*what* the design is trying to achieve) and the design features (*how* the design goals are to be satisfied). The design matrix describes the relationship between the characteristic vectors of the domains and forms the basis for functional analysis of the design in order to identify acceptable designs. Uncoupled and decoupled designs are shown to satisfy the independence Axiom and thus are acceptable. Coupled designs do not satisfy the independence Axiom and thus are unacceptable.

The Independence Axiom states that the FRs must always be maintained independent of one another by choosing appropriate DPs. To be able to satisfy the FRs, the designer must always think in terms of FRs before any solution is sought. Robust design is a design that satisfies the FRs easily, although large tolerances are given to DPs and PVs. Decomposition of FRs and DPs can be done by zigzagging between the functional and the physical domains to deal with complex designs and complex systems.

The Information Axiom deals with information content, the probability of satisfying the FRs, and complexity. Information content is defined in terms of the probability of success and is the additional information required to satisfy the FR. Complexity is related to information content, since it is more difficult to meet the design objectives when the probability of success is low. Computing the information content in a design is facilitated by the notion of the design range and the system range. The design range is specified for each FR by the designer, whereas the system range is the resulting actual performance of the design embodiment.

Notes

1. Many commercial products have been developed based on axiomatic design. Some of these can be viewed at the following web sites: www.trexel.com, www.axiomaticdesign.com, www. tribotek-inc.com. Axiomatic design has been used in designing large systems with several hundred or more functional requirements (FRs) and design parameters (DPs).

2. In natural systems, FRs are functions of a natural system and DPs are the physical (chemical or biological) entities that perform the functions. PVs are the physical processes that create the physical entitites.

3. www.axiomaticdesign.com.

4. Although the mathematical formula for information is the same as that used in information theory, the information content in axiomatic design and that in information theory have different significance (see Appendix).

5. These theorem numbers correspond to those given in Suh (2001) (see Appendix).

References

Carrascosa, M., Eppinger, S.D., and Whitney, D.E. 1998. "Using the design structure matrix to estimate product development time," ASME Design Automation Conference, Atlanta, GA, DETC98-6013, September.
Suh, N.P. 1990. *The Principles of Design*, Oxford University Press, New York.
Suh, N.P. 1995. "Design and operation of large systems," *Journal of Manufacturing Systems*, Vol. 14, No. 3, pp. 203–213.
Suh, N.P. 2001. *Axiomatic Design: Advances and Applications*, Oxford University Press, New York.

Appendix: Corollaries and Theorems

Some of these theorems are derived in this book as well as in the references given. For those theorems not derived in this book, the readers may consult the original references.

1. Corollaries

COROLLARY 1 (Decoupling of Coupled Designs)

Decouple or separate parts or aspects of a solution if FRs are coupled or become interdependent in the designs proposed.

COROLLARY 2 (Minimization of FRs)

Minimize the number of FRs and constraints.

COROLLARY 3 (Integration of Physical Parts)

Integrate design features into a single physical part if the FRs can be independently satisfied in the proposed solution.

COROLLARY 4 (Use of Standardization)

Use standardized or interchangeable parts if the use of these parts is consistent with the FRs and constraints.

COROLLARY 5 (Use of Symmetry)

Use symmetrical shapes and/or components if they are consistent with the FRs and constraints.

COROLLARY 6 (Largest Design Ranges)

Specify the largest allowable design range in stating FRs.

COROLLARY 7 (Uncoupled Design with Less Information)

Seek an uncoupled design that requires less information than coupled designs in satisfying a set of FRs.

COROLLARY 8 (Effective Reangularity of a Scalar)

The effective reangularity R for a scalar coupling "matrix" or element is unity. [Note: Reangularity is defined in Suh (1990).]

2. Theorems of General Design

THEOREM 1 (Coupling Due to Insufficient Number of DPs)

When the number of DPs is less than the number of FRs, either a coupled design results or the FRs cannot be satisfied.

THEOREM 2 (Decoupling of Coupled Design)

When a design is coupled because of a larger number of FRs than DPs (i.e., $m > n$), it may be decoupled by the addition of new DPs so as to make the number of FRs and DPs equal to each other if a subset of the design matrix containing $n \times n$ elements constitutes a triangular matrix.

THEOREM 3 (Redundant Design)

When there are more DPs than FRs, the design is a redundant design, which can be reduced to an uncoupled design or a decoupled design, or a coupled design.

THEOREM 4 (Ideal Design)

In an ideal design, the number of DPs is equal to the number of FRs and the FRs are always maintained independent of each other.

THEOREM 5 (Need for New Design)

When a given set of FRs is changed by the addition of a new FR, by substitution of one of the FRs with a new one, or by selection of a completely different set of FRs, the design solution given by the original DPs cannot satisfy the new set of FRs. Consequently, a new design solution must be sought.

THEOREM 6 (Path Independence of Uncoupled Design)

The information content of an uncoupled design is independent of the sequence by which the DPs are changed to satisfy the given set of FRs.

THEOREM 7 (Path Dependency of Coupled and Decoupled Design)

The information contents of coupled and decoupled designs depend on the sequence by which the DPs are changed to satisfy the given set of FRs.

THEOREM 8 (Independence and Design Range)

A design is an uncoupled design when the designer-specified range is greater than

$$\left(\sum_{\substack{i \neq j \\ j=1}}^{n} \frac{\partial FR_i}{\partial DP_j} \Delta DP_j \right)$$

in which case the nondiagonal elements of the design matrix can be neglected from design consideration.

THEOREM 9 (Design for Manufacturability)

For a product to be manufacturable with reliability and robustness, the design matrix for the product, [A] (which relates the FR vector for the product to the DP vector of the product), times the design matrix for the manufacturing process, [B] (which relates the DP vector to the PV vector of the manufacturing process), must yield either a diagonal or a triangular matrix. Consequently, when either [A] or [B] represents a coupled design, the independence of FRs and robust design cannot be achieved. When they are full triangular matrices, either both of them must be upper triangular or both lower triangular for the manufacturing process to satisfy independence of functional requirements.

THEOREM 10 (Modularity of Independence Measures)

Suppose that a design matrix [DM] can be partitioned into square submatrices that are nonzero only along the main diagonal. Then the reangularity and semangularity for [DM] are equal to the product of their corresponding measures for each of the nonzero submatrices. [Note: See Suh (1990).]

THEOREM 11 (Invariance)

Reangularity and semangularity for a design matrix [DM] are invariant under alternative orderings of the FR and DP variables, as long as the orderings preserve the association of each FR with its corresponding DP.

THEOREM 12 (Sum of Information)

The sum of information for a set of events is also information, provided that proper conditional probabilities are used when the events are not statistically independent.

THEOREM 13 (Information Content of the Total System)

If each DP is probabilistically independent of other DPs, the information content of the total system is the sum of the information of all individual events associated with the set of FRs that must be satisfied.

THEOREM 14 (Information Content of Coupled versus Uncoupled Designs)

When FRs are changed from one state to another in the functional domain, the information required for the change is greater for a coupled design than for an uncoupled design.

THEOREM 15 (Design–Manufacturing Interface)

When the manufacturing system compromises the independence of the FRs of the product, either the design of the product must be modified or a new manufacturing process must be designed and/or used to maintain the independence of the FRs of the products.

THEOREM 16 (Equality of Information Content)

All information contents that are relevant to the design task are equally important regardless of their physical origin, and no weighting factor should be applied to them.

THEOREM 17 (Design in the Absence of Complete Information)

Design can proceed even in the absence of complete information only in the case of a decoupled design if the missing information is related to the off-diagonal elements.

THEOREM 18 (Existence of an Uncoupled or Decoupled Design)

There always exists an uncoupled or a decoupled design that has less information than a coupled design.

THEOREM 19 (Robustness of Design)

An uncoupled design and a decoupled design are more robust than a coupled design in the sense that it is easier to reduce the information content of designs than to satisfy the Independence Axiom.

THEOREM 20 (Design Range and Coupling)

If the design ranges of uncoupled or decoupled designs are tightened, they may become coupled designs. Conversely, if the design ranges of some coupled designs are relaxed, the designs may become either uncoupled or decoupled.

THEOREM 21 (Robust Design When the Design Range Has a Nonuniform pdf)

If the probability distribution function (pdf) of the FR in the design range is nonuniform, the probability of success is equal to 1 when the system range is inside the design range.

THEOREM 22 (Comparative Robustness of a Decoupled Design)

Given the maximum design ranges for a given set of FRs, decoupled designs cannot be as robust as uncoupled designs in that the allowable tolerances for the DPs of a decoupled design are less than those of an uncoupled design.

THEOREM 23 (Decreasing Robustness of a Decoupled Design)

The allowable tolerance and thus the robustness of a decoupled design with a full triangular matrix diminish with an increase in the number of functional requirements.

THEOREM 24 (Optimum Scheduling)

Before a schedule for robot motion or factory scheduling can be optimized, the design of the tasks must be made to satisfy the Independence Axiom by adding decouplers to eliminate coupling. The decouplers may be in the form of a queue or of separate hardware or buffer.

THEOREM 25 ("Push" System versus "Pull" System)

When identical parts are processed through a system, a "push" system can be designed with the use of decouplers to maximize productivity, whereas when irregular parts requiring different operations are processed, a "pull" system is the most effective system.

THEOREM 26 (Conversion of a System with Infinite Time-Dependent Combinatorial Complexity to a System with Periodic Complexity)

Uncertainty associated with a design (or a system) can be reduced significantly by changing the design from one of serial combinatorial complexity to one of periodic complexity.

3. Theorems Related to Design and Decomposition of Large Systems

THEOREM S1 (Decomposition and System Performance)

The decomposition process does not affect the overall performance of the design if the highest-level FRs and Cs are satisfied and if the information content is zero, irrespective of the specific decomposition process.

THEOREM S2 (Cost of Equivalent Systems)

Two "equivalent" designs can have substantially different cost structures, although they perform the same set of functions and they may even have the same information content.

THEOREM S3 (Importance of High-Level Decisions)

The quality of design depends on the selection of FRs and the mapping from domain to domain. Wrong selection of FRs made at the highest levels of design hierarchy cannot be rectified through the lower-level design decisions.

THEOREM S4 (The Best Design for Large Systems)

The best design for a large flexible system that satisfies *m* FRs can be chosen among the proposed designs that satisfy the Independence Axiom if the complete set of the subsets of FRs that the large flexible system must satisfy over its life is known *a priori.*

THEOREM S5 (The Need for a Better Design)

When the complete set of the subsets of FRs that a given large flexible system must satisfy over its life is not known *a priori,* there is no guarantee that a specific design will always have the minimum information content for all possible subsets and thus there is no guarantee that the same design is the best at *all times.*

THEOREM S6 (Improving the Probability of Success)

The probability of choosing the best design for a large flexible system increases as the known subsets of FRs that the system must satisfy approach the complete set that the system is likely to encounter during its life.

THEOREM S7 (Infinite Adaptability versus Completeness)

A large flexible system with infinite adaptability (or flexibility) may not represent the best design when the large system is used in a situation where the complete set of the subsets of FRs that the system must satisfy is known *a priori.*

THEOREM S8 (Complexity of a Large Flexible System)

A large system is not necessarily complex if it has a high probability of satisfying the FRs specified for the system.

THEOREM S9 (Quality of Design)

The quality of design of a large flexible system is determined by the quality of the database, the proper selection of FRs, and the mapping process.

4. Theorems for Design and Operation of Large Organizations (Suh, 1995)

THEOREM M1 (Efficient Business Organization)

In designing large organizations with finite resources, the most efficient organizational design is the one that specifically allows reconfiguration by changing the organizational structure and by having flexible personnel policy when a new set of FRs must be satisfied.

THEOREM M2 (Large System with Several Subunits)

When a large system (e.g., organization) consists of several subunits, each unit must satisfy independent subsets of FRs so as to eliminate the possibility of creating a resource-intensive system or a coupled design for the entire system.

THEOREM M3 (Homogeneity of Organizational Structure)

The organizational structure at a given level of the hierarchy must be either all functional or product-oriented to prevent duplication of effort and coupling.

5. Theorems Related to Software Design

THEOREM SOFT 1 (Knowledge Required to Operate an Uncoupled System)

Uncoupled software or hardware systems can be operated without precise knowledge of the design elements (i.e., modules) if the design is truly an uncoupled design and if the FR outputs can be monitored to allow closed-loop control of FRs.

THEOREM SOFT 2 (Making Correct Decisions in the Absence of Complete Knowledge for a Decoupled Design with Closed-Loop Control)

When the software system is a decoupled design, the FRs can be satisfied by changing the DPs if the design matrix is known to the extent that knowledge about the proper sequence of change is given, even if precise knowledge about the elements of the design matrix may not be known.

6. Theorems Related to Complexity

THEOREM C1 (Complexity of an Uncoupled System with Many Interconnected Parts)

Complexity of an uncoupled system with many interconnected parts is not necessarily greater than that of a system with fewer interconnected parts unless the interfaces between the interconnected parts of the uncoupled system increase uncertainty by reducing the overlap between the system range and the design range.

THEOREM C2 (Complexity of a Decoupled System with Many Interconnected Parts)

Complexity of a decoupled system with many interconnected parts is not necessarily greater than that of a system with fewer interconnected parts unless the interfaces between the interconnected parts of the decoupled system increase uncertainty by reducing the overlap between the system range and the design range.

THEOREM C3 (Complexity of a Coupled System with Many Interconnected Parts)

Complexity of a coupled system with many interconnected parts is greater than that of a system with fewer interconnected parts since any variation at the interfaces between the interconnected parts of the coupled system increases uncertainty by reducing the overlap between the system range and the design range.

THEOREM C4 (Complexity of an Uncoupled System with Complicated Arrangement of Parts)

Complexity of an uncoupled system with complicated arrangement of parts is not necessarily greater than that of a system with less complicated arrangement of parts unless the interfaces between the parts of the uncoupled system increase uncertainty by reducing the overlap between the system range and the design range.

THEOREM C5 (Complexity of a Decoupled System with Complicated Arrangement of Parts)

Complexity of a decoupled system with complicated arrangement of parts is not necessarily greater than that of a system with less complicated arrangement of parts unless the interfaces between the parts of the decoupled system increase uncertainty by reducing the overlap between the system range and the design range.

THEOREM C6 (Complexity of a Coupled System with Complicated Arrangement of Parts)

Complexity of a coupled system with complicated arrangement of parts is greater than that of a system with less complicated arrangement of parts since any variation at the interfaces between the parts of the coupled system increases uncertainty by reducing the overlap between the system range and the design range.

THEOREM C7 (Imaginary Complexity of a Decoupled System with Complicated Arrangement of Parts)

The time-independent imaginary complexity of a decoupled system with complicated arrangement of parts can be large if the design parameters (DPs) are not changed in the sequence given by the design matrix.

THEOREM C8 (Complexity of Sociopolitical–Economic Systems)

The complexity of sociopolitical–economic systems increases with the number of entities (i.e., organizations or individuals) that can affect the ultimate outcome.

THEOREM C9 (Reduction of Complexity of Sociopolitical–Economic Systems)

If all the constituents of a social system can agree on the common set of FRs and if the FRs can be satisfied independently, the complexity of the decision-making process can be reduced when the final decision is made by a single entity after understanding and taking into account the uncertainties introduced by other constituents of the system.

THEOREM C10 (Reduction of Complexity of Sociopolitical–Economic System through Reinitialization or Redesign)

When a sociopolitical–economic system is moving into a chaotic state because of time-dependent combinatorial complexity, the system should be reinitialized or redesigned to reduce complexity.

Exercises

2.1. Prove that if each information content term of the right-hand side of equation (2.9) is multiplied by a weighting factor k_i, the total information content will not be equal to information.

2.2. With the result of exercise 2.1 in mind, prove Theorem 16 (Equality of Information Content): All information contents that are relevant to the design task are equally important regardless of their physical origin, and no weighting factor should be applied to them.

2.3. Consider the design of a hot and cold water tap. The functional requirements are the flow rate and the temperature of the water. If we have a faucet that has one valve for hot water and another valve for cold water, the design is coupled since the temperature and flow rate cannot be controlled independently. We can design an uncoupled faucet that has one knob only for temperature control and another knob only for the flow-rate control. Design such an uncoupled faucet by decomposing the FRs and DPs. Integrate the DPs to reduce the number of parts.

2.4. Professor Smith of the University of Edmonton raised the following question about the water-faucet design. If we take the coupled design (i.e., the design with two valves, one for cold water and the other for hot water) and then add a servocontrol mechanism, we may be able to control the flow rate and the temperature independently. Therefore, Professor Smith says that a coupled design is as good as the uncoupled design.

How would you answer Professor Smith's question? Analyze the design proposed by Professor Smith by establishing FRs and DPs, by creating a design hierarchy through zigzagging, and by constructing the design matrices at each level. Is Professor Smith's design coupled, uncoupled, or decoupled?

2.5. In some design situations, we may find that we have to make design decisions in the absence of sufficient information. In terms of the Independence Axiom and the Information Axiom, explain when and how we can make design decisions even when we do not have sufficient information. What kinds of information can we do without and what kinds of information must we have in design? Illustrate your argument using a design task with three FRs as an example.

2.6. Prove Theorem 18, which states that there is always an uncoupled design that has a lower information content than coupled designs.

2.7. Prove Theorems 2, 6, 15, and 16.

2.8. A surgical operating table for hospitals is to be designed. The position of the table must be adjustable along the horizontal and the vertical directions as well as the inclination of the table. Design a mechanism that can satisfy these functional requirements.

If the functional requirements of the table are modified so that the table has to change from one fixed position (i.e., fixed horizontal, vertical, and inclination) to another fixed position, how would you design the mechanism?

2.9. One of the major problems in the automobile business is the warranty cost associated with the weatherstrip. It is typically made of extruded rubber to prevent dust, water, and noise from coming into the vehicle. The weatherstrip also affects the force required to close the door. One of the problems identified is that the gap between the door and the body can vary from about 10 to 20 mm. Design the weatherstrip.

2.10. Compare elements of axiomatic design theory to those of other design methodologies, specifically Quality Function Deployment (QFD), Robust Design (Taguchi methods), and Pugh Concept Selection. Where do they agree and where do they differ?

2.11. The two linear equation sets below describe two designs. Each of the design matrices can be made uncoupled or decoupled depending on whether the variable x is set to 0 or 1, respectively. Analytically compute the probability of success of each of the designs for each value of x (four cases in total). All distributions are uniform. What can you conclude about the relationship between information content and coupling?

$$\begin{Bmatrix} -1 < FR_1 < 1 \\ -1 < FR_2 < 1 \end{Bmatrix} = \begin{bmatrix} 1 & 0 \\ x & 1 \end{bmatrix} \begin{Bmatrix} 0 < DP_1 < 2 \\ 0 < DP_2 < 2 \end{Bmatrix}$$

$$\begin{Bmatrix} -1 < FR_1 < 1 \\ 1.5 < FR_2 < 3.5 \end{Bmatrix} = \begin{bmatrix} 1 & 0 \\ x & 1 \end{bmatrix} \begin{Bmatrix} 0 < DP_1 < 2 \\ 0 < DP_2 < 2 \end{Bmatrix}$$

2.12. The equation below describes a design with two DPs and two FRs. The first DP has a uniform distribution and the second DP has a normal $(2, 0.8)$ distribution. Write a short program (in MATLAB, for example) that numerically computes the probability of success of this design, and plot $FR(DP_1, DP_2)$.

$$\begin{Bmatrix} 2 < FR_1 < 5 \\ 1.5 < FR_2 < 3.5 \end{Bmatrix} = \begin{bmatrix} 0.7 & 1.6 \\ 1.1 & 0.5 \end{bmatrix} \begin{Bmatrix} 1 < DP_1 < 3 \\ DP_2 = 2, \ \sigma = 0.8 \end{Bmatrix}$$

2.13. Given a system with m independent events with probability of success P_i, prove that the total information contents is the sum of individual information contents of these events.

2.14. Prove that the variance of decoupled design is always less than that of a coupled design (see equation (2.18)).

3

▆▆ Complexity Theory Based on Axiomatic Design

3.1   Introduction: A Functional View of Complexity

This chapter is a treatise on the fundamental principles of the complexity theory presented in this book. The concepts and theory presented in this chapter form the foundation for the subsequent chapters. In chapters 4 through 10, the basic ideas presented in this chapter are illustrated, amplified, and applied to a variety of different problems, ranging from engineering and biology to socioeconomic–political issues.

Complexity is measured in the functional domain rather than in the physical domain, which distinguishes this theory from other complexity theories and yields a unique and coherent perspective on complexity. Complexity is narrowly defined as a measure of uncertainty in achieving the functional requirements (FRs) of a system. Therefore, it is a relative measure—there is no absolute complexity. As a result of this definition and the relative nature of complexity, many different types of complexity in both engineered systems and natural systems can be generally treated using the same theory.

The complexity that follows directly from the Information Axiom of axiomatic design is time-independent real complexity, which is the same as the information content defined in axiomatic design. In addition to the time-independent real complexity, there are three other types of complexity: time-independent imaginary complexity, time-dependent combinatorial complexity, and time-dependent periodic complexity.

An important result of this complexity theory is the discovery that there exists functional periodicity in stable systems. Functional periodicity is defined by a set of FRs that repeats on a periodic basis. There are many different kinds of functional periodicity: temporal, geometric, material biological, chemical, electrical, manufacturing/ processing, thermal, and informational functional periodicity.

When a functional periodicity is introduced to an engineered system, it reduces the time-dependent complexity of the system and makes the system stable. Many natural systems—atoms, biological systems, and so on—also have a functional periodicity, which appears to be a fundamental requirement for the long-term stability of natural systems.

Based on these observations, it is concluded that a system—both engineered and natural systems—must be either at an equilibrium state or must have a functional periodicity for long-term stability. Therefore, functional periodicity is a fundamental requirement for the long-term stability of both natural and engineered systems. It is a *basic property* of a stable system.

The stability argument presented in this chapter may provide a basis for explaining the underlying assumptions of classical and modern physics. An attempt is made to explain the equilibrium-based Newtonian mechanics and thermodynamics, the particle/ wave duality assumed in quantum mechanics, and the fundamental assumption of string theory based on the complexity theory, which stipulates the existence of functional periodicity in a stable system.

In the next section, the implications of the design axioms on complexity are presented, since the notion that uncertainty must be measured in the functional domain (not the physical domain) through the mapping of FRs and DPs is derived from axiomatic design.

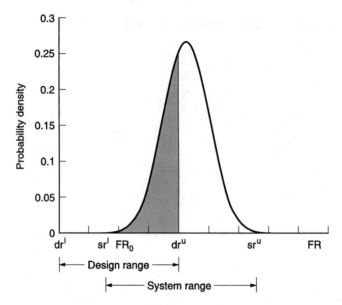

Figure 3.1 Design range, system range, and system pdf. The probability of satisfying the FR is given by the shaded area under the system pdf in the design range. dr^l and dr^u are the lower and upper bounds of the design range; sr^l and sr^u are the lower and upper bounds of the system range.

3.2   Axiomatic Design and Complexity

In axiomatic design, the design process is described in terms of the mapping between domains (Suh, 1990, 2001). The design goals for a product (software, systems, etc.) are described in terms of FRs (i.e., "what we want to achieve") in the functional domain. In the case of natural systems, FRs are "what we want to know."

The design task is to achieve the set of specified FRs by mapping these FRs to DPs in the physical domain (see figure 2.1, page 21 above). Thus, the selection of DPs determines the probability of satisfying the FRs. The Independence Axiom specifies that DPs must be chosen so that the resulting design is either uncoupled or decoupled, which are characterized by a diagonal and a triangular design matrix, respectively.

When the FR is defined, its desired target value FR_0 and its design range are specified, as shown in figure 3.1. However, the actual probability density function (pdf) of the resulting design embodiment determines the system range, which may be different from the design range, as discussed in chapter 2. The portion of the design range overlapped by the system range is called the common range.

If the system pdf for a given FR_i is denoted $p_s(FR_i)$, then the probability P_i of satisfying FR_i is given by

$$P_i(dr^l \leq FR_i \leq dr^u) = \int_{dr^l}^{dr^u} p_s(FR_i)d(FR_i) \tag{3.1}$$

where dr^l and dr^u are the lower and upper limits of the design range, respectively.

Information content I_i is defined in terms of the probability P_i of satisfying a given FR$_i$ as

$$I_i = -\log_2 P_i \tag{3.2}$$

In the general case of m FRs, the information content for the entire system I_{sys} is

$$I_{\text{sys}} = -\log_2 P_{\{m\}} \tag{3.3}$$

where $P_{\{m\}}$ is the joint probability that all m FRs are satisfied.

When all FRs are statistically independent, as is the case for an uncoupled design, $P_{\{m\}} = \Pi_i P_i$. Therefore I_{sys} may be expressed as

$$I_{\text{sys}} = \sum_{i=1}^{m} I_i = -\sum_{i=1}^{m} \log_2 P_i \tag{3.4}$$

When all FRs are not statistically independent, as is the case for a decoupled design,

$$I_{\text{sys}} = -\sum_{i=1}^{m} \log_2 P_{i|\{j\}} \qquad \{j\} = \{1, 2, \ldots, i-1\} \tag{3.5}$$

where $P_{i|\{j\}}$ is the conditional probability that FR$_i$ is satisfied given that the other relevant (i.e., correlated) $\{\text{FR}_j\}, j = 1, \ldots, i-1$, are also satisfied.

In some systems such as software systems, the probability of satisfying the FRs is binary, either zero or 1. When it is zero, we do not get any useful output. When it is 1, it satisfies the FR. In that case, the probability P is given by the number of successful outcomes (1's) divided by the number of trials (Suh, 2001).

Because the system has a fixed number of FRs, complexity is unrelated to the number of FRs, but instead is related to the probability that a system will achieve all of its FRs. A system with low total I (high probability of satisfying all FRs) is less complex than a system with the same number of FRs and DPs, but with high total I (low probability of satisfying all FRs). This leads us to a specific definition of complexity.

Information content as a function of the absolute scale
In some systems, the probability of success decreases for a given design range when the absolute magnitude of the FR increases (Wilson, 1980; Suh, 1990). For example, the probability of measuring the distance between two points within a given tolerance decreases as the distance (i.e., absolute scale or range) increases if a traditional measuring instrument is used (rather than a satellite-based system). This becomes an issue in dealing with nanomanufacturing, where nanoscale objects must be connected to macroscale systems. Also the probability of creating an error message increases as the length of the message increases. In many of these situations, when the total range (e.g., length) is only one or two orders of magnitude larger than the design range (i.e., allowable tolerance), the system range can be easily made to lie in the design range. However, when the range is six or more magnitudes larger than the design range, the common range may be difficult to establish, increasing the complexity of the task.

3.3   Definition of Complexity

Complexity is defined as a measure of uncertainty in achieving the specified FRs.

Uncertainty may be a result of poor design of engineered systems or a result of not understanding the system. In the former case, complexity is a function of the relationship between the design range and the system range just as information content is. When there are many FRs that a system must satisfy at the same time, the quality of design in terms of the independence of FRs affects the uncertainty of satisfying the FRs. The uncoupled design is likely to be least complex. However, the complexity of a decoupled design can be high because of imaginary complexity if we do not understand the system—it is not really complex, but appears to be complex because of our lack of understanding.

Complexity can be a function of time or can be completely independent of time, depending on whether or not the system range changes as a function of time (Suh, 1999). Therefore, complexity can be classified into the following two kinds: *time-dependent complexity* and *time-independent complexity*.

Time-independent complexity consists of two orthogonal components: time-independent *real complexity* and time-independent *imaginary complexity*. Their vector sum is called *total time-independent complexity*. As the terms imply, real complexity is a result of not satisfying the FR at all times. On the other hand, a system with imaginary complexity is designed to satisfy the FRs at all times, but we lack understanding of the system and thus are not able to make the system satisfy the FRs.

In general, the real complexity of coupled designs is larger than that of uncoupled or decoupled designs. The presence and the magnitude of the off-diagonal element (i.e., the coupling terms) of the design matrix will increase the uncertainty if the system range is not inside the design range. The uncoupled design does not have any imaginary complexity since such a system does not depend on an understanding of the relationship among the FRs (i.e., no coupling terms). On the other hand, a decoupled design can have imaginary complexity. Imaginary complexity can be reduced when the design matrix is known. When FRs are not clearly stated or when design equations are not clearly written, one can waste time and effort trying to figure out how the system should work. Many engineers waste time because of the imaginary complexity created by a nonsystematic approach to design.

When the system range moves as a function of time, there are two types of time-dependent complexity: time-dependent *combinatorial complexity* and time-dependent *periodic complexity*. Time-dependent combinatorial complexity can lead to a chaotic situation if the number of combinations continues to explode as a function of time or if the underlying physical phenomenon continues to move the system range away from the design range. On the other hand, time-dependent periodic complexity reduces the number of combinations to a finite set and may reduce the complexity problem to a deterministic one.

The introduction of functional periodicity into a system that has time-dependent combinatorial complexity may substantially reduce the uncertainty of satisfying the FR. Functional periodicity simplifies the design by making it deterministic, which requires much less information. Whenever a system with time-dependent combinatorial complexity is converted to a system with time-dependent periodic complexity, uncertainty is reduced and the design is simplified. Nature, through the process of evolution, consists of stable systems that possess a variety of functional periodicity.

3.4 Time-Independent Complexity: Real Complexity, Imaginary Complexity, and Total Complexity

Time-independent complexities—*real complexity* and *imaginary complexity*—are defined to deal with *real* uncertainty and *imaginary* uncertainty, respectively. Imaginary complexity and real complexity are orthogonal to one another. *Total time-independent complexity* is defined as the vector sum of these two orthogonal components of time-independent complexity.

3.4.1 Real complexity

Real complexity is defined as a measure of uncertainty when the probability of achieving the FR is less than 1.0 because the system range is not identical to the design range. In figure 3.1, the uncertainty is given by the unshaded area under the system pdf. Real uncertainty in design exists because the actual embodiment of the design does not quite satisfy the desired FR within the design range at all times.

The probability of achieving a given FR is determined by the overlap between the design range and the system range, called the common range (figure 3.1). Therefore, real uncertainty exists, even when the Independence Axiom is satisfied, if the system range is not the same as the design range. Thus, real complexity can be related to the information content, which was defined in terms of the probability of success of achieving the desired set of FRs (see equation (3.3)). If we denote real complexity as C_R, then we will define real complexity to be equal to the information content as

$$C_R = I \tag{3.6}$$

The information content is a measure of uncertainty and thus is related to real complexity.[1]

Time-dependent real complexity may be reduced when the design is either uncoupled or decoupled, that is, when the design satisfies the Independence Axiom. For uncoupled designs, the system range for each FR can be shifted horizontally by changing the DPs until the information content is at a minimum, since other FRs are not affected by such a change. Therefore, the mean value of FR provided by the system can be determined by adjusting the corresponding DP until the information is at a minimum. For decoupled designs, the system range can be shifted to seek the minimum information point by changing the DPs in the sequence given by the design matrix. The best values of DPs can be obtained by finding where the value of real complexity reaches its minimum when the following two conditions are satisfied:

$$\sum_{j=1}^{n} \frac{\partial C_R}{\partial \mathrm{DP}_j} = 0 \tag{3.7}$$

$$\sum_{j=1}^{n} \frac{\partial^2 C_R}{\partial DP_j^2} > 0 \qquad (3.8)$$

When the design is uncoupled, the solution to equations (3.7) and (3.8) can be obtained for each DP without regard to any of the other DPs, that is, each term of the series must be equal to zero. In the case of decoupled designs, these equations must be evaluated in the sequence given by the design equation because the design matrix is triangular.

In the case of a coupled design, real complexity can also be changed, but the minimum information point for each FR is no longer meaningful, because when one of the DPs is changed in order to affect only one FR, all other FRs may change. Therefore, the minimum information point is defined only for the entire set of DPs where the information for the entire set of FRs is the minimum. This corresponds to an "optimum" point, which is often sought in operations research. However, this is a poor design solution since all of the FRs can be satisfied exactly in the design space if the Independence Axiom is satisfied. In many cases of coupled design, equation (3.7) may never be satisfied. Coupled designs have larger real complexity than uncoupled or decoupled designs for the same set of FRs.[2]

EXAMPLE 3.1: COMPLEXITY OF A SYSTEM WITH MANY INTERCONNECTED PARTS

According to Webster's dictionary, the word "complex" is defined as:

1. Composed of many interconnected parts.
2. Characterized by a very complicated or involved arrangement of parts, units, etc.
3. So complicated or intricate as to be hard to understand or deal with.

Are these definitions given in the dictionary and the definition given in this theory of complexity consistent?

Solution

In the complexity theory presented in this book, complexity is defined as a measure of uncertainty in satisfying FRs. Therefore, time-independent real complexity is determined by the overlap between the design range and the system range. If the system range is always completely inside the design range, FRs are satisfied at all times, and therefore the real complexity is zero. Based on this understanding of complexity, are all systems with many connecting parts more complex?

If the interface between the interconnected parts adds additional uncertainty in satisfying the FRs, the Webster's dictionary and the definition of complexity used in this complexity theory are consistent. However, the mere presence of many interconnected parts does not necessarily make a system more complex, if the interconnected parts do not add any additional uncertainty.

If the interconnecting parts are additional DPs, the effect of these additional DPs can be compensated by using one of the DPs to cancel out the uncertainty introduced by all other DPs.[3] If the interconnecting parts are simply the source of noises, they may or may not be compensated by taking active measures to eliminate the source of uncertainty.

One of the definitions of complexity given by Webster's dictionary is that a complex system has "a very complicated or involved arrangement of parts and units." Such a complicated arrangement may contribute to uncertainty in satisfying the FRs, which increases the real complexity. However, if the uncertainty does not increase, complexity is not greater, in spite of the fact that the system has a complicated arrangement of parts. Also if a given system is "complicated or intricate so as to be hard to understand or deal with," the complexity may increase, but the complexity that increases may be time-independent imaginary complexity.

When a system is a coupled design, the complexity is expected to be greater when there are interconnected parts, since any variation in any one of the connections is likely to move the system range out of the design range.

When a system is an uncoupled or decoupled design, the uncertainties introduced by the interconnections can be compensated by changing one of the DPs to correct for the uncertainties introduced by others. This will be discussed further in chapter 4.

Based on these observations, we may state the following theorems:

THEOREM C1 (Complexity of an Uncoupled System with Many Interconnected Parts)

Complexity of an uncoupled system with many interconnected parts is not necessarily greater than that of a system with fewer interconnected parts unless the interfaces between the interconnected parts of the uncoupled system increase uncertainty by reducing the overlap between the system range and the design range.

THEOREM C2 (Complexity of a Decoupled System with Many Interconnected Parts)

Complexity of a decoupled system with many interconnected parts is not necessarily greater than that of a system with fewer interconnected parts unless the interfaces between the interconnected parts of the decoupled system increase uncertainty by reducing the overlap between the system range and the design range.

THEOREM C3 (Complexity of a Coupled System with Many Interconnected Parts)

Complexity of a coupled system with many interconnected parts is greater than that of a system with fewer interconnected parts since any variation at the interfaces between the interconnected parts of the coupled system increases uncertainty by reducing the overlap between the system range and the design range.

THEOREM C4 (Complexity of an Uncoupled System with Complicated Arrangement of Parts)

Complexity of an uncoupled system with a complicated arrangement of parts is not necessarily greater than that of a system with a less complicated arrangement of parts unless the interfaces between the parts of the uncoupled system increase uncertainty by reducing the overlap between the system range and the design range.

THEOREM C5 (Complexity of a Decoupled System with Complicated Arrangement of Parts)

Complexity of a decoupled system with a complicated arrangement of parts is not necessarily greater than that of a system with a less complicated arrangement of parts unless the interfaces between the parts of the decoupled system increase uncertainty by reducing the overlap between the system range and the design range.

THEOREM C6 (Complexity of a Coupled System with Complicated Arrangement of Parts)

Complexity of a coupled system with a complicated arrangement of parts is greater than that of a system with a less complicated arrangement of parts since any variation at the interfaces between the parts of the coupled system increases uncertainty by reducing the overlap between the system range and the design range.

THEOREM C7 (Imaginary Complexity of a Decoupled System with Complicated Arrangement of Parts)

The time-independent imaginary complexity of a decoupled system with complicated arrangement of parts can be large if the design parameters (DPs) are not changed in the sequence given by the design matrix.

When a system is a coupled design, the only way to reduce time-independent real complexity is by designing an uncoupled design that satisfies the same set of FRs. The following example involving an internal-combustion (IC) engine illustrates how one might develop a decoupled engine design.

EXAMPLE 3.2: REAL COMPLEXITY OF SPARK-IGNITION IC ENGINE

Consider a conventional four-stroke cycle spark-ignition internal-combustion engine that most of us drive, which is shown in figure E3.2a. It typically has a manifold port injection system (MPI), which injects gasoline to the manifold just outside of the intake valve of the power cylinder. When the valve opens and the piston moves downward, the fuel and air mixture is drawn into the chamber by the low pressure in the cylinder. When the piston is about to reach the bottom dead center (BDC), the intake valve closes and the piston compresses the mixture as the piston moves upward with the exhaust valve closed. Just before the piston reaches the top dead center (TDC), a spark ignites the fuel/air mixture, which is further compressed. With the combustion of the fuel/air mixture when the piston is near the TDC, the piston is pushed downward by the pressure of the combustion product. When the piston almost reaches the BDC, the exhaust valve opens and the combustion product is exhausted, which carries with it the small amount of fuel vapor that did not go into the cylinder when the intake valve first opened.

When the engine is cold, the gasoline does not evaporate completely and some goes into the cylinder as liquid droplets, which coat the surface of the cold cylinder wall. These liquid droplets do not combust completely since they evaporate near the end of the combustion cycle. They are exhausted as partially oxidized hydrocarbon. When the engine gets hot, the hydrocarbon exhausted by the engine decreases, which is further reduced by the catalytic converter.

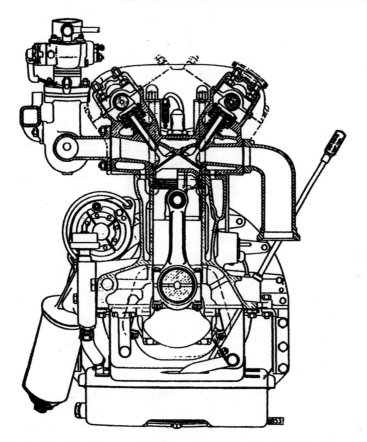

Figure E3.2a A conventional four-stroke cycle spark-ignition internal-combustion engine. (Courtesy of Ford Jaguar Cars.)

To minimize the emission of the hydrocarbon, we use a palladium catalyst in a catalytic converter, which is attached close to the exhaust port of the engine. The catalytic converter is not effective when it is cold. Therefore, the emission of hydrocarbon is greatest during the first 3 minutes after the engine is started.

The allowable nonmethane hydrocarbon emission mandated by the U.S. government is about 0.25 grams per mile. The Federal government has also established the fuel efficiency of passenger cars, but it cannot be directly related to the performance of an engine because fuel efficiency is a function of vehicle weight, shape, tire pressure, and many other factors.

Determine the real complexity of the engine in meeting the emission standard of the U.S. government at 1 minute after the engine is turned on.

Solution

To determine the time-independent real complexity, we have to establish the design range of the FR for hydrocarbon emission and compare that against the actual emission level by establishing the system range under a variety of operating conditions. The real complexity can be computed by determining the overlap (common range)

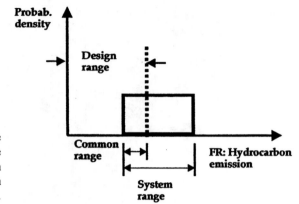

Figure E3.2b Design range
and system range
of a spark-ignition
internal-combustion
engine (hypothetical case).

between the design range and the system range, as illustrated in figure E3.2b. The real complexity is given by

$$C_R = -\log_2 A_c$$

where A_c is the area of the system pdf over the common range and the total area of the system pdf is equal to 1.

We should note the following:

1. As long as the design does not change, the system range is not going to change. Therefore, the real complexity will not be reduced.
2. The real complexity will increase if the emission standard lowers the allowable emission level.
3. When there are more FRs that we have to satisfy, the total complexity of the engine will be the sum of the complexity for each FR.
4. The real complexity of the engine will depend on the temperature of the engine, which changes as a function of time. However, it may be assumed that the emission of hydrocarbon is reproducible under identical conditions.

Real complexity and the scale issue
During the last decade, nanotechnology has become a topic of intense interest in academia and industry because of its potential impact on various technologies and engineered systems. However, the interest in nanoscience and nanotechnology has been around for many decades, since many natural systems such as biological systems consist of nanoscale elements. Until recently, however, we have not had the tools with which to construct or investigate nanoscale systems.

As the scale becomes smaller, the corresponding design range also becomes smaller.[4] Conversely, in dealing with nanoscale entities, the range or scale of FRs and DPs must be made small, since the probability of achieving the FRs or DPs decreases as the ratio (range/tolerance) increases. Therefore, the information content and thus the time-independent real complexity of nanotechnologies can be large with the magnitude of the scale unless the range (i.e., scale) is made correspondingly small, that is, the ratio (range/tolerance) is maintained within a limit.

The practical implication of real complexity associated with nanotechnology is that the overall scale must be made small. Therefore, macro- or microdevices that use nanoscale components must be made up of an integrated system consisting of a large number of microscale entities to be able to satisfy the FR within the design range. In engineering, the ratio of (range/tolerance) that could be determined using typical engineering instruments was of the order of 10^4 in typical manufacturing operations to 10^6 in semiconductor and microelectronics processing (Kim, 2004). This observation leads to Theorem C8:

THEOREM C8 (Complexity of a System that Crosses Many Scales)

When a system must be integrated across many length (or time) scales, it must be divided into smaller subunits, which are then integrated, to minimize the information content and reduce the time-independent real complexity.

3.4.2 Imaginary complexity

Imaginary complexity is defined as uncertainty that is not real uncertainty, but arises because of the designer's lack of knowledge and understanding of a specific design itself. Even when the design is a good design, consistent with both the Independence Axiom and the Information Axiom, imaginary (or unreal) uncertainty exists when we are ignorant of what we have. For example, a combination lock is easy to open once we know the sequence of numbers we have to activate, but in the absence of the information on the combination, it would appear to be complex. This uncertainty, which is not real but associated with the lack of knowledge, is defined as the imaginary complexity.

To understand the distinction between real and imaginary uncertainty, consider a decoupled design with m FRs and n DPs given by the triangular matrix in equation (3.9) where $m = n$:

$$\begin{Bmatrix} FR_1 \\ FR_2 \\ FR_3 \\ \cdots \\ \cdots \\ \cdots \\ FR_m \end{Bmatrix} = \begin{bmatrix} X & 0 & 0 & 0 & \cdots & 0 \\ X & X & 0 & 0 & \cdots & 0 \\ X & X & X & 0 & \cdots & 0 \\ \cdots & \cdots & \cdots & \cdots & \cdots & 0 \\ \cdots & \cdots & \cdots & \cdots & \cdots & 0 \\ \cdots & \cdots & \cdots & \cdots & \cdots & 0 \\ X & X & X & X & \cdots & X \end{bmatrix} \begin{Bmatrix} DP_1 \\ DP_2 \\ DP_3 \\ \cdots \\ \cdots \\ \cdots \\ DP_n \end{Bmatrix} \tag{3.9}$$

This may be generally written as

$$\{FR\} = [A^{LT}]\,\{DP\} \tag{3.10}$$

where $[A^{LT}]$ is a lower triangular matrix.

The design represented by equation (3.9) satisfies the Independence Axiom. Thus it can be implemented because there is no uncertainty associated with it if the DPs are changed in the order indicated in equation (3.9) and if each of the system ranges is inside its associated design range.[5] If the system range is the same as the design range for all FR_i, then the real complexity is equal to zero. If the system range is not the same as the design range for any of the FR_i, there is real uncertainty and real complexity. This real complexity cannot be removed unless the system range and the design range are made to be the same for all FRs by choosing new DPs or by making the design more robust so as to remove uncertainty.

The decoupled design given by equation (3.9) can be a source of *imaginary complexity*, despite the fact that the design does satisfy the Independence Axiom and the real complexity is zero. Imaginary complexity exists whenever the perceived complexity is not entirely due to real complexity. This imaginary uncertainty exists only in the mind of the designer because the designer does not know that the design represented by equation (3.9) is a good design or when the designer does not write the design equation.

Suppose the designer does not recognize that the design is decoupled, although it is represented by equation (3.9), and thus does not know that the DPs must be changed in a proper order to make the design achieve the given set of m FRs. Then the designer resorts to trial-and-error methods of evaluation, trying many different sequences of DPs to satisfy the FRs. There are $n!$ distinct sequences of DPs, of which only one is correct. Then the probability of finding the right sequence of n DPs to satisfy the entire set of m FRs is given by

$$P = \frac{1}{n!} \qquad (3.11)$$

The probability of finding the right sequence through a random trial-and-error process goes down rapidly with an increase in the number of DPs, as shown by the table below. When n is 5, the probability of finding the right sequence is 0.008, which is quite low. Therefore, this design appears to be very complicated and one would say that it is very complex because the uncertainty is large. However, it is not a case of real uncertainty; this uncertainty is artificially created by lack of understanding of the system designed. Therefore, this kind of uncertainty is defined as *imaginary uncertainty*. Because of lack of fundamental understanding of axiomatic design theory, this imaginary uncertainty leads to the erroneous conclusion that a design is complex, although it may not be. When this type of mistake is made, the cost and the time taken for new product development are expected to be greater than when the imaginary complexity is absent.

n	$n!$	$P = 1/n!$
1	1	1
2	2	0.5
3	6	0.1667
4	24	0.04167
5	120	0.8333×10^{-2}
6	720	0.1389×10^{-2}
7	5,040	0.1984×10^{-3}

If we denote imaginary complexity as C_I, then the maximum imaginary complexity may be related to the probability of finding the right sequence given by equation (3.11) as

$$(C_I)_{max} = \log n! \qquad (3.12)$$

For very large n ($n > 100$), equation (3.12) may be written as

$$C_I \approx n(\log n - 1) \qquad (3.13)$$

The imaginary complexity given by equation (3.12) is the maximum imaginary complexity. If the design matrix is such that there is more than one possible sequence of n DPs that can equally satisfy the m FRs, then the probability of finding an appropriate sequence is given by

$$P = z/n! \tag{3.14}$$

where $z =$ the number of sequences that will satisfy the FRs. Therefore, as z increases, the design will appear to be less complex because the imaginary uncertainty decreases. However, the real uncertainty does not change with z.

Time-independent imaginary complexity exists only in a decoupled system. When the system is uncoupled, there is no imaginary complexity because the FRs can be satisfied in any order. When the system is coupled, it is difficult to satisfy the FRs when there is also a random variation of DPs.

In natural sciences, interactions at the physical and biological entities are studied to understand the behavior of nature. Even in these situations, imaginary complexity can exist when a set of DPs is chosen for investigation without identifying the FRs that correspond to the DPs if the specific nature behaves as a decoupled design. Sometimes, the relationship represented by an off-diagonal element between an FR and a DP is studied extensively, although it is not the main DP for the given FR, which may add to imaginary complexity.

EXAMPLE 3.3: XEROGRAPHY-BASED PRINTING MACHINE

HG Company, one of the leading printing-press manufacturers in the world, has just developed a commercial label-printing machine based on a xerography technique. This machine can quickly print commercial labels as soon as the original copy is inserted into the machine because it is based on the xerography principle. The design of the machine is schematically illustrated in figure E.3.3a.

The optical image of the label is transmitted to the surface of the selenium-coated aluminum cylinder using light. The cylinder rotates at a constant speed. When the charged section of the cylinder passes by the toner box, the oppositely charged liquid toner transfers to the charged part of the selenium surface. To control the thickness of the toner layer on the selenium drum, the wiper roll removes

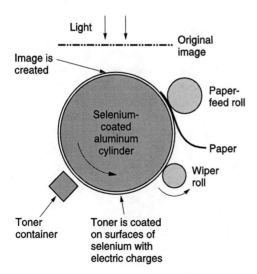

Figure E3.3a Schematic drawing of the xerography-based printing machine. The image is transmitted to the selenium-coated aluminum cylinder using light. When the charged section of the cylinder passes by the toner box, the oppositely charged toner liquid transfers to the charged part of the selenium surface. The wiper roll removes the extra-thick toner layer from the surface of the cylinder. Paper comes in contact with the selenium surface under the light pressure exerted on the paper by the paper-feed roll.

the extra-thick toner layer from the surface of the cylinder. Paper is fed into the gap between the main selenium cylinder and the paper-feed roll. When the paper comes in contact with the selenium surface under the light pressure exerted on the paper by the paper-feed roll, the image is firmly printed on the paper.

The Advanced Engineering Division of HG Co., which was developing this printing machine, ran into trouble. They found that sometimes the selenium coating is badly scratched, creating poor images and damaging the expensive selenium-coated rolls (about $4000 per cylinder, which was about 18 inches in diameter). Since the beta machine had to be shipped in a few months, they assigned many scientists and engineers to figure out the problem and solve it.

The scientists and engineers came to the conclusion that the scratch marks (in the form of lines) must have been the result of abrasive wear. They attributed the source of abrasion to unknown abrasive particles that somehow got into the toner tank. This reasoning received much internal support from everyone in the Advanced Engineering Division, since the machine (which was about 30 feet long) was being assembled at a corner of a large machine shop. They conjectured that tiny metal chips from the machining operation somehow got into the tank, occasionally scratching the selenium drum.

To make sure that the toner was free of any abrasive particles, they installed special filters that would remove all particles greater than a few microns and put a plastic sheet around the machine to create a clean environment. However, the despicable scratch marks would not go away! The high-level managers of the company became uneasy about the situation and decided to consult a tribologist at MIT about this abrasive wear problem. The tribologist told them to read a reference book on tribology to learn all about the things that affect abrasive wear.

After a few months, the tribologist received an urgent call from HG Co. They said that they have to ship the beta machine to a customer's factory in a week and yet the scratch marks were still there—apparently the reference book did not do any good! The tribologist was asked to hop on an airplane right way and visit the factory where the machine was being tested. So he went.

What do you think the tribologist found at the HG Company?

Solution

The tribologist, who also knows something about axiomatic design, listened to the HG engineers and scientists, who explained all the things they had done and their theory of the cause of the problem. They were sure that somehow devilish small particles were getting into the printing machine and the toner box and that it was these particles that caused the scratches on the surface. Indeed, the examination of the surface and micrographs indicated that the scratch marks were typical scratches caused by abrasive particles. However, the tribologist was not convinced that the explanation given by the HG engineers and scientists was correct.

The FRs of the machine, assuming that abrasive particles somehow got into the toner box, may be chosen to be:

FR_1 = Create electrically charged images.
FR_2 = Coat the charged surface with toner.

$FR_3 =$ Wipe off the excess toner.
$FR_4 =$ Make sure that abrasive particles do not cause abrasion.
$FR_5 =$ Feed the paper.
$FR_6 =$ Transfer the toner to the paper.
$FR_7 =$ Control throughput rate.

The tribologist reasoned that the DPs used by HG personnel (although they did not use axiomatic design) in their trial-and-error approach were:

$DP_1 =$ Optical system with light on selenium surface.
$DP_2 =$ Electrostatic charges of the selenium surface and the toner.
$DP_3 =$ Wiper roller.
$DP_4 =$ Filter.
$DP_5 =$ Paper-feeding mechanism.
$DP_6 =$ Mechanical pressure.
$DP_7 =$ Speed of the cylinder.

Since there are seven FRs and seven DPs, there are more than 5000 sequences of DPs to consider if they try to run the tests by trying different sequences of DPs. The probability of success using a trial-and-error method is quite small. Even if they devised an orthogonal array experiment, there are still too many tests to determine the cause. Furthermore, if the design is a decoupled design, simply identifying important DPs through the orthogonal array experiment will not yield the answer. Indeed their extensive tests did not yield any solution!

The design matrix that represents the thinking of the HG engineers may be represented as

	DP_1	DP_2	DP_3	DP_4	DP_5	DP_6	DP_7
FR_1	X	0	0	0	0	0	0
FR_2	X	X	0	0	0	0	0
FR_3	0	0	X	0	0	0	0
FR_4	0	0	X	X	X	0	0
FR_5	0	0	0	0	X	0	0
FR_6	0	0	0	0	0	X	0
FR_7	0	0	0	0	X	0	X

According to the above design matrix, the order of FR_4 and FR_5 as well as DP_4 and DP_5 should be changed to obtain a triangular matrix. What the matrix is saying is that if the paper-feeding mechanism or process creates particles, filtering the toner outside the machine will not do any good. The filter must also remove the particles generated by the paper-feeding mechanism. This is not easy to achieve.

Another solution is to prevent large particles from ever approaching the interface by means of controlling the fluid motion. For abrasion to occur, kinematic considerations indicate that somehow the abrasive particle, whatever it may be made of, must be stationary at the interface between the main cylinder and the wiper roll. If the particle goes through, then, at most, the selenium surface will be indented rather than scratched. Then FR_4 (make sure that abrasive particles do not cause abrasion) may be decomposed as:

FR_{41} = Prevent the abrasive particle from being anchored at the interface between the main cylinder and the wiper roll.

FR_{42} = Prevent the particles from approaching the interface.

At this point, it is instructive to consider the kinematics and fluid mechanics of the toner motion near the entrance between the wiper roll and the main roll.[6] When the machine is first started, if the main cylinder rotates first before the wiper roller is rotated, the toner will be dragged along and any particle in the toner will anchor at the narrow section of the opening between the roller and the main cylinder. Furthermore, if the surface speed of the main cylinder is greater than that of the counterrotating wiper roller, the pressure at the narrow gap will be greater and the tendency to squeeze in the abrasive particle at the interface between the main cylinder and the wiper roller will be greater.

On the other hand, if the wiper roller starts turning first and if the surface speed of the wiper roller is greater than and opposite to the surface speed of the main cylinder (as indicated in figure E3.3b), then the pressure at the entrance will be less. It will reduce the tendency for large particles to come into the narrow gap. Furthermore, the vortex motion in the toner will prevent the large particles from approaching the main cylinder/wiper interface, as shown in figure E3.3b.

Then DPs may be chosen as

DP_{41} = The order of rotation of the wiper roller and the main cylinder (wiper roller rotates first).

DP_{42} = The surface speed of the wiper roller greater than and opposite to the surface speed of the main cylinder.

The tribologist made the suggestion that DP_{41} and DP_{42} be implemented. The machine had a digital control system, and therefore they could be implemented immediately. He also asked the HG engineers to put abrasive particles into the toner intentionally. When the machine was turned on, there were no more scratch marks!

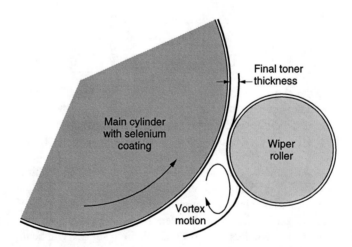

Figure E3.3b The vortex motion of the toner and the rotational direction of the main cylinder and the wiper roller.

The tribologist happily hopped on an airplane and returned to Boston. He had spent six working hours at HG Company to solve the problem, while many months using the trial-and-error approach prior to his visit produced no success.

If a design is uncoupled with a diagonal design matrix and zero information content, both the real uncertainty and the imaginary uncertainty are equal to zero. In this case, both those who do and those who do not understand axiomatic design may come to the same conclusion on complexity and uncertainty.

3.4.3 Total time-independent complexity

The total time-independent complexity C_T is defined as

$$C_T = C_R + C_I \tag{3.15}$$

C_R, the time-independent real complexity, and C_I, the time-independent imaginary complexity, may be plotted in a two-dimensional complex plane, as shown in figure 3.2. The vertical axis is the axis of the imaginary complexity, that is, the axis of "ignorance," since it is caused by lack of knowledge, which yields the perception that the design is more complex than it really is. The horizontal axis represents real uncertainty as a result of the design and/or unknown behavior of nature. C_I and C_R are orthogonal to each other because the imaginary complexity has no relationship to the real complexity and vice versa. The total complexity C_T is shown as the vector sum of C_R and C_I because they are orthogonal to each other.

It is difficult to predict the exact values of C_R and C_I *a priori* if the design is coupled or decoupled. However, a bound for C_I can be estimated if the design is decoupled using equation (3.12). When the design is uncoupled, the imaginary component of complexity is equal to zero and only real complexity may exist if the system range is not inside the design range for all FRs. In the case of a coupled design, the magnitude of the imaginary complexity can be very large and dominate the real complexity.

Based on the foregoing discussion of real complexity and imaginary complexity, we may adopt the following definition of time-independent complexity:

> Time-independent complexity is a measure of uncertainty in achieving a given set of FRs, which does not change as a function of time. It consists of two orthogonal components—real complexity and imaginary complexity. Real complexity is defined

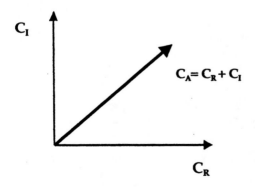

Figure 3.2 Time-independent real and imaginary components of complexity. The vertical axis is the axis of ignorance, and the horizontal axis represents real uncertainty as a result of the design and/or unknown behavior of nature.

as a measure of real uncertainty in achieving a given set of FRs and thus is equal to the information content given by equation (3.5). Imaginary complexity—perceived uncertainty—is caused by the designer's lack of knowledge about the system designed.

3.5   Time-Dependent Complexity: Combinatorial Complexity and Periodic Complexity

The time-independent complexity discussed in the preceding section dealt with two different kinds of complexities involved in making design decisions: the real complexity associated with uncertainties inherent in the system designed and the imaginary complexity associated with uncertainties caused by lack of design knowledge, in other words, ignorance. In this section, time-dependent complexity will be defined. Time-dependent complexity may also viewed as "event-driven" complexity.

Time-dependent complexity occurs because future events occur in unpredictable ways and thus cannot be predicted. Often this results in a time-varying system range, that is, the system range moves away from the design range as shown in figure 3.3. There are two types of time-dependent complexity, which are defined as follows:

The *combinatorial complexity* is defined as the complexity that increases as a function of time due to a continued expansion in the number of possible combinations with time, which may eventually lead to a chaotic state or a system failure. Time-dependent combinatorial complexity arises because the future events occur in unpredictable ways and thus cannot be predicted, although the future events depend on the current state.

The *periodic complexity* is defined as the complexity that only exists in a finite time period, resulting in a finite and limited number of probable combinations.

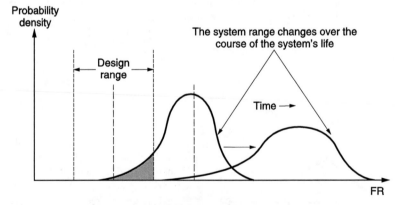

Figure 3.3 Movement of the system range over time. When the system range moves away from the design range over time, we have time-dependent combinatorial complexity.

To reduce combinatorial complexity, we have to devise a means of preventing the system range from moving out of the design range. In the next section and in chapter 5, the means of reducing time-dependent complexity through the transformation of a system with time-dependent combinatorial complexity into a system with time-dependent periodic complexity is discussed.

Time-dependent complexity arises because in many situations, future events cannot be predicted *a priori*. Many of these problems are combinatorial problems that can grow more complicated indefinitely as a function of time because the future events depend on the decisions made in the past, but in an unpredictable way. In some cases, this unpredictability is due to violation of the Independence Axiom. An example is the problem associated with scheduling a job shop. Job shops are typically engaged in machining a variety of parts that are brought to them by their customers. In this case, the future scheduling—which parts are produced using which machines—is affected by the decisions made earlier and its complexity is a function of the decisions made over its past history. This type of time-dependent complexity will be defined as time-dependent *combinatorial complexity*.

▓ EXAMPLE 3.4: BAD WEATHER AND THE AIRLINE SCHEDULE

Suppose that we have a snowstorm around the Detroit area so that airplanes cannot land and take off during the day. Then the airplanes for Boston and other cities cannot take off from Detroit. As time goes on, the flights from Boston to other cities will be disrupted since there will not be enough airplanes to dispatch them according to the original schedule. Therefore, the airlines will not be able to satisfy their FRs of sending airplanes on schedule. The situation is going to get worse as time passes and the snowstorm continues. This is an example of time-dependent combinatorial complexity.

There is another kind of time-dependent complexity, *periodic complexity*. Consider the problem of scheduling airline flights. Although airlines develop their flight schedules, there exist uncertainties in actual flight departures and arrivals because of unexpected events, such as bad weather or mechanical problems. The delayed departure or arrival of one airplane will affect many of the subsequent connecting flights and arrival times. However, since the airline schedule is periodic each day, all of the uncertainties introduced during the course of a day terminate at the end of a 24-hour cycle, and hence this combinatorial complexity does not extend to the following day. That is, each day, the schedule starts all over again; in other words, it is periodic and thus uncertainties created during the prior period are irrelevant. However, during a given period there are uncertainties due to combinatorial and other complexities. This type of time-dependent complexity will be defined as time-dependent *periodic complexity*.

What are the differences between time-independent complexities and time-dependent complexities?

In addition to those characteristics of the four complexities, the following aspects of these complexities should be noted:

(a) Both combinatorial complexity and periodic complexity are real complexities.
(b) In the case of time-independent complexity, the complexity of a system is governed by the given set of FR and DP relationships, since the real complexity

measures the relationship between the design range and the system range and since the imaginary complexity is a particular aspect of decoupled designs. This is in contrast to the case of time-dependent complexity, which sometimes depends on the initial conditions, but unless the system goes back to the same set of initial conditions cyclically, the distant future behavior is totally unpredictable. In the case of combinatorial complexity, the initial condition has little control over the long-term behavior of the system, that is, the initial condition is not distinguishable from the state of the system at any other time in terms of its ability to control long-term behavior. However, in a system with periodic complexity, the initial condition at the beginning of each cycle determines the behavior of the system during the period and thus forever, since the functional period repeats itself.

3.6 Reduction of Complexity

To make a system robust and reliable by satisfying the FRs and Cs throughout the system's life cycle, we must reduce the complexity of the system, starting from the design stage. There are several ways of reducing complexity.

3.6.1 Elimination of time-independent real complexity

Two things must be done to reduce real complexity: satisfy the Independence Axiom and eliminate the bias and reduce the variance shown in figure 2.3.

To reduce real complexity, we must put the system range inside the design range and reduce variance of the system range. In one-FR design, it is easy to achieve this goal. Since the FR is a function of a DP, varying the magnitude of the DP may eliminate bias. Also, lowering the stiffness (i.e., coefficient) of the FR/DP relationship can reduce the variance.

If there are many FRs, we must make sure that the design satisfies the Independence Axiom. If the design is coupled, it will be difficult to make the system range lie inside the design range, because the bias and the variance of the FR cannot be adjusted independently of other FRs. That is, if we try to eliminate the bias of one of the FRs, the bias of other FRs may be adversely affected. Similarly, in a coupled design, the variance cannot be changed without affecting other FRs.

3.6.2 Elimination of time-independent imaginary complexity

Imaginary complexity is a result of not knowing the exact relationship between the FRs and the DPs of a system, if the system has more than one FR. By writing down the design equation, the need for a trial-and-error approach to system development can be eliminated, which will eliminate the imaginary complexity.

It should be also recalled that any coupling terms (i.e., the off-diagonal element of the design matrix) reduce the allowable design range and thus increase the real complexity.

3.6.3 Reduction of time-dependent combinatorial complexity

If a system with combinatorial complexity runs for a long time,[7] it is most likely to fail either because the physical phenomena that control the FR will lead to its failure or because the system generates more possible combinations of FRs that eventually lead to a chaotic state. To reduce such complexity, we may transform a system with a time-dependent combinatorial complexity to a system with time-dependent periodic complexity. The idea is to disrupt a process by reinitializing the FRs on a periodic basis, that is, replacing a combinatorial complexity with a periodic complexity. The period is a *functional period*, not necessarily a temporal period.

The first step in introducing a *functional periodicity* is to decouple a coupled system to make sure that the FRs can be satisfied independently so that the system obeys the Independence Axiom. Then a set of FRs that repeats cyclically must be identified. Among these FRs, we must then identify the presence of FRs that may have a characteristic of a combinatorial complexity. The initial states of these FRs must be then established at the beginning of each period through *reinitialization*. Reinitialization is a key step in establishing a functional periodicity. Means of transforming a combinatorial complexity to a periodic complexity are discussed in later chapters in great detail.

3.7 Functional Periodicity

3.7.1 Functional periodicity in engineering and science

Natural systems have functional periodicity. As discussed in section 3.9, it appears that for long-term stability, functional periodicity is a basic requirement. Since what we see in nature are the ones that have survived, they must have a functional stability at a system level or even at a subatomic level.

In designing engineered systems, functional periodicity can be introduced into a system by various means. To achieve this goal, two requirements must be satisfied. First, there must be a repeating set of functions. Second, we must identify the function that we want to prevent from going into a random and eventually chaotic state.

Some functional periodicities are:

1. Temporal periodicity.
2. Geometric periodicity.
3. Biological periodicity.
4. Manufacturing process periodicity.
5. Chemical periodicity.
6. Thermal periodicity.
7. Information process periodicity.
8. Electrical periodicity.
9. Circadian periodicity.
10. Material periodicity.

Some of the functional periodicities identified above will be explained in great detail in the following chapters. In this section, they will be briefly explained.

(a) *Temporal periodicity*
The most common periodicity encountered in engineering is the temporal period. A set of functions repeats itself on a regular time interval, like a pendulum that oscillates about an equilibrium position. For example, trains (in some countries) are operated on a regular schedule. They have a fixed schedule for a set of trains, which depart and

arrive exactly at the same time every day. Even watches can be set by the departure time of trains. The characteristic of temporal periodicity is that the same set of FRs is repeated over and over again at the same time. Another example is the function performed by clocks built into a computer to synchronize the computation. The period can be one hour, two days, one month, or one year. The circadian cycle is an example of a temporal period.

In nature, the solar system and the planetary motion have a temporal periodicity. The circadian cycle in turn sets the functional periodicity of biological systems.

(b) Geometric periodicity

Geometric functional periodicity is introduced by a repeating set of geometry such as between two sliding surfaces to disrupt the process of wear-particle generation and accumulation so as to maintain low friction and reduce wear (Suh, 1986). Fibers and fabrics also incorporate geometric functional periodicity. As will be shown in chapter 7, "woven" electrical connectors were invented to make use of geometric functional periodicity. The functional periodicity is introduced by means of a patterned surface topography.

In natural systems, the crystallinity of metals is an example of geometric functional periodicity. This crystallinity is a means of minimizing the free energy of the system, which also provides long-term stability.

(c) Biological periodicity

Biological systems are natural systems that have evolved for a long time and acquired biological functional periodicity. For example, cells divide on a periodic basis. Some cells undergo mitosis on a regular time interval (circadian cycle), but many divide when they are ready, which could be governed by the completion of polymerization of proteins and DNA molecules that populate different organelles of a cell. In neurons (another type of cell), the charge decay may control the period. Plants also have a biological periodicity. They survive by creating seeds, which repeat the functions of a given plant. These living biological systems are the ones with long-term stability.

(d) Manufacturing process periodicity

In manufacturing operations, parts must undergo a set of processes before they are completed. Each one of the processes takes a finite time, but often variation is introduced by tool wear, and so on. In this case, the functional period may be treated as the completion of all the processes, that is, functions, for a given part, which may vary in terms of time. Transport robots used in manufacturing cells can be scheduled based on manufacturing process functionality. The productivity of manufacturing systems can be maximized when the functional period is based on the recurrence of a set of FRs (Suh and Lee, 2004).

Biological systems may be considered "natural manufacturing systems" with functional periodicity, ranging from the germination of seeds to death of plants. Insects are "robots" that fertilize ovules and create seeds.

(e) Chemical periodicity

In nature, the periodic chart of elements is the most important chemical functional periodicity. The electron around the nucleus has a functional periodicity.

Some chemical processes depend on the concentration of certain chemicals, which may vary during a given process. The functional period for these chemical processes can be set by the concentration gradient or the concentration of a given chemical component.

(f) Thermal periodicity

When the temperature of a system cycles between a peak temperature and a minimum temperature, the temperature cycle may set the functional period. The temperature of the Earth has a thermal functional periodicity because of the existence of temporal periodicity.

Among engineered systems, the thermodynamic cycles (e.g., the Carnot cycle) of a heat engine are an example of thermal functional periodicity. In these thermal systems, when the thermal cycle is lost, the system may no longer function as intended.

(g) Information process periodicity

In many digital devices, information processing requires functional periodicity. To process information, computers use precision clocks (made of either crystals or integrated circuits) to control the flow of information. In some software programs, the error accumulates with the length of time that the software is used, which ultimately results in system failure. This is overcome by rebooting the computer. It is possible to reinitialize the software to overcome this problem. Another example of the information process functional periodicity is the overflowing of the memory space in computers when the program does not periodically trash the stored information that is no longer used in computation.

(h) Electrical periodicity

A most important electrical functional periodicity is the alternating cycle current (AC), which provides a number of advantages over the direct current (DC). Another example is the alternating electric field used in LCD. When LCD displays are used, they must be subjected to an opposing voltage periodically to make them function for a long time. Otherwise, charges accumulate at the interfaces, leading to system failure. This is an example of a functional periodicity.

In nature, the weather exhibits an electrical functional periodicity. When electric charges build up in clouds and exceed a critical limit, they are discharged to the Earth during a thunderstorm.

(i) Circadian periodicity

Many things are regulated by the change in day and night as the Earth revolves around the sun. All living beings are subjected to this circadian cycle. Even weather may be regarded as a representative of a phenomenon that is governed by this circadian functional periodicity.

(j) Material periodicity

Some natural materials such as metals and ceramics have a periodic structure in the form of crystallinity. Often the structure represents the lowest free energy formation.

Among engineered systems, some synthetic materials such as fibers and fabrics have a functional periodicity, which imparts desired properties such as higher strength, flexibility and toughness. Another example is the periodic annealing done in wire drawing to eliminate the dislocations generated during the plastic deformation.

Another important example of the material functional periodicity is the quantum behavior of electrons in matter. In quantum mechanics, it is assumed that electrons have a particle/wave duality—behaving both as discrete particles and as waves. It may be conjectured that the electrons must have a functional periodicity in the time-space domain to have a long-term stability.

3.7.2 Three dimensions in functional periodicity

Functional periodicity can have many dimensions. In dealing with geometry and materials, functional periodicity can extend along three orthogonal directions. In one-dimensional (1-D) functional periodicity, the functional periodicity extends along only one dimension, whereas in two-dimensional (2-D) functional periodicity, it extends to orthogonal directions forming a plane. In three-dimensional (3-D) functional periodicity, the functional periodicity extends to three orthogonal directions. Depending on the dimensionality of the material functional periodicity, materials can have significantly different properties.

3.7.3 Functional periodicity in sociology, economics, and politics

As it will be shown in chapter 10, the complexity theory presented in this chapter applies equally well to economics, politics, and organizations. The four types of complexity are present in the socioeconomic–political arena just as they exist in science, engineering, and technology. One major difference is that in the socioeconomic–political arena, there tend to be many extra DPs, which are difficult to control, especially in a free society. However, that is the price we have to pay to enjoy basic human rights, and political and economic freedom. The redundant DPs may also protect the society from making major mistakes that a totalitarian system tends to commit.

The time-independent real complexity is difficult to measure in some of these non-scientific fields, because it is difficult to establish FRs and their design ranges. Furthermore, time-independent imaginary complexity may play a major part in many of these fields, consuming resources and energy. Many of these socioeconomic–political fields are burdened with time-dependent combinatorial complexity, which is stopped when a major disaster occurs. As will be shown, a judicious introduction of functional periodicity and a new set of FRs can have a major effect in these fields.

The functional periodicity that applies to socioeconomic–political fields is the following:

1. Organizational periodicity.
2. Economic functional periodicity.
3. Political functional periodicity.
4. Academic functional periodicity.

The functional periodicity stated above has the following characteristics:

(*a*) *Organizational periodicity*
Organizations need organizational functional periodicity. When organizations do not renew themselves by resetting and reinitializing their functional requirements, they can

become an entity that wastes resource and regurgitates issues that do not benefit the members of the organization.

(b) Economic functional periodicity

The economy can deteriorate when it is governed by a combinatorial complexity. Overly regressive taxation or overly progressive taxation has created havoc in many parts of the world, when it is not checked in time to deal with the negative consequences of the system. Equally important is the greater rate of injection of currency into the economic system—overstimulation of the economy—than the ability by the system to absorb it using productive mechanisms that can add value to society, which leads to inflation. Underinvestment, on the other hand, can lead to depression. Economic functional periodicity is a mechanism that enables the reinitialization of the economic system.

(c) Political functional periodicity

Political corruption often follows when the system does not have a functional periodicity and is not able to renew itself. Political terms and elections play a major beneficial role by introducing political functional periodicity.

(d) Academic functional periodicity

Schools have academic terms, which let students recharge and professors regain their objectivity. Academic institutions must also reinitialize their curricula and other activities to introduce innovation to teaching and learning. Often this happens when the administration changes.

3.8   General Comparison of the Complexities of Engineered Systems and Natural Systems

Engineered systems are designed by human beings. Therefore, they can be very complex if the designers do not perform their design tasks well. In principle, engineered systems should have zero time-independent real and imaginary complexities and no time-dependent combinatorial complexity. Furthermore, if the system range must change as a function of time, the designer should be able to introduce time-dependent periodic complexity. These observations should hold regardless of the size of the system and the number of FRs and DPs of the system. At this time, the time-dependent imaginary complexity is a major factor in the cost and effort required in developing and operating large systems, ranging from the space shuttle to healthcare systems.

In living systems, which are perhaps the most important natural systems, it appears that the natural process of selection has eliminated those with time-dependent combinatorial complexity and only those with time-dependent periodic complexity have survived. It is difficult to assess the time-independent real complexity, since we do not know the design range of various functions that a living system must satisfy. Instead, the system range of living beings defines the limits of their capability. System range makes us what we are! Nature renews itself by letting living beings have a finite period of life and reproduce to sustain the species.

Table 3.1, which was also presented in chapter 1, shows examples of functional periodicity that exist in nature and engineered systems. The functional periodicity of

Table 3.1   Examples of functional periodicity in nature and engineered systems

Examples of functional periodicity	Examples in nature	Examples in engineered systems
Temporal periodicity	Planetary system, solar calendar	Airline/train schedules, computers, pendulum
Geometric periodicity	Crystalline solids, surfaces of certain leaves	Undulated surface for low friction, "woven" electric connectors
Biological periodicity	Cell cycle, life–death cycle, plants, grains	Fermentation processes such as wine making
Manufacturing/processing periodicity	Biological systems	Scheduling a clustered manufacturing system
Chemical periodicity	Periodic table of chemical elements/atoms	Polymers
Thermal periodicity	Temperature of Earth	Heat cycles (e.g., Carnot cycle)
Information processing periodicity	Language	Reinitialization of software systems, music
Electrical periodicity	Thunderstorm	LCD, alternating current
Circadian periodicity	Living beings	Light-sensitive sensors
Material periodicity	Wavy nature of matter, atomic structure, crystallinity	Fabric, wire drawing, microcellular plastics

nature has enabled the existence of what constitutes the universe at this time. The functional period of nature ranges from very short to extremely long periods relative to the typical lifespan of living beings. Once we recognize the importance of functional periodicity, we may use the idea to create many innovative solutions to achieve the aspirations of humans. With the advances in biological sciences, the biological functional periodicity may be manipulated to increase the longevity of living beings.

3.9 On Functional Periodicity, Long-Term Stability, Equilibrium States, Quantum States, and Superstring Theory

3.9.1 Summary

Starting from the postulate that there must exist fundamental axioms that characterize a good design, axiomatic design theory was advanced. Following axiomatic design theory, a complexity theory was presented in this chapter, which showed that there are four types of complexity and that the introduction of functional periodicity can transform a system with time-dependent combinatorial complexity, which eventually becomes chaotic, to a stable system that has a time-dependent periodic complexity. For long-term stability, both natural systems and engineered systems must be either at a stable *equilibrium state* or have a *functional periodicity*. Classical physics such as Newtonian mechanics and thermodynamics are based on the existence of equilibrium states, whereas modern physics such as quantum mechanics and superstring theory suppose the existence of a functional periodicity. The particle/wave duality of matter that forms the basis of quantum mechanics is consistent with the stability argument presented in this section.[8]

3.9.2 Stability as the basic driving force of nature

There are two basic characteristics of natural matter (e.g., electrons around an atom, molecules, and assembly of biological molecules) and engineered systems: they must be *stable* for a long-time and serve specific *functions*. Functions are "what the system is designed to achieve." These functions are satisfied by physical or biological entities. For example, biological cells that are made up of various constituents achieve a set of functions. Even electrons spinning around an atom also satisfy certain functions, for instance, provide a charge neutrality of the atom. Furthermore, the fact that these systems have survived billions of years of evolution proves that they are stable. Similarly, an engineered system that performs a desired set of functions must be stable throughout its lifetime.

If a system has a time-dependent combinatorial complexity, the system will become chaotic and eventually fail. For long-term stability and survival, a system—both natural and engineered—must be at an equilibrium state or have a functional periodicity. The long term of an engineered system is its life cycle and for a natural matter, it is the life cycle of living beings or the elements.

3.9.3 Hypothesis on long-term stability of natural systems or engineered systems

For long-term stability of natural and engineered systems, the system must be either in an equilibrium state with its surroundings or have a functional periodicity. Systems without stability because of the lack of functional periodicity have a transitory existence or are in a chaotic state and eventually disappear or mutate into another system or matter which is either in equilibrium or is periodic.

3.9.4 Equilibrium state for long-term stability

If an engineered system or a natural matter is at an equilibrium state with its surroundings, it is going to be stable until its equilibrium state is disturbed. If it is at a metastable equilibrium state, it will go into an equilibrium state or unstable equilibrium state or another metastable state, if the system is disturbed by the application of external energy that is larger than energy barriers surrounding the local minimum. Classical physics such as Newtonian mechanics and thermodynamics have been developed based on the postulate that equilibrium states exist. In Newtonian mechanics, forces are assumed to be in equilibrium: the first and third Newtonian laws are based on static stable equilibrium and the second law is based on a dynamic stable equilibrium relationship. In thermodynamics, the equilibrium condition forms the basis for both the first and second law of thermodynamics (Hatsopolous and Keenan, 1981).

3.9.5 Functional periodicity for long-term stability

Some natural and engineered systems, or natural matter, are not at an equilibrium state at all times. However, they are stable if they possess a functional periodicity. That is, an absolute static equilibrium is not a requirement for long-term stability and survival if and only if the system can renew itself through functional periodicity. Modern physics such as quantum mechanics and superstring theory assumes the existence of a functional periodicity in natural matter such as atoms, electrons, and subatomic particles.

(a) Quantum states

It is generally accepted that quantum mechanics provides the best description of matter and energy on a tiny scale. Quantum mechanics assumes that basic particles such as electrons have the characteristics of both a particle and a wave and that the position and velocity of the particle can be treated in terms of probability of the particle occupying a region in position and momentum spaces. Furthermore, Heisenberg's uncertainty principle states that the position and velocity of an object cannot be determined accurately at the same time. The Schroedinger equation was constructed by seeking a wave function that represents the complex amplitude of displacement as a function of time and space, and thereby satisfies the conservation of energy and the quantum nature of energy and matter. The square of the wave function represents the probability of finding the particle in a given region. Although the predictions made by quantum mechanics are the basis for modern electronic devices and modern physics, for many people the particle/wave duality has been a difficult concept to understand (Giancoli, 2000; Merzbacher, 1998). The generally accepted "Copenhagen" interpretation applies the duality principle in a statistical sense, that is, when large numbers of particles are involved (or a large number of observations are made on a particle). Still unsolved paradoxes such as the famous "Schroedinger cat" dilemma illustrate the difficulty of wave/particle duality in the limit of a single particle.

Based on the hypothesis presented in this section, the particle/wave duality of electrons, for example, can be rationalized. Electrons will be stable if they are at equilibrium states. When electrons are at an equilibrium state, they may be identified as a particle. Even when electrons are not at an equilibrium state, such as when they are moving around a nucleus of an atom, they can be stable if they have a functional periodicity. When electrons seek a stable state by having a functional periodicity, they will appear to have the characteristics of a wave with a functional periodicity. This view is consistent with the quantum mechanical view, which assumes that the electrons in a crystal exist in the form of waves, obeying the Schroedinger equation. When a stable electron interacts with other particles, it may begin as a particle and then acquire a functional periodicity, which may be the basis for Young's double slit experiment with electrons.

(b) String theory

Superstring theory is the latest theory that is believed to unite all fundamental laws of nature. String theory can describe the behavior of elementary particles by representing them as vibrating strings. The length of the string is varied to represent various particles and the strings can interact. Superstring theory can derive Einstein's equation by demanding that the string move self-consistently in space-time (Kaku, 1994). An interesting aspect of the superstring theory is that it involves functional periodicity, indicating that the concept of functional periodicity may be fundamental for stability of any existing matter, including materials and living beings. This particular use of strings may be one of many different ways of representing a functional periodicity of matter.

3.9.6 Equilibrium state versus functional periodicity

Based on complexity theory, it was argued that for long-term stability, natural matter or engineered systems must be at an equilibrium state or have a functional periodicity. In some systems with functional periodicity, there may not be a sharp demarcation between an equilibrium state and functional periodicity because of limitations to knowledge or technology.

A system that is stable as a result of functional periodicity could appear to an observer as operating at an equilibrium state if the cycling period is short relative to the observation period. Therefore, the laws of Newtonian mechanics and thermodynamics may be applicable even when the underlying system, such as that described by super-string theory, possesses a functional periodicity with a functional period much shorter than the time over which the system changes from one equilibrium state to another equilibrium state.

3.10 Complexity Theory and Design Axioms

The basic idea for the complexity theory presented in this book was generated as a result of the design axioms. However, having developed the complexity theory, it is also possible to develop the two design axioms from the complexity theory, using the rationale given in this section.

From the definition of complexity that complexity is a measure of uncertainty in satisfying FRs, we may state the following axiom:

THE COMPLEXITY AXIOM

Reduce the complexity of a system.

The Complexity Axiom is equivalent to a combination of the Independence Axiom and the Information Axiom, except that the Complexity Axiom is less explicit than the Independence Axiom on the functional independence and more explicit on the means of achieving robustness than the Information Axiom.

3.11 Summary

This chapter examined the issue of complexity, information, and uncertainty based on the Independence Axiom and the Information Axiom. It was shown that there are four different kinds of complexities.

In the time-independent situation, there are two kinds of complexity, *real complexity* and *imaginary complexity*, which are orthogonal to each other. Total complexity is defined to be the vector sum of the real and the imaginary complexities.

In the time-dependent complexity arena, there are two kinds of complexity, *combinatorial complexity* and *periodic complexity*. In a system that is subject to combinatorial complexity, the uncertainty of the future outcome continues to grow over time, and as a result, the system cannot have long-term stability and reliability. In the case of systems with periodic complexity, the system is deterministic and can renew itself over each period. Therefore, a stable and reliable system must be periodic. Starting from the application of the Independence Axiom, it was shown how a coupled system was decoupled through design changes and how a combinatorial complexity problem could be changed into a periodic complexity problem.

A system with time-dependent combinatorial complexity can be changed to a system with time-dependent periodic complexity. The time-dependent periodic complexity requires that a set of functions repeat periodically. At the beginning of each period, the initial state of the system (i.e., the FRs) must be determined to reinitialize the system. The functional periodicity can be obtained by many different means: temporally, geometrically, biologically, chemically, thermally, and electrically. Also they can be controlled by manufacturing processes, information processes, and circadian cycles.

The consistency between nature and Theorem 26 (Conversion of a System with Infinite Time-Dependent Combinatorial Complexity to a System with Periodic Complexity) was discussed. It was shown that many things in nature are periodic, consistent with the need to change a combinatorial complexity design to one of periodic complexity to reduce uncertainty. It was argued that the theorem may apply to political and societal systems as well. Periodic renewal of political systems and societal systems is essential for long-term sustainability of the system.

One of the fundamental requirements for long-term survivability is the stability of the system. For a system to be stable, it must be either at an equilibrium state or must possess functional periodicity. The particle/wave duality of electrons, quantum states, and superstring theory may all be different characterization of the functional periodicity of natural matters that are stable and have existed since the creation of the universe.

Notes

1. In section 3.2, it was stated that we have to know the joint probability to compute the information content of decoupled and coupled designs. When the design is a decoupled design, we have to know the conditional probabilities in determining the system range. When the design is a fully coupled design, the joint probability of all the FRs must be known to compute the information content. Unfortunately, the computation of the joint probability and, thus, determination of the information content for decoupled or coupled designs is difficult because the joint probability and the conditional probability may not be available during the design stage.

2. This is consistent with Theorem 18 (Existence of an Uncoupled Design): There always exists an uncoupled design that has less information than a coupled design.

3. Theorem 4 (Ideal Design) states that the number of FRs and DPs must be the same. When the design has only one FR and many DPs, all DPs except one chosen FR may be fixed. Then all the uncertainties introduced by all the fixed DPs can be compensated by varying the one chosen DP.

4. This issue was pointed out by my colleague Professor Sang-Gook Kim.

5. When the system ranges of the FRs shown in equation (3.9) are not inside their respective design ranges, we have to know the conditional probability of satisfying the FRs to compute the real complexity, which may not be readily available without doing extensive testing. In some

cases, we may be able to determine the system range through simulation. This is known as a "tradeoff study," which is conducted to determine how the choice of DPs and their values affect the performance of the system. When the design is fully coupled with many FRs, the tradeoff study is difficult to conduct. Furthermore, when the tradeoff study is done without listing all of the FRs, the result may provide misleading information.

6. In selecting DPs, the designer's knowledge of associated physics and engineering is obviously indispensable. Axiomatic design cannot make up for lack of fundamental understanding of physics, mathematics, and other associated knowledge bases.

7. "Time" in the time-dependent complexity is used in a general sense. Here, "time" signifies progression of events.

8. This section is based on the research note by Suh (2004).

References

Giancoli, D.C. 2000. *Physics for Scientists and Engineers*, Prentice-Hall, Upper Saddle River, NJ.

Hatsopoulos, G.N. and Keenan, J.H. 1981. *Principles of General Thermodynamics*, Krieger Publishing Co., Melbourne, FL.

Kaku, M. 1994. *Hyperspace*, Oxford University Press, New York.

Kim, S.-G. 2004. "Axiomatic design of multiscale systems," presented at the Third International Conference on Axiomatic Design, Seoul, Korea.

Merzbacher, E. 1998. *Quantum Mechanics*, Wiley, New York.

Shannon, C.E. and Weaver, W. 1949. *The Mathematical Theory of Communication*, University of Illinois Press, Urbana.

Suh, N.P. 1986. *Tribophysics*, Prentice-Hall, Englewood Cliffs, NJ.

Suh, N.P. 1990. *The Principles of Design*, Oxford University Press, New York.

Suh, N.P. 1999. "A theory of complexity, periodicity, and design axioms," *Research in Engineering Design,* Vol. 11, pp. 116–131.

Suh, N.P. 2001. *Axiomatic Design: Advances and Applications,* Oxford University Press, New York.

Suh, N.P. 2004. "On functional periodicity as the basis for long-term stability of engineered and natural systems and its relationship to physical laws," *Research in Engineering Design*, Vol. 15, pp. 72–75.

Suh, N.P. and Lee, T. 2004. "System integration based on time-dependent periodic complexity," U.S. Patent 6,701,205 B2.

Wilson, D.R. 1980. "An exploratory study of complexity in axiomatic design," Ph.D. thesis, MIT, August.

Exercises

3.1. Consider the design and manufacture of a car. Estimate the order of magnitude of real and imaginary complexities that the engineers at Toyota Motor Co. may be dealing with by analyzing how they design the vehicle.

3.2. Estimate the real and the imaginary complexity an engineer must deal with when the products are evaluated by a trial-and-error method.

3.3. State the conditions under which a combinatorial complexity can be changed into a periodic complexity.

3.4. Compare the quantitative definition of complexity provided by axiomatic design to other quantitative definitions of complexity (for example, Kolmogorov complexity and Shannon's information entropy). What is the fundamental difference?

3.5. The stability criterion for systems is one of the most important criteria for both engineered systems and natural systems. It was shown that for a system to be stable, it must be either at an equilibrium state or have a functional periodicity.

   (a) Identify a mechanical system that is stable and has a functional periodicity because of the balance between kinetic energy and potential energy.
   (b) Identify natural systems that are stable because of the presence of functional periodicity.

3.6. Identify a socioeconomic–political system that does not have long-term stability and describe the causes of its instability. How should the system be made more stable?

3.7. If a system has time-dependent combinatorial complexity and yet cannot be transformed into a system with time-dependent periodic complexity, how should the system be made stable?

3.8. Is a democratic system inherently stable? Describe your reasoning.

3.9. An engineered system can be made stable by constantly seeking an equilibrium position by using a feedback control loop or by introducing functional periodicity. If you were to develop a humanoid robot, how would you give it stability as it walks on its own?

3.10. If we have to machine a steel rod to a diameter of 20 mm $\pm$ 0.001 mm using a general-purpose lathe, there is a finite probability that it cannot be done without the input of a skilled machinist. Thus, it has a finite time-independent real complexity. Describes means of eliminating the real complexity.

4

Reduction of Time-Independent
Complexity

4.1 Introduction

One of the goals of design and scientific endeavor is to reduce the complexity associated with the design of artifacts or scientific understanding to zero.

In chapter 3, it was stated that there are four types of complexities: time-independent real complexity, time-independent imaginary complexity, time-dependent combinatorial complexity, and time-dependent periodic complexity. To make a system perform its intended functions within their specified design ranges, we must establish the FRs and constraints of the system correctly and then satisfy them with zero complexity. Any performance problem that arises at the beginning of the system operation after it is designed is usually a result of poor design. Such a system has time-independent real complexity. Long-term performance, reliability, and safety issues arise when the design is encumbered with time-dependent combinatorial complexity problems.

In this chapter, we will consider how time-independent complexity can be reduced. In chapter 2 on Axiomatic Design, it was shown how the FRs can be satisfied at all times by making the system range to be in the design range. When the overlap between the design range and the system range is not complete, there is a finite amount of uncertainty in achieving the FRs. This fact is represented by the time-independent *real* complexity, which is equal to the information content. In addition to the real complexity, there is also time-independent imaginary complexity—the combination lock problem. The time-independent *imaginary* complexity is not a measure of real complexity and should be zero if the designer knows the FR/DP relationships of a decoupled design.

When the FRs are *always* satisfied within their design range by a system (i.e., product, process, etc.), the real complexity and the time-dependent combinatorial and periodic complexities are equal to zero. The least complex design is the one that always satisfies the FRs even when the design range is made smaller and smaller, eventually approaching the ultimate limit of zero. A design with zero complexity is the ideal design, provided that it meets all the constraints imposed on the design. When the complexity of two different designs is equal to zero, one may choose one design over the other based on some other criterion such as cost.

As the allowable variation of FRs decreases (or the desired accuracy of what we want to know increases), the real complexity of an uncoupled design will become less than that of a decoupled or coupled design. Similarly, the real complexity of a decoupled design will be less than that of a coupled design. This is true because the allowable tolerance of the DPs decreases when the coupling terms—off-diagonal terms—are added to the design matrix. In this case, the real complexity may increase. The larger the magnitude of the off-diagonal terms of the design matrix, the greater the real complexity.[1] The real complexity of uncoupled or decoupled design can be decreased by eliminating the *bias* and reducing the *variance* of FRs, which was discussed in chapter 2.

The purpose of this chapter is to show how time-independent real complexity of a coupled system can be decreased by replacing it with a decoupled or uncoupled design. Similarly, when a decoupled design can be uncoupled, the real complexity may be reduced. Three case studies will be used to illustrate this process of reducing the real complexity by uncoupling or decoupling. One is the design of a knob. Another case study is based on injection molding of precision parts. The third case study is the design of spark-ignition internal-combustion (IC) engines.

These case studies are intended to show that when the design is uncoupled or decoupled, time-independent real complexity can be reduced since it is often possible to put the system range inside the design range. Therefore, the robustness and reliability of uncoupled or decoupled designs increase with a decrease in real complexity. Although, without extensive data,[2] it is difficult to prove definitively that the cost of the product should also be less when real complexity decreases, the limited data available indicate that the development cost and human effort are less since we can allow greater tolerance for DPs and still satisfy the FRs.

4.2   Reduction of Time-Independent Complexity

How do we determine the real complexity?
One of the major goals of design should be to reduce the time-independent real complexity to zero. The real complexity is a consequence of the system range being outside

of the design range. Therefore, the real complexity may increase when the design range of an FR is made smaller. A robust design is a design that has no time-independent real complexity and no time-dependent combinatorial complexity.

To determine the real complexity, we must take the following actions:

1. Define FRs with their design ranges.
2. State the constraints at each level.
3. Develop designs that satisfy the FRs and the Independence Axiom.
4. Determine the real complexity by comparing the system range with the design range of each FR. When the information content is finite, the complexity is also finite.
5. Adjust the system range to reduce bias and variance so that it will be inside the design range.
6. Decompose the FRs and DPs, and reduce the real complexity following steps 1 through 5.

Since the time-independent real complexity is equal to the information content, the real complexity of a design is zero when the information content is zero. To reduce the real complexity, we must satisfy the Independence Axiom by creating an uncoupled or a decoupled design, and reduce the bias and the variance of the system range or make the design range large, as discussed in chapter 2. This process of reducing real complexity is relatively straightforward. Readers are referred to Suh (2001).

The more difficult case of reducing the real complexity is when the design is so fully coupled that the bias cannot be removed since FRs are dependent on each other. In this case of coupled design, even the stiffness of an FR cannot be reduced lest it adversely affect the stiffness of other FRs. In this case, the best recourse is to replace the coupled design with an uncoupled or decoupled design, which means developing a new system.

If the FRs of a system are coupled, the real complexity may be large because the FRs cannot be controlled independently from one other. The reasons are as follows:

1. The FRs of some designs may not be satisfied by an existing design no matter what we do with the design, if it is coupled. For example, it will be impossible to reduce the hydrocarbon emission from the conventional spark-ignition IC engines, since there is no way that the functions of this engine can be decoupled or uncoupled without the use of a catalytic converter. That is, the desired solution lies outside of the available design window of current commercial IC engine designs. To satisfy more stringent emission standards, a new engine must be designed.
2. Although a coupled design may have a unique set of DPs for a given set of FRs, its real complexity increases faster than that of an uncoupled or decoupled design when the system is subject to random variation. Therefore, it is difficult to reduce the bias and variance to put the system range inside the design range at all times.
3. The allowable tolerance for DPs decreases with every coupling term that exists in the design matrix (Suh, 2001). The larger the off-diagonal terms, the smaller is the allowable tolerance of DPs. That is, coupled designs are not robust.
4. When a design that is coupled at high levels must be decomposed, it is difficult to determine and vary the low-level DPs without adversely affecting some of the highest-level FRs and thus to satisfy the highest-level FRs within the design range.
5. We may not be able to express the elements of the design matrix with precise mathematical expressions to determine a precise set of FRs when the design is coupled with many leaf-level FRs and DPs.

Reduction of complexity of a system with many interconnected parts
In chapter 3, complexity with many interconnected parts was discussed and several theorems were given. In chapter 10, complexity issues related to social systems will be examined in light of these theorems. In this section, we will explore the question of how we may further reduce the complexity of a system with many connected parts.

If each interconnected part of the system is equivalent to a DP and thus the system is one with many extra DPs, then uncertainties introduced by all DPs except one can be compensated by using the last DP if the design is either uncoupled or decoupled. In this case, the FR may be expressed as

$$FR = f[DP^a, DP^b, \ldots, DP^x] \tag{4.1}$$

The random variation of FR may be expressed as

$$\delta FR = \frac{\delta FR}{\delta DP^i}\delta DP^i + \sum_{\text{all other DPs}} \frac{\delta FR}{\delta DP^j}\delta DP^j \tag{4.2}$$

DP^i is the DP selected to compensate for the effect of all other interconnecting parts represented by the second term on the RHS of equation (4.2). To reduce uncertainty in satisfying FR, the random variation of FR should be made equal to zero by setting the first term on the RHS of equation (4.2) to be equal in magnitude but opposite in sign to the second term on the RHS of the equation.

EXAMPLE 4.1: COMPLEXITY IN DECISION-MAKING IN AN ORGANIZATION

Let us assume that every professor in the Department of Mechanical Engineering at MIT agrees on FRs and that the FRs can be satisfied independently. However, each one of the 60 professors in the Department has different views on the best DP that can satisfy the FR. Because of this diverse opinion of the faculty, the decision-making can be complex if each one of the professors can affect the outcome, since the FRs may not be satisfied within their design ranges. What is the best decision-making process that will enable the Department to achieve the FR?

Solution

The best solution is to gather all the opinions of the faculty and understand their implications. Then the person in charge (normally the Department Head) makes the final decision to be sure that the uncertainty and thus the complexity is minimized. This process works when the FRs are accepted by every faculty member.

Reduction of time-independent imaginary complexity
Designers should not waste their time worrying about imaginary complexity, but many do because they think design can be done intuitively based solely on experience without clearly establishing the FR/DP relationship in the form of the design matrix. In some scientific inquiries, where we lack detailed knowledge, we fumble through different possibilities, although the solution may be very simple if we again establish the FR/DP relationship.

Time-independent imaginary complexity can arise from many different sources. One of the primary causes is a consequence of not knowing the exact relationship between the FRs and DPs of a decoupled design, as shown in chapter 3. This complexity can be reduced by changing the DPs of a decoupled design in the order dictated by the triangular matrix.

Other causes of imaginary complexity may be:

1. FRs are not clearly stated, and the design embodiment is created by a trial-and-error process. The design may be good, but it is difficult to assess whether or not the design is good because the design equation cannot be written in the absence of clearly stated FRs and DPs.
2. Many designers work on wrong FRs either because the customer has no clearly stated FRs or because the designer has not asked for clear FRs.
3. Some designers confuse physical parts with DPs. (For example, how many DPs does an airplane wing have?)
4. In scientific inquiries, we may not have sufficient knowledge to define FRs or DPs.

Uncoupled designs have many desirable characteristics as discussed in previous chapters. One of the advantages of an uncoupled design is its lack of time-independent imaginary complexity.

4.3   Case Study: Reduction of Time-Independent Real Complexity of a Knob

We want to determine the time-independent real complexity of a simple design to illustrate how the real complexity can be decreased by uncoupling FRs. Consider the knob for grasping and turning a shaft shown in figure 4.1.[3] The knob is to be made of a polyamide plastic by injection molding. It was designed to satisfy the following two FRs:

FR_1 = Grasp the end of the shaft tightly with axial force of 30 N $\pm$ 3 N.
FR_2 = Turn the shaft by applying 15 N-m $\pm$ 2 N-m of torque.

What is the complexity of this knob? The real complexity is determined by the overlap between the design range and the system range of the FRs. The DPs for the design shown in figure 4.1 are:

DP_1 = Interference fit between the shaft and the inside diameter of the knob.
DP_2 = The flat surface.

The design equation may be written as

$$\begin{Bmatrix} FR_1 \\ FR_2 \end{Bmatrix} = \begin{bmatrix} X & X \\ x & X \end{bmatrix} \begin{Bmatrix} DP_1 \\ DP_2 \end{Bmatrix} \qquad (4.3)$$

where the lower-case x is used to signify the fact that the effect of DP_1 on FR_2 is much less than the other effects indicated by upper-case X. In this case, x is almost equal to zero.

Equation (4.3) states that the interference fit (DP_1) between the shaft and the inside diameter of the knob provides the gripping force on the shaft and also affects the ability to turn the shaft. Similarly, DP_2 (the flat surface on the shaft and the knob) satisfies not only FR_2 (turn the shaft) but also affects the gripping force because the turning

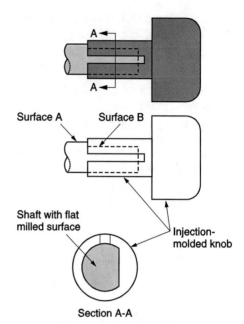

Figure 4.1 Design of an injection-molded knob to hold and turn the shaft.

action opens the slot. For instance, when the torque is applied by the knob, the grip force on the shaft may decrease with increase of the slot opening as a result of the normal load acting on the flat surface, depending on where the reaction force exerted on the inside surface of the knob is applied.[4] If the grip force is less than the required force to keep the knob on the shaft, the knob will slide off the shaft.

Given the exact magnitude of forces and material properties, we should be able to come up with a unique set of DP_1 and DP_2 that may work reasonably well, provided that the dimensions of the injection-molded parts are accurate enough to maintain the interference fit. However, it is better to come up with a design that avoids this kind of problem. The design shown in figure 4.1 is a coupled design or, at best, a decoupled design (depending on the design ranges of the FRs) and thus cannot be the best design.

As shown in section 2.6, the off-diagonal terms reduce the allowable tolerance of DPs. Therefore, it is more difficult to satisfy the DPs within the required tolerances using the manufacturing processes selected for manufacture of the knob. If we choose the dimensions of the knob without considering the effect of the coupling terms, the system range for the FRs, especially FR_1, is most likely to be outside of the design range. Therefore, there is a finite real complexity.

How do we eliminate this real complexity?
Some will suggest that the solution to this coupled design problem is to make the outer diameter of the knob shaft (i.e., the thickness of the cylindrical section) thicker, which will make the slot open up less and thus minimize the reduction of the gripping force. However, this solution has its cost; not only does it require more materials, but it also has higher information content, which may result in greater real complexity.

The process of increasing the thickness of the cylindrical section is equivalent to making the information content larger and thus the real complexity greater. This is

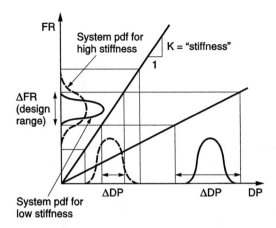

Figure 4.2 Dependence of system pdf on "stiffness." The design range of the gripping force is given by ΔFR. The manufacturing capability is given by the bell-shaped pdf in the physical domain (the DP axis). Depending on the stiffness, the same DP distribution translates into very different system distributions in the functional domain. The real complexity decreases with the lowering of the stiffness.

illustrated in figure 4.2. The design range for the force is given by ΔFR, which is specified in the functional domain. The manufacturing capability is given by the bell-shaped distribution of DPs in the physical domain (i.e., along the DP axis). The figure shows two bell-shaped distribution curves in the physical domain, which are assumed to be the same because the same manufacturing process is used. Depending on the stiffness of the plastic cylindrical section, the same bell-shaped distribution along the DP axis translates into very different system distributions in the functional domain, as shown along the FR axis. When the stiffness is lower, the system pdf (the solid bell-shaped curve along the FR axis) fits in the design range, but when the stiffness increases, the system pdf (the dashed bell-shaped curve) is outside the design range.

When the thickness of the cylinder wall increases, the stiffness increases, and thus the interference between the shaft and the inner diameter of the cylinder (DP_1) must be manufactured to a tighter tolerance to be within the design range. Therefore, the allowable tolerance between the shaft and the inner diameter of the knob must be made smaller since small changes in the tolerance will result in a large difference in the gripping force. The overlap between the system range and the design range will become smaller with an increase in wall thickness, thus increasing the information content and the real complexity. The probability of success will decrease and the real complexity will increase as the tolerance becomes tighter.

To eliminate the real complexity, the manufacturing operations must be made increasingly accurate. This probably requires that the machining of the shaft diameter be done more accurately and the tolerance of the injection-molded part (if the knob is made by injection molding) be kept tight. The latter may require special molds and molding cycles or measuring all of the parts to select those that meet the specification. Thus, the manufacturing process becomes more complex when the design is coupled. If we do not change the manufacturing process, then the random variation of DPs will be the same, which may yield unacceptable parts, and the real complexity will be finite.

Design with zero real complexity
We can reduce the time-independent real complexity, possibly to zero, by making a slight modification to the original design shown in figure 4.1. An improved design is shown in figure 4.3. In this design, the slot terminates where the flat part of the knob

Figure 4.3 A new uncoupled design. In this design, the flat surface used for turning is away from the slot. The bottom corners of the slot should be rounded to reduce the stress concentration.

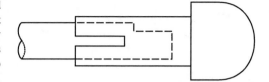

begins. Since the flat surface is completely away from the slot, the turning action does not force the slot to open, and therefore the axial grip is not affected. This is a completely uncoupled design. The information content is zero since the system range for each FR can be made to lie inside its design range. It is much *less complex* to make than the original design because the thickness of the cylinder wall can be made to provide the right gripping force at the knob/shaft interface and thus the manufacturing tolerance can be made as large as possible. The design shown in figure 4.3 may be made to have zero real complexity (i.e., the system range is always inside the design range) by proper choice of manufacturing processes and DPs.

How do we actually determine the wall thickness and the desired interference? The gripping force can be controlled either by changing the stiffness (thickness) of the slotted section of the knob or by adjusting the interference fit between the shaft and the knob. What we want to do is to make the thickness as thin as possible to reduce cost and to make the interference as large as possible. The thickness must be determined by considering two limiting factors: manufacturability and failure of the knob under stress. Can it be manufactured by injection molding? Does the maximum stress at the bottom corner of the slit, which is the stress concentration point, cause either fracture or plastic deformation? (Obviously, we should use a fillet to round the sharp inside corners.)

What we want to do is to maximize the grip force and, at the same time, minimize the stress.

4.4   Case Study: Precision of Injection-Molded Parts

Certain injection-molded plastic parts, such as the "air-flow meter" used to monitor the air going into an internal-combustion engine, must be precise in their dimensions for accurate measurement of the flow rate. However, the dimensions and the shape of injection-molded plastic parts cannot be controlled precisely because the FR for dimensional accuracy is coupled to the FR for injection of the molten plastic into the mold and to the FR for cooling of the injected molten plastic in the mold.

The injection-molding process works as follows. Plastic is molten and mixed in the plasticating part (the rotating screw that moves axially) of the injection-molding machine. The plastic is injected into the mold by the pumping action of the screw when it moves forward, pushing the molten plastic into the mold. During the injection process, the plastic continues to cool as it flows into the mold and develops a complicated temperature profile, in some cases freezing the plastic at the mold surface as it flows into the mold. Once the mold is almost filled, extra pressure known as the "packing pressure" is applied to partially compensate for the shrinkage that occurs during cooling and to ensure the replication of the mold cavity shape throughout the part. The pressure distribution is different throughout the mold, being highest at the gate of the mold. The temperature of the plastic is decreased by removing the thermal energy in the molten plastic through the surface of the cold mold. Finally, the part acquires the shape of the mold cavity.

The molded part is different from the mold cavity shape, often having warpage. The variation of the part dimensions of injection-molded parts is largely due to the coupling of the FRs of injection molding. The process couples the flow and cooling of plastics during the injection process. When the molten plastic is cooled to freeze it in the mold cavity, the specific volume of the plastic shrinks. The shrinkage is a function of the amount of material present at a given location, which is a function of the local pressure. Furthermore, the mold is rigid and does not change its dimensions. Owing to these two effects, the differential pressure that exists in the mold and the partial solidification during flow, there is differential shrinkage. As a result, residual stresses are developed in the solidified plastic part. When the part is taken out of the mold, it distorts to minimize the strain energy by reducing residual stress and strain. Therefore, the molded part dimensions are different from the mold cavity shape. In conventional injection molding, to minimize the shrinkage during cooling, molten plastic is cooled under "packing pressure." Therefore, in the conventional injection-molding process, the total mass of plastic injected is coupled to the final geometry.

To overcome the problems associated with conventional injection molding, the mold-filling process is simulated, using a model that includes the rheological behavior of polymers and the transport processes, to determine the mold shape, process conditions and gate geometry that can best yield the desired shape of the molded part. However, this optimization technique does not eliminate the coupled nature of the injection molding to part dimensions.

How would you uncouple the injection-molding process so as to get accurate part geometry regardless of the injection and cooling process in the mold?

Solution

One way of decoupling the part geometry from other FRs of the injection-molding process, especially during cooling, is to generate a large number of tiny bubbles in the plastic that expand in the mold. This can be done by creating a thermodynamic instability that results when the thermodynamic state of a polymer with a large amount of dissolved gas at high pressure is exposed to a new thermodynamic state at low pressure. A large amount of gas (such as N_2 or CO_2) that is dissolved in the molten plastic must form a separate phase when the pressure is lowered. When it is done suddenly, inducing a rapid phase change through the rapid depressurization, a large number of cells nucleate because the large driving force (due to the large pressure change) activates a large number of nucleation centers. These nucleated cells grow to microscale bubbles of fairly uniform size, expanding to fill the mold and thus eliminating the need to apply the packing pressure (Suh, 1996; Okamoto, 2003).

When the polymer with dissolved gas is injected under high pressure at a fast injection rate (pressure drop rate unit time of at least 10 GPascals/second), a large number of tiny gas bubbles (with a diameter in the order of 10 microns) are present in the injection-molded parts. In this process, commercially known as the MuCell[5] process, no "packing pressure" is applied, which reduces differential shrinkage and residual stress. The growth and expansion of nucleated voids controls the local pressure and the expansion of the plastic. As a result, the pressure in the mold is much more uniform than in conventional molding processes. Because of the much more uniform pressure in the mold, the residual stress is less in the molded part, which results in less dimensional distortion. It is also possible that the presence of bubbles relaxes any residual stress as the part is ejected from the mold.

The MuCell injection-molding process decouples the part dimensions from the amount of material injected and the pressure under normal processing conditions, because the bubble expansion controls the final geometry in the mold. The MuCell parts shrink but more uniformly, because the packing pressure is essentially constant across a wide weight range, which means that changes to the part weight do not result in dimensional changes. This is unlike traditional molding where changes to the cavity pressure will result in changes to shrinkage (Kishbaugh, 2004).

The MuCell process was originally developed to reduce the consumption of plastics, but this geometric accuracy, together with some other advantages such as low injection pressure and short cycle time, was obtained as a bonus.

4.5   Case Study: Reduction of Time-Independent Real Complexity of Spark-Ignition Internal-Combustion (IC) Engine

4.5.1   Introduction to spark-ignition IC engine

How does a typical four-stroke spark-ignition IC engine work?
A conventional four-stroke cycle spark-ignition internal-combustion (IC) engine is shown in figure 4.4.[6] In the IC engine that uses a manifold port injection system (MPI), the fuel is injected right outside the intake valve of the cylinder before the valve opens and the piston in the cylinder begins to move down. Air also comes in through the manifold. When the valve opens, the fuel and air mixture is drawn into the cylinder because of the pressure difference created by the piston motion toward the bottom dead center (BDC), that is, the bottom end of the downward stroke.

After the transfer of the fuel and air into the cylinder through the intake valve is completed, the intake valve closes. During this process, the exhaust valve remains closed. During this intake process, the fuel and vapor also mix in the cylinder. This is the first stroke.

As the piston reverses its direction and moves up (pushed by the connecting rod connected to the crankshaft), the fuel/air mixture is compressed. During this second stroke, both the intake and exhaust valves are closed. When the piston almost reaches the top dead center (TDC), that is, the top end of the piston stroke, the compressed fuel/air mixture is ignited by the spark plug. The mixture is normally ignited at roughly 10° of crankshaft rotation before TDC.

After a finite delay, the flame propagates. By the time the piston reaches TDC, the combustion accelerates and the combustion products rapidly expand, pushing the piston downward under very high pressure, which turns the crankshaft via the connecting rod. This is the third stroke.

When the piston almost reaches the BDC, the exhaust valve opens and the combustion products are exhausted as the piston moves back up toward TDC during this fourth stroke. This process repeats.

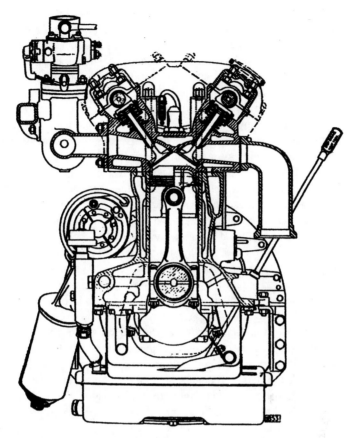

Figure 4.4 Typical four-stroke cycle spark-ignition internal-combustion (IC) engine. (Courtesy of Ford Jaguar Cars.)

Coupled nature of conventional IC engines

The engine development process has often involved making decisions between competing engine characteristics, including fuel efficiency, power output, physical size, emission characteristics, reliability, and durability, to name a few. In particular, emission characteristics are one criterion that is often evaluated by governmental organizations like the Environmental Protection Agency (EPA) in the United States. For instance, if some emission levels, such as nitrous oxides (NOx), hydrocarbons (HC), carbon monoxide (CO), or particulate matter, are too high for an engine, the engine may require expensive exhaust treatments using special devices such as a catalytic converter. In other instances, the engine might not be certified for operation or sale if it has poor emission characteristics. As a result, engine emissions should be carefully considered during the engine development process.

Carbon monoxide and NOx emissions (including both NO and NO_2) are formed during combustion. Carbon monoxide generally results when combustion occurs with an air and fuel mixture that has more fuel than the stoichiometric reaction requires (also known as a "rich" mixture). To address carbon monoxide concerns, most engines attempt to operate with stoichiometric or lean (less fuel than stoichiometric) air and fuel mixtures. However, some pockets of fuel-rich zones will typically still exist in the air and fuel mixtures of conventional engines. These pockets can result in carbon monoxide production. Conversely, NOx emissions are high when the air and fuel mixtures are lean or near stoichiometric values. Techniques used to address NOx formation include the recirculation of exhaust gases (EGR) into fresh air and fuel mixtures.

Among other causes, HC emissions can result from incomplete combustion or unburned fuel passing through a power cylinder during a period of intake and exhaust valve overlap. Cylinders of conventional engines often provide areas where it is difficult to sustain combustion, such as in the crevices between a piston and a cylinder wall. Additionally, most fuel-injection systems cannot provide fuel that is completely evaporated before combustion begins. Fuel may also cling to the walls of a cylinder after it has been injected, forming a wet sheet of fuel that does not burn. This often leads to incomplete combustion in at least portions of a combustion chamber, thus resulting in HC emissions. Hydrocarbon emissions are often worse when an engine is first started, as the engines are typically cold and complete evaporation of fuel is difficult to support. In recent studies, the ignition time is delayed when the engine is cold to minimize the emission of partially combusted hydrocarbons.

Both in diesel and spark-ignition engines, the fuel-to-air ratio is not the same throughout the cylinder—thus not stoichiometric—in part as a result of poor mixing. Some part of the fuel/air mixture is fuel-rich and some part is oxygen-rich (i.e., lean). The crown (top) of the piston, the injection angle, and valve size and location, and so on, are varied to control the flow of injected fuel/air mixture, but the problem still persists. This nonstoichiometric ratio may limit the maximum compression ratio of the engine, which controls the flame-propagation speed and the combustion chemistry.

Another problem of the conventional four-stroke spark-ignition engine is the knocking of the engine. This knocking problem limits the maximum compression ratio of conventional IC engines and, thus, the power efficiency of the engines. This limited compression ratio, in turn, determines the volume of the cylinder that still contains the hot combustion product when the piston is at the highest position of its compression stroke. Knocking is a result of self-ignition or autoignition. To prevent knocking, the most desirable combustion process in the power cylinder of spark-ignition IC engines is the one where a flame sheet propagates from the ignition point outward at a high compression ratio. Because of the expansion of the gas behind the flame front, the unburned fuel vapor and air mixture experiences high pressure and temperature before the flame front reaches the unburned region. When the pressure and temperature of the unburned fuel vapor/air mixture are high enough, the mixture can self-ignite (i.e., autoignition), causing a rapid rise in pressure, which induces vibration of the cylinder walls and audible knocks. This process is accelerated when there is enough time for sufficient autoignition precursors to form.

Two mechanisms control "knocking": the formation of precursors and the temperature rise that accelerates the flame-propagation rate. At high engine speeds, knocking may not be a problem since there is less time available for the precursors to form. On the other hand, as engine speed increases, there is less heat loss from the gases so that gas temperatures will be higher. This accelerates the precursor formation rate so that less time is required to form a concentration high enough for autoignition to occur. As a result of these two competing effects, some engines show knocking at high speeds, and some at low speeds.

Knocking can be severe when the fuel vapor/air mixture is at its stoichiometric ratio. This problem has been solved in current engines in two expensive ways: the use of anti-knock additives and the lowering of the compression ratio. To prevent autoignition, high-octane fuel—a mixture of many hydrocarbons with high-octane additives—is used in high-compression engines. If knocking persists even with the use of high-octane gaso-line, it is eliminated by changing the ignition timing to ignite the fuel vapor/air mixture at a lower pressure (thus at a lower compression ratio) when the piston has moved downward from its highest position. This lowers fuel efficiency.

Conventional methods of developing products, and specifically, internal-combus-tion engines, often lead to lengthy development cycles and consequently high cost because of the iterative nature of such methods. For example, an engine designer may make a modification to one component of an engine which, in turn, requires him to make many other modifications in other already designed and tested components of the engine. Making such a change may require reevaluating the previously tested compo-nents, thereby adding cost and time to the development process.

4.5.2 Design of a decoupled spark-ignition IC engine

The conventional IC engine is a coupled design that has large time-independent *real* complexity, since it cannot meet the emission standard of the U.S. government without using a catalytic converter. The purpose of this section is to show how the real com-plexity might be reduced by replacing a coupled IC engine, which has been used for more than a century, with a decoupled IC engine.[7]

The spark-ignition IC engine should satisfy the following FRs:

FR_1 = Measure the right amount of fuel for each combustion cycle.
FR_2 = Evaporate fuel.
FR_3 = Measure the right amount of air (oxidizer) for each combustion cycle.
FR_4 = Mix the vaporized fuel with the oxidizer.
FR_5 = Inject the mixture into the combustion chamber at a preset pressure.
FR_6 = Ignite the fuel/oxidizer mixture.
FR_7 = Deliver the power.
FR_8 = Exhaust the combustion product.
FR_9 = Minimize frictional loss.
FR_{10} = Control the emission of NOx, HC, and CO.

To satisfy these FRs, a decoupled engine has been proposed, which has two kinds of cylinders: a power cylinder and a fuel/air mixing and conditioning cylinder. The mix-ing cylinder is used to satisfy FR_2, FR_3, FR_4, and FR_5. Its function is to prepare the fuel/air mixture for the power cylinder, in which combustion takes place. This design employs the separation of functions using one cylinder—the mixing cylinder—to meter the fuel and air, and then mix the fuel vapor with air, and the other cylinder—the power cylinder—to combust the mixture and deliver power. This arrangement, together with other features, can minimize the emission of NOx, HC, and CO, and increase fuel effi-ciency. For the purpose of illustration, only a two-cylinder engine will be considered.

Figure 4.5 shows the design of a decoupled engine (Suh and Cho, 2002).[8] The engine comprises a mixing cylinder for mixing and compressing air and fuel, and a second cylinder for combusting the fuel and air and exhausting it from the engine. A conduit

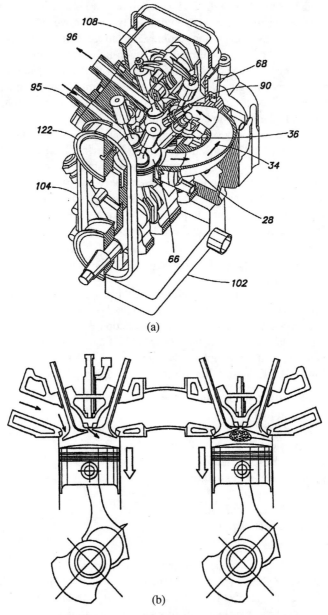

Figure 4.5 (a) Cut-off view of the engine and (b) cross-section of the pistons and cylinders. The first cylinder is the mixing cylinder and the second cylinder is the power cylinder. The two cylinders are connected by a conduit. The fuel and air flow into the mixing cylinder, where they are mixed and compressed. The mixture from the mixing cylinder flows into the conduit. The pressurized mixture in the conduit flows into the power cylinder, where it is combusted. There is a single connecting rod (the orientation of the connecting rod in the figure is not correctly shown). Part (b) shows when the air is coming into the mixing cylinder with fuel injection by a separate fuel injector. Ignition is taking place in the power cylinder. Note that the fuel may also be injected using a manifold port injection system.

provides fluid communication between the mixing cylinder and the power cylinder for delivering the air and fuel mixture from the mixing cylinder to the power cylinder. One or more valves control the delivery of the air and fuel mixture between the cylinders. In one embodiment, the valves are adapted to open and close out of phase with one another. In another embodiment, the swept volume of the power cylinder is smaller than that of the mixing cylinder. In some embodiments, the engine is adapted to prevent liquefied fuel that may exist in the mixing cylinder from entering the power cylinder.

This design has the feature of a "two and four stroke" engine. The engine has a cylinder block having a mixing cylinder and a power cylinder. The piston in the mixing cylinder reciprocates to complete an intake stroke and a compression stroke. During the first stroke, it forms a homogeneous air and fuel charge. The piston in the power cylinder reciprocates to complete a power stroke and an exhaust stroke within the second cylinder. The engine also has a crankshaft mounted within the cylinder block for rotation about its axis and connecting rods, which connect pistons to the crankshaft such that the second end of the first connecting rod is adapted to rotate with the crankshaft about the axis of rotation.

A conduit for fluid communication exists between the mixing cylinder and the power cylinder. The conduit has an opening into the mixing cylinder and the power cylinder for delivering all of the air and fuel charge from the mixing cylinder to the power cylinder. The conduit is selectively closable for closing fluid communication between the mixing cylinder and the conduit, and also between the conduit and the power cylinder.

The engine also has an exhaust passage for exhaust gas from the power cylinder. The opening and closing of the passageways can be done using either conventional valves or specially designed valves. Gases are isolated in the conduit between the compression and power strokes. Furthermore, the conduit is positioned above the mixing cylinder, whereby any volume of liquefied fuel transferred from the mixing cylinder to the transfer port and the power cylinder is minimized.

The mixing cylinder of the engine receives air and fuel to be mixed in the cylinder and compressed within the cylinder by the piston driven by a connecting rod attached to a crankshaft, thereby creating a compressed mixture of air and fuel charge. The conduit is designed to receive substantially all of the compressed air and fuel charge while retaining any liquid fuel in the mixing cylinder. The conduit also contains the compressed air/fuel charge to maintain the compressed air/fuel mixture at an elevated operating-pressure range. The power cylinder receives a portion of the compressed air/fuel mixture through the conduit, which combusts the compressed air and fuel charge to drive its piston connected to a connecting rod and the crankshaft.

Since the fuel and air must come into the mixing cylinder before being transferred to the power cylinder, all liquid fuel remains in the mixing cylinder. This retention of the liquid fuel in the mixing cylinder eliminates the possibility of liquid fuel going into the power cylinder, which is a major source of partially oxidized hydrocarbon emission when the engine is cold. The air and the fuel are compressed within the mixing cylinder by the piston to create a compressed air/fuel charge. Substantially all the compressed air/fuel charge is delivered to a conduit chamber while retaining any liquid fuel in the mixing cylinder. The compressed air/fuel charge is combusted within the power cylinder to drive its piston.

After combustion in the power cylinder, the exhaust products are expelled through an exhaust aperture while a new compressed mixture of air and fuel is delivered to the power cylinder. The inlet and exhaust apertures of the power cylinder are adapted to remain open concurrently for a period of time so that the incoming mixture of air and fuel can assist in expelling the exhaust products. The exhaust aperture is also adapted to close, leaving a portion of the exhaust products within the power cylinder.

The conduit acts as a pressure accumulator while providing fluid communication between a mixing cylinder and a power cylinder. The accumulator is adapted for retaining an air and fuel mixture within an elevated pressure range while the engine is in operation, thus allowing the air and fuel mixture to be delivered to the power cylinder at desired times and/or pressures.

The engine may have multiple power cylinders that are adapted to receive a portion of an air and fuel mixture delivered from one mixing cylinder. Conduits provide fluid communication between the mixing cylinder and each of the power cylinders for delivering the portions of the air and fuel mixture from the mixing cylinder to the power cylinders. This arrangement also provides more power per engine weight than conventional four-stroke engines, as more than half of its cylinders provide power during each crankshaft revolution.

The number of cylinders can be either in even numbers, that is, two, four, six, and so on, or in odd numbers. In the case of an even number of cylinders, one of the two cylinders is the mixing cylinder, and the other cylinder is the power cylinder. The diameter of the mixing cylinder is normally larger than that of the power cylinder by about 10–15% to increase the volumetric efficiency of the engine without the use of turbochargers. The large displacement volume of the mixing cylinder will increase the amount of air brought into the power cylinder, resulting in more efficient combustion.

In the engine with an odd number of cylinders, one mixing cylinder supplies the mixture into two power cylinders (see figure 4.6). The first conduit provides fluid communication between the mixing cylinder and one of the two power cylinders. A second conduit provides fluid communication between the mixing cylinder and the second power cylinder. Valves exist for closing and opening fluid communication through the conduits between the mixing cylinder and each one of the power cylinders. The diameter of the mixing cylinder is larger than the combined volume of the power cylinders to supply a stoichiometric ratio of air for the fuel to be combusted without the use of special equipment such as turbochargers.

Design parameters of the proposed decoupled engine
Based on the foregoing description, the FRs and DPs at the highest level of a spark-ignition IC engine may be summarized as follows:

FR_1 = Measure the right amount of fuel for each combustion cycle.
FR_2 = Evaporate fuel.
FR_3 = Measure the right amount of air (i.e., oxidizer) for each combustion cycle.
FR_4 = Mix the vaporized fuel with the oxidizer.
FR_5 = Inject the mixture into the combustion chamber at the preset pressure.
FR_6 = Ignite the fuel/oxidizer mixture.
FR_7 = Deliver the power.
FR_8 = Exhaust the combustion product.
FR_9 = Minimize frictional loss.
FR_{10} = Control the emission of NOx, HC, and CO.

DP_1 = Injection time of fuel injector at constant pressure.
DP_2 = Geometry of fuel injector/atomizer.
DP_3 = Volume of the mixing cylinder.
DP_4 = Air injector/fuel injector arrangement in the mixing cylinder.

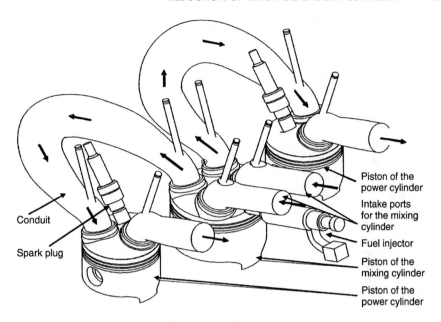

Figure 4.6 A three-cylinder option. The mixing cylinder in the middle has two intake ports and two exhaust ports like a pent-roof shape combustion chamber for a typical four-valve DOHC (double overhead camshaft) gasoline engine. The fuel injector can be installed under the intake ports as in this diagram if a SOHC (single overhead camshaft) is used; otherwise, it can be installed at the center of the combustion chamber with a DOHC, which is better for fast flame propagation. The bore size of the middle mixing cylinder can be bigger than those of the power cylinders to increase the volumetric airflow. Cylinder phase angle can be 180° for better mass balance, and this would not decrease engine performance.

DP_5 = Upward stroke of the piston in the mixing cylinder and fuel vapor/air supply line.

DP_6 = Spark plug.

DP_7 = Downward stroke of the piston in the power cylinder.

DP_8 = Upward stroke of the piston in the power cylinder and other parts of the exhaust system.

DP_9 = Undulated surfaces inside the cylinder/lubrication.

DP_{10} = Emission control systems.

The design equation at this highest-level design is:

$$
\begin{Bmatrix} FR_1 \\ FR_2 \\ FR_3 \\ FR_4 \\ FR_5 \\ FR_6 \\ FR_7 \\ FR_8 \\ FR_9 \\ FR_{10} \end{Bmatrix}
=
\begin{bmatrix}
X & 0 & 0 & 0 & 0 & 0 & 0 & 0 & 0 & 0 \\
x & X & 0 & 0 & 0 & 0 & 0 & 0 & 0 & 0 \\
0 & 0 & X & 0 & 0 & 0 & 0 & 0 & 0 & 0 \\
0 & x & 0 & X & X & 0 & 0 & 0 & 0 & 0 \\
0 & x & x & 0 & X & 0 & 0 & 0 & 0 & 0 \\
0 & 0 & 0 & 0 & 0 & X & 0 & 0 & 0 & 0 \\
0 & 0 & 0 & 0 & 0 & 0 & X & 0 & 0 & 0 \\
0 & 0 & 0 & 0 & 0 & 0 & 0 & X & 0 & 0 \\
0 & 0 & 0 & 0 & 0 & 0 & 0 & 0 & X & 0 \\
X & 0 & X & x & 0 & 0 & 0 & x & 0 & X
\end{bmatrix}
\begin{Bmatrix} DP_1 \\ DP_2 \\ DP_3 \\ DP_4 \\ DP_5 \\ DP_6 \\ DP_7 \\ DP_8 \\ DP_9 \\ DP_{10} \end{Bmatrix}
\qquad (4.4)
$$

Equation (4.4) has a triangular matrix if the order of FR_4 and FR_5 is reversed. Thus this is a good design that satisfies the independence of FRs when the DPs of the decoupled designs are changed in the order shown. These FRs and DPs may be further decomposed to develop detailed design embodiments, but any decomposition that satisfies these highest FRs and DPs will equally satisfy the design intentions described here.

FR_1 (Measure the right amount of fuel for each combustion cycle) and DP_1 (Injection time of fuel injector at constant pressure) may be further decomposed:

$FR_{1.1}$ = Measure the temperature of the mixing cylinder.
$FR_{1.2}$ = Measure the pressure of the fuel in the fuel pump.
$FR_{1.3}$ = Measure the speed of the engine.
$FR_{1.4}$ = Determine the right amount of fuel per cycle based on the temperature of the mixing cylinder.
$FR_{1.5}$ = Control the injector time.

The corresponding DPs are:

$DP_{1.1}$ = Temperature sensor.
$DP_{1.2}$ = Pressure sensor.
$DP_{1.3}$ = Speed sensor.
$DP_{1.4}$ = Algorithm for fuel amount.
$DP_{1.5}$ = Duration of the electric power on the fuel injector solenoid.

The design equation is:

$$
\begin{Bmatrix} FR_{1.1} \\ FR_{1.2} \\ FR_{1.3} \\ FR_{1.4} \\ FR_{1.5} \end{Bmatrix} = \begin{bmatrix} X & 0 & 0 & 0 & 0 \\ 0 & X & 0 & 0 & 0 \\ 0 & 0 & X & 0 & 0 \\ X & X & X & X & 0 \\ 0 & 0 & 0 & 0 & X \end{bmatrix} \begin{Bmatrix} DP_{1.1} \\ DP_{1.2} \\ DP_{1.3} \\ DP_{1.4} \\ DP_{1.5} \end{Bmatrix} \tag{4.5}
$$

This results in a decoupled design.

FR_4 (Mix the vaporized fuel with the oxidizer) and DP_4 (Air injector/fuel injector arrangement in the mixing cylinder) may also be decomposed to show the details of the design that promote the mixing of fuel vapor and the injected air:

$FR_{4.1}$ = Supply air through many nozzles distributed over the mixing cylinder.
$FR_{4.2}$ = Open/close the air supply line.
$FR_{4.3}$ = Mix air with the vaporized fuel vapor.

$DP_{4.1}$ = Air supply line and nozzles.
$DP_{4.2}$ = Valve.
$DP_{4.3}$ = Fuel injector position.

The design equation is:

$$
\begin{Bmatrix} FR_{4.2} \\ FR_{4.1} \\ FR_{4.3} \end{Bmatrix} = \begin{bmatrix} X & 0 & 0 \\ X & X & 0 \\ 0 & 0 & X \end{bmatrix} \begin{Bmatrix} DP_{4.2} \\ DP_{4.1} \\ DP_{4.3} \end{Bmatrix} \tag{4.6}
$$

FR$_5$ (Inject the mixture into the combustion chamber at a preset pressure) and DP$_5$ (Upward stroke of the piston in the mixing cylinder and fuel vapor/air supply line) may also be decomposed as:

FR$_{5.1}$ = Compress the mixture to a preset pressure.

FR$_{5.2}$ = Transport the pressurized mixture to the mixing cylinder at constant pressure.

DP$_{5.1}$ = Timing of the opening of the exhaust valve of cylinder C and the intake valve of cylinder P at the preset pressure.

DP$_{5.2}$ = Conduit and piston motions in the mixing cylinder and the power cylinder.

The design equation is:

$$\begin{Bmatrix} FR_{5.1} \\ FR_{5.2} \end{Bmatrix} = \begin{bmatrix} X & 0 \\ x & X \end{bmatrix} \begin{Bmatrix} DP_{5.1} \\ DP_{5.2} \end{Bmatrix} \tag{4.7}$$

FR$_{10}$ (Control the emission of NOx, HC, and CO) and DP$_{10}$ (Emission control systems) may be decomposed as:

FR$_{10.1}$ = Control the emission of NOx.

FR$_{10.2}$ = Control the emission of HC.

FR$_{10.3}$ = Control the emission of CO.

DP$_{10.1}$ = Injection of the extra fuel near the end of the compression and injection cycle of the mixing cylinder.

DP$_{10.2}$ = Cylinder C and the screen.

DP$_{10.3}$ = Stoichiometric fuel/air ratio.

The formation of NOx is a sensitive function of temperature. In one embodiment, the fuel may be injected twice into the mixing cylinder. The first injection occurs during the intake stroke to bring in air and fuel vapor during the downward stroke of the piston in the mixing cylinder to create a nearly stoichiometric mixture. The second injection occurs near the end of the compression-transfer period of the mixing cylinder to enrich the fuel vapor/air mixture that will be ignited in the power cylinder to prevent the formation of NOx by reducing the relative amount of oxygen.

To reduce the injection of liquid fuel droplets, a screen in front of the first transfer valve in the mixing cylinder and the presence of the mixing cylinder will control the HC emission, especially when the engine is cold. The emission of CO is reduced when the stoichiometric ratio of the fuel and air is maintained.

The design equation for FR$_{10.x}$ and DP$_{10.x}$ has a diagonal matrix as shown below:

$$\begin{Bmatrix} FR_{10.1} \\ FR_{10.2} \\ FR_{10.3} \end{Bmatrix} = \begin{bmatrix} X & 0 & 0 \\ 0 & X & 0 \\ 0 & 0 & X \end{bmatrix} \begin{Bmatrix} DP_{10.1} \\ DP_{10.2} \\ DP_{10.3} \end{Bmatrix} \tag{4.8}$$

Engine cycle

Figure 4.7 shows a first piston 20 in a mixing cylinder (not shown) through which it reciprocates.[9] It also shows a second piston 24 within a power cylinder (not shown)

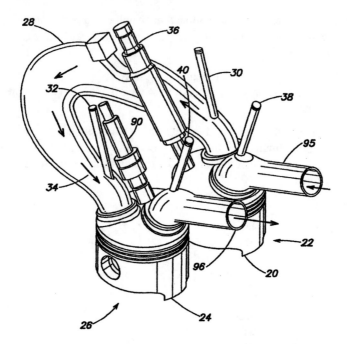

Figure 4.7 Detailed drawing of the mixing cylinder (only the piston 22 is shown) and the power cylinder (only the piston 24 is shown). 28 is the conduit. 30 and 32 are the valves that close and open the conduit. 38 is the intake valve of the mixing cylinder, and 40 is the exhaust valve of the power cylinder.

through which it reciprocates. A conduit 28 provides fluid communication between the mixing cylinder and the power cylinder. Transfer valves 30 and 32 are located at each end of the conduit to control the flow of the air and fuel mixture 34 between the conduit and each of the cylinders. A fuel injector 36 is adapted to provide fuel into the mixing cylinder.[10] An intake valve 38 is arranged to provide air to the mixing cylinder, and an exhaust valve 40 is arranged to allow combustion products to escape from the power cylinder.

The operating cycle of the present design differs from the four-stroke and two-stroke operating cycles that define most engines. In a four-stroke cycle, each cylinder of the engine is used to accomplish four different functions with four separate strokes of a piston within the same cylinder, including intake, compression, power, and exhaust. The intake stroke involves drawing air and/or fuel into the cylinder as the piston moves downward. The air and fuel mixture is then compressed within the cylinder as the piston moves upward. Typically, just before the piston reaches top dead center (TDC), a spark ignites the compressed air/fuel mixture, thereby beginning the combustion process. The combusting air and fuel mixture drives the piston downward, thereby providing useful mechanical work through a rotating crankshaft that is typically connected to the piston via a connecting rod. Combustion ends as the piston nears bottom dead center (BDC) and begins moving upward. At this point, an exhaust aperture is opened, allowing the combustion products to be removed from the cylinder by the piston as it travels toward TDC. The intake valve opens again, either before or after the exhaust valve closes, and the cycle repeats itself.

In a two-stroke engine, the four functions described above are accomplished in two strokes. There is first an intake/exhaust stroke, which occurs when the piston is near BDC. Here, an intake valve or other type of aperture is opened, allowing a pressurized air and fuel mixture into the cylinder. The new air and fuel mixture displaces any gases that previously existed within the cylinder such as exhaust products from a previous cycle. These gases are expelled through an open exhaust valve or other type of aperture. Once the new air and fuel mixture is located in the cylinder and the previous gases are displaced, the intake and exhaust valves are closed as the piston moves upward toward TDC, thereby compressing the air and fuel mixture. Combustion then begins as a spark ignites the air and fuel mixture when the piston nears TDC. The combusting air and fuel mixture drives the piston downward, thereby providing useful mechanical work through a rotating crankshaft that is typically connected to the piston via a connecting rod. Once the piston nears BDC, the intake and exhaust apertures open, and a new air and fuel mixture is introduced to the cylinder. The new air and fuel mixture then displaces the exhaust products of the previous cycle so that the cycle may repeat.

The general operating cycle of the present design accomplishes the four different functions described above in four separate strokes. Two of these strokes occur in a mixing cylinder with a mixing cylinder piston, and the other two strokes occur in a power cylinder with a power cylinder piston. The intake stroke involves drawing air and/or fuel into the mixing cylinder as the mixing cylinder 22 piston moves downward. The air and fuel mixture is then compressed within the cylinder as the mixing cylinder 22 piston moves upward. Sometime before the piston reaches TDC, a first transfer valve 30 opens fluid communication to a conduit 28. Substantially all of the air and fuel mixture 34 is then transferred to the conduit in a pressurized state. The transfer valve 30 closes as the mixing cylinder piston 22 nears TDC. The intake valve 38 opens after the mixing cylinder piston reaches TDC and begins on its downward stroke, allowing the intake and compression strokes of the cycle to be repeated within the mixing cylinder.

At a desired time, a second transfer valve 32 opens fluid communication between the conduit 28 and the power cylinder 26. The compressed air and fuel mixture 34 is then transferred from the conduit to the power cylinder as the power cylinder piston is on its upward stroke. This allows the air and fuel mixture to remain within an elevated operating-pressure range as it is transferred to the power cylinder. The valve 32 between the conduit and the power cylinder closes before the power cylinder piston reaches TDC, and then a spark 46 ignites the compressed air/fuel mixture, thereby beginning the combustion process. The combusting air and fuel mixture drives the power cylinder piston 24 downward, thereby providing useful mechanical work through a rotating crankshaft 48 that is typically connected to the piston with a connecting rod 50. Combustion ends as the power cylinder piston nears BDC and then begins moving upward. Near BDC, an exhaust aperture 40 is opened, allowing the combustion products to be removed from the power cylinder by the piston as it travels toward TDC. The transfer valve 32 between the conduit and power cylinder opens again, either before or after the exhaust valve closes. The power and exhaust strokes of the cycle are then repeated within the power cylinder.

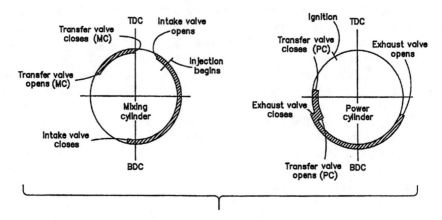

Figure 4.8 An engine cycle.

Steps of engine operating cycle

The engine cycle and the engine structures that are involved with the respective cycle are now described in more detail with respect to the particular embodiment of the engine cycle represented in figure 4.8. In particular, figure 4.8 describes the motions of the pistons and valves associated with both the mixing cylinder and the power cylinder, according to one aspect of the invention.

Figure 4.9 shows the motions of the pistons, valves, and the air and fuel mixture at various points throughout the cycle defined in figure 4.8. In this particular embodiment, the pistons of both the mixing cylinder and the power cylinder move in phase with one another, although other arrangements are possible. It is noted that figure 4.9 shows the mixing cylinder piston and the power cylinder piston as being attached to separate crankshafts. However, the pistons are connected to the same crankshaft via connecting rods.

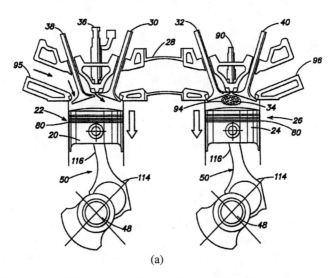

(a)

Figure 4.9 The motions of the pistons, valves, and the air and fuel mixture at various points throughout the cycle defined in figure 4.8. See text for explanation of parts (a)–(e).

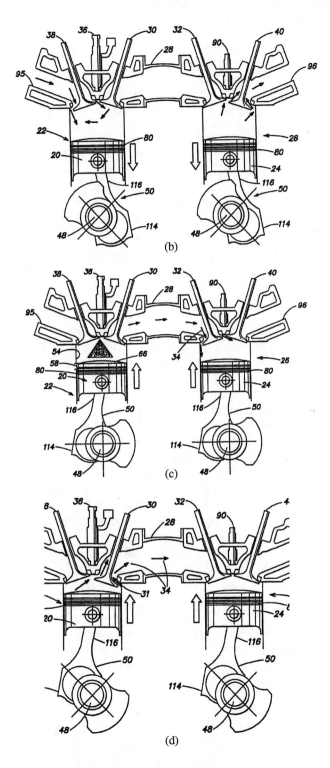

Figure 4.9 *(continued)*

109

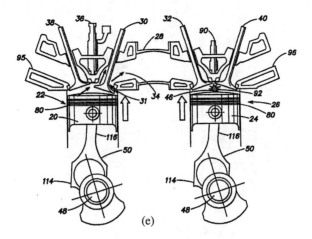

(e)

Figure 4.9 *(continued)*

Intake

The operating cycle of figures 4.8 and 4.9 is now described, beginning with the motions of the mixing cylinder. The mixing cylinder piston, as shown in figure 4.9(a), is approximately 45° of crank angle after it has descended from its TDC position. At this point, the downward motion of the piston has created a reduced pressure zone within the mixing cylinder. This reduced pressure allows air to be drawn into the mixing cylinder through the intake valve that opens at approximately 30° after TDC. The mixing cylinder will draw in a substantially similar volume of air during each engine cycle.

In some embodiments, the mixing cylinder swept volume may be increased to improve the volumetric efficiency of the engine. In particular, it may be larger than the swept volume of the power cylinder. Alternatively, in other embodiments, air could be pushed into the cylinder by peripheral components, such as turbochargers, superchargers, ram air devices, or other suitable means, as the invention is not limited in this respect. In these scenarios, the amount of air drawn into the mixing cylinder may vary between cycles. Air continues to enter the mixing cylinder, as is shown in figure 4.9(b), until the mixing cylinder piston nears BDC. In particular, the figure 4.8 embodiment has the intake of air continuing until 10° (crank angle) past BDC when the intake valve closes.

Fuel delivery

Fuel may be injected into the mixing cylinder (or into the intake manifold) during the air intake process with a low-pressure fuel injector. Fuel injection is shown to begin between 40° and 60° after TDC in the cycle diagram of figure 4.8. However, figure 4.9(c) shows fuel being delivered with a high-pressure fuel injector well after the mixing cylinder piston has reached BDC and is returning toward TDC, as the invention is not limited in this respect. To deliver fuel, as shown in figures 4.7 and 4.9, a fuel injector is used to directly deliver fuel into the mixing cylinder, although other embodiments may incorporate different types of fuel delivery systems such as carburetors, port fuel injectors, indirect fuel injectors, gaseous fuel injectors, or other suitable fuel-delivery systems, as the invention is not limited in this respect. In some embodiments, fuel is injected substantially orthogonally into air that is flowing into the mixing cylinder. Injecting fuel in this manner helps promote evaporation and mixing. In other embodiments, multiple fuel injections may be used as well.

Fuel delivery continues until the desired amount of fuel has been injected into the mixing cylinder. Operating conditions of the engine at any given moment may determine how much fuel is required. For instance, if more air is delivered to the mixing cylinder, then more fuel will be required to maintain a similar air/fuel ratio within the mixing cylinder. In many embodiments, more air and fuel is allowed into the cylinder when the engine requires more power. The amount of air provided to the cylinder may be controlled by a throttling device within the intake system of the engine. In other embodiments, peripheral devices such as turbochargers, superchargers, and/or ram air devices may also affect the amount of air provided to the mixing cylinder and thus affect the amount of fuel required.

While the strategy behind the present invention is generally to operate with an air/fuel mixture near the stoichiometric value, there may be certain scenarios where altering the air/fuel ratio is desired, as the present invention is not limited in this respect. For instance, some embodiments of the invention may regularly draw substantially the same amount of air into the mixing cylinder during every engine cycle. In such embodiments as well as others, the torque output of the engine and/or the operating speed of the engine can be changed by altering the air/fuel ratio. Operating the engine with a rich air and fuel mixture may increase the engine torque and/or engine speed, while operating the engine with a lean air and fuel mixture may decrease the engine torque and/or engine speed.

Fuel and air mixing

Fuel and air homogenization is promoted by various features and aspects of the mixing cylinder, as unevaporated fuel or nonhomogenized air and fuel mixtures can cause incomplete combustion and hydrocarbon emissions. A fuel-delivery system that atomizes most of the fuel as it is delivered into the mixing cylinder helps evaporate fuel and homogenize the mixture. However, some of the injected fuel may impinge the walls 54 of the cylinder and form a liquid fuel film. Liquid fuel may also be trapped between the outer cylindrical walls 58 of the piston and the cylinder walls 54. Such liquid fuel typically causes incomplete combustion and hydrocarbon emissions in a conventional engine. However, if liquid fuel resides within the mixing cylinder of the present invention, it will remain there until it evaporates. Some embodiments of the invention may include a receptacle in the piston crown for retaining liquid fuel until it can evaporate. Furthermore, the environment of the mixing cylinder is maintained at a temperature that promotes the rapid evaporation of fuel within the mixing cylinder. For one embodiment operating at 3500 revolutions per minute, a temperature of 500 K accomplishes this effect.

The mixing cylinder may also include other features, such as turbulators, placed at various positions within the cylinder that promote the evaporation and homogenization of the air and fuel mixture through turbulent air motions within the cylinder. These turbulators may include structures placed near the valve port, on the crown 66 of the piston, on the firedeck of the cylinder head, or in any other suitable location, as the invention is not limited in this respect. The fact that combustion does not occur within the mixing cylinder provides a wide degree of freedom in designing turbulators, which are often designed to endure the rigors of a combustion environment in conventional engines.

The mixing cylinder may also incorporate mixing features that might otherwise be subject to combustion pressures and temperatures in a conventional engine. Active mixing devices, such as a mixing fan disposed in the crown of a piston or on the firedeck of the cylinder head, may be included within the mixing cylinder to promote fuel evaporation and mixture homogenization. Such a mixing fan may comprise a rotor that actively moves air and fuel about the mixing cylinder. The active mixing fan can be driven by fluids directed to a separate drive rotor that is disposed outside of the mixing cylinder via a shaft. Fluids, such as engine oil, engine coolant, or any other suitable fluids, may serve to rotate the drive rotor, which in turn rotates the mixing fan. Alternatively, the reciprocating motion of the piston, an electric drive system, or even a magnetic drive system between the fan and the walls of the cylinder may serve to drive the active mixing device. In some embodiments, the mixing fan may be heated by various engine fluids, or even electrically, to improve fuel evaporation. Other suitable drive means may be used, as the present invention is not limited in this respect.

Compression

Return now to figure 4.9(e), where the air and fuel mixture is shown to be compressed after the intake valve 38 closes and the piston 20 begins moving upward toward TDC. The compression stroke continues until a first transfer valve 30 opens fluid communication between the mixing cylinder 22 and a conduit 28. This occurs from approximately 60° before TDC until TDC, although other opening times, closing times, and delivery durations are possible, as the invention is not limited in this respect. Substantially all of the air/fuel mixture 34 is then transferred to the conduit through the aperture 31 of the conduit, as is depicted in figure 4.9(d) and (e). This transfer, as depicted, is timed to substantially prevent any backflow of gases into the mixing cylinder. Substantially complete transfer of the homogenized air and fuel mixture is possible in embodiments with very little clearance volume in the mixing cylinder.

It is noted that the aforementioned aspects and features that promote evaporation and homogenization within the mixing cylinder 22 also reduce the likelihood of transfer of liquid fuel to the conduit 28. However, should any portion of the fuel not evaporate before the air and fuel mixture is delivered to the conduit 28, the fact that the aperture 31 to the conduit is located near the top of the mixing cylinder will further prevent the liquid fuel from entering the conduit. Additionally, injected liquid fuel droplets will tend to contact the mixing cylinder walls and the piston because of their greater weight and thus greater momentum. Then the liquid droplets will likely stick to the wall or piston because of surface tension. Furthermore, some embodiments may include additional features in or near the entrance to the conduit to insure that liquid fuel is retained in the mixing cylinder. One such feature is a mesh screen placed near the aperture 31 between the mixing cylinder 22 and the conduit 28. Should any liquid fuel be carried toward the conduit, it will likely impact the screen and be removed from the air before it passes into the conduit. A tortuous passageway can also be placed between the mixing cylinder and conduit to serve a similar function. Additionally, other features to further insure that liquid fuel does not enter the conduit may also be incorporated into the engine, as the invention is not limited in this respect.

The mixing cylinder 22 of the various embodiments of the invention is not required to contain hot, combusted gases. As a result, numerous advantageous features can be incorporated into the mixing cylinder. For instance, the sealing mechanisms 80 that typically exist between the outer cylindrical surface of the piston 50 and the inner wall 54 of the cylinder do not have to contain hot, extremely high-pressure gases during combustion. Therefore, they can be manufactured from materials that are less expensive and that present less frictional resistance to the movement of the engine. Additionally, the surfaces of the cylinder wall may comprise undulated surfaces to reduce frictional drag between the piston and cylinder. Such surfaces reduce the work required of the engine to compress air and fuel within the mixing cylinder, and/or ultimately to create a more efficient engine.

Another benefit realized by the use of a separate mixing cylinder 22 is that less heat needs to be removed from the mixing cylinder environment. Many embodiments of the invention include features such as an engine coolant jacket that surrounds the mixing cylinder to help maintain its temperature. It does not need to remove as much heat as it would in a conventional engine. As a result of lower temperatures, the cylinder may be made of a much lighter-weight material, and/or a material that does not need to withstand extremely high temperatures typically associated with combustion, such as some aluminum alloys.

The mixing cylinder 22 may also have a much higher compression ratio than that of a typical cylinder. Compression ratio is defined as the volume within the cylinder when the piston is at BDC over the volume in the cylinder when the piston is at TDC. Most compression ratios of typical engines cannot be too high because an air and fuel mixture may autoignite if compressed too much in the hot environment that exists in a cylinder that supports combustion. Such autoignition can cause "knocking" in a spark-ignition engine, as is discussed later.

Accumulation
The pressure level in the conduit is raised as the air and fuel mixture is delivered from the mixing cylinder 22. The conduit 28 is typically maintained within an elevated operating pressure range except for certain conditions where the conduit is under substantially atmospheric pressure, such as during initial engine starting or during some transient operation modes. The pressure levels of both the mixing cylinder and the conduit are depicted in figure 4.10 (see page 118) for an embodiment of the engine operating at 3500 revolutions per minute with the mixing cylinder piston and a power cylinder piston moving in phase. This embodiment also has a base diameter of 158 mm and a stroke of 42 mm; the compression ratio of the mixing cylinder is 20:1 and the compression ratio of the power cylinder is 9:1. In this embodiment and at this engine speed, the conduit maintains an elevated operating pressure between 4 and 6 bars, although other suitable pressures may be employed, as the present invention is not limited in this respect.

The air and fuel mixture 34 delivered to the conduit 28 may exist in the conduit along with a portion of an air and fuel mixture that was delivered in a previous cycle or cycles. In this sense, the conduit 28 can act as an accumulator that collects homogenized air and fuel mixtures 34 and holds them in the accumulator within a substantially elevated operating pressure range. In one embodiment, the conduit 28 defines a volume substantially equal to the swept volume of the mixing cylinder 22. This allows the conduit to retain several times the amount of air delivered during one cycle of the engine, if desired. However, conduits defining larger or smaller volumes may be employed, as the present invention is not limited in this respect.

Valves found in conventional engines typically only have to hold a pressurized gas within a cylinder. However, the valves 30 and 32 at each end of the conduit 26 in the present invention are required to hold a pressurized gas within the conduit, as well as within their respective cylinders. Although the pressure within the conduit is generally lower than the pressure within the mixing cylinder 22, and substantially lower than the peak pressures witnessed in a power cylinder 26, some modifications may be made to the valves to help them close fluid communication. These changes may include increasing the valve spring strength to provide a greater closing force and/or making the valves out of a much lighter material such as titanium. Lighter materials such as titanium may also improve valve train dynamics and even help prevent valve surge in some embodiments. This can be particularly helpful in embodiments that have rapid valve motions.

The presence of the conduit 28 between mixing cylinder 22 and power cylinder 26 allows the engine to effectively have a variable compression ratio. In a conventional engine, the compression ratio determines what pressure the air/fuel mixture 34 will have when it is in a fully compressed state near the beginning of combustion. This is generally a fixed value in a conventional engine. However, the conduit 28 of the present invention acting as an accumulator can take on various different pressure levels as desired by the engine controller. In some embodiments, particularly those with solenoid-actuated valves or other valves that may be adjusted during operation, the compression ratio or effectively the pressure at which the air/fuel mixture is delivered to the power cylinder 26 prior to combustion may be varied according to the engine operating parameters.

Delivery of air and fuel mixture to power cylinder

The embodiment represented by figure 4.9(c) shows an air and fuel mixture 34 being delivered from the conduit 28 to the power cylinder when a second transfer valve 32 opens at the opposite end of the conduit 28. This occurs at approximately 120° before TDC (in the power cylinder) and continues for approximately 40° until the valve closes in the embodiment represented by figure 4.8, although other valve-opening times and durations may also be suitable for other embodiments. The air and fuel mixture is delivered to the power cylinder 26 as the piston 24 in the power cylinder is on its upward stroke, allowing the transfer of the air and fuel mixture to occur within the elevated operating pressure range. This is represented in the plot in figure 4.11 (see page 118) of pressure in the conduit 28 and power cylinder 26 versus crank position. The opening and closing of the second transfer valve 32 is generally timed to prevent flow from occurring in a reverse direction, that is, from the power cylinder to the conduit. However, such flow may occur under some scenarios, such as during engine starting. It is noted that the conduit pressure is shown scaled ×10 in figure 4.11. The accumulating aspect of the conduit 28 may allow the air and fuel mixture 34 to be delivered to the power cylinder 26 at a time desired for the particular engine operating conditions. Additionally, the accumulating aspect of the conduit 28 may also allow control of the pressure level at which the air and fuel mixture 34 is delivered to the power cylinder. Control over these variables can greatly assist in tuning the engine to provide improved emission characteristics.

Some embodiments of the conduit 28 may include a fuel delivery device 36 adapted to inject a small portion of atomized or otherwise gaseous fuel into the air and fuel mixture 34 as it enters the power cylinder 26. Such a portion of fuel is intentionally designed to create a fuel-rich portion of an otherwise homogenized air and fuel mixture 34. This fuel-rich portion is adapted to reside near an ignition device in the power cylinder to aid in initiating combustion. It may also be used in conjunction with an air and fuel mixture 34 that is otherwise lean of fuel. This strategy can be used to lower emissions of NOx and/or hydrocarbons under some circumstances.

The power cylinder piston 24 may continue on its upward stroke for approximately 80° of crank angle after the air and fuel mixture 34 has been delivered and the transfer valve closes, as is depicted in figures 4.8 and 4.9(d). However, the timing of the delivery of the air and fuel mixture 34 to the power cylinder 26 may differ in other embodiments, as the invention is not limited in this respect. In some embodiments, the second transfer valve 32 between the conduit 28 and the power cylinder 26 can even vary according to particular engine operating parameters, such as engine speed, engine power, and emission characteristics, to name a few.

Combustion

After the air and fuel mixture 34 is delivered to the power cylinder and the second transfer valve 32 closes, the air and fuel mixture is ignited to begin the combustion process. In the embodiments represented by figures 4.8 and 4.9(e), this occurs when the power cylinder piston 24 is between 30° and 10° before TDC. A spark plug 90 protruding through the firedeck 64 of the cylinder head 68 is typically used to initiate combustion, although other suitable devices may be used as well. Ignition of the air and fuel mixture 34 starts adjacent to the protruding end of the spark plug 90, where it forms a flame kernel 92. As the piston nears TDC, the kernel 92 rapidly spreads until a flame front that extends to the cylinder walls 54 is created. This flame front progresses through the cylinder, combusting the air and fuel mixture 34 as it moves through the power cylinder 26, pushing the piston 24 on its downward stroke.

As the air and fuel mixture 34 is burned, the temperature and pressure within the power cylinder 26 rapidly increase. The rapidly increased pressure drives the power cylinder piston 26 downward, thereby creating useful mechanical work. This work is transferred from the piston 26 to the crankshaft 48 of the engine via a connecting rod 50 as shown in figure 4.9(a).

As was previously discussed, the air and fuel mixture 34 enters the power cylinder free of liquid fuel and in a homogenized state (except for embodiments that intentionally have a fuel-rich area for ignition). Having such a homogenized, liquid-free air and fuel mixture allows the flame front 94 to burn the air and fuel mixture 34 substantially completely as it propagates through the cylinder 26, which can improve the hydrocarbon emission characteristics of the engine. Furthermore, an air and fuel mixture free of liquid fuel will make it difficult for any liquid fuel to become trapped in the crevices between the piston and the cylinder wall or on the cylinder walls, where it can be difficult to combust. Uncombusted fuel in such crevices and on the cylinder walls can cause hydrocarbon emissions.

Furthermore, a homogenized air and fuel mixture helps prevent knocking from occurring in the power cylinder. As combustion progresses through the cylinder, the pressure and temperature increase dramatically. The pressure and temperature may become great enough to cause any unburned fuel-rich areas of the air and fuel mixture 34 to autoignite at secondary locations in the cylinder. If this occurs, an additional flame front may be created that can disrupt the combustion process. The additional flame front can cause incomplete combustion of the air and fuel mixture 34, leading to hydrocarbon emission problems. Also, the additional flame front may cause shock waves that can propagate through the engine, causing damage to it.

While knocking can be caused by a nonhomogenized mixture, it can also be caused by hot spots within a cylinder. Deposits left on the power cylinder surfaces by incomplete combustion of previous cycles may remain hot after combustion has occurred. If they remain hot long enough, they can ignite the air and fuel mixture delivered to a power cylinder during a subsequent engine cycle, thus causing secondary ignition and the afore-mentioned knocking phenomenon. By providing a homogenized mixture to the power cylinder, embodiments of the present invention promote complete combustion of the air and fuel mixture. This also prevents the formation of deposits on the surfaces of the power cylinder, thereby reducing the likelihood of the knocking phenomenon occurring.

In some cases, unwanted autoignition can occur during the compression stroke of an engine cycle. This is not the case for embodiments of the present invention as substantially all compression takes place in the mixing cylinder 22. The mixing cylinder does not sustain combustion and therefore should not contain any deposits where autoignition can begin. Furthermore, the mixing cylinder 22 is not subjected to high combustion temperatures and can therefore remain at a temperature that will help prevent autoignition, as was previously discussed.

Exhaust

The combustion process is shown to terminate at approximately 70° before BDC in the embodiment represented by figures 4.8 and 4.9(b). At this point, an exhaust valve 40 opens fluid communication with an exhaust port 96 disposed outside of the power cylinder 26. This allows the still pressurized combustion products within the cylinder to escape through the exhaust port. As the power cylinder piston begins moving upward toward TDC, it helps expel the remaining combustion products from the power cylinder 26.

In some embodiments, substantially complete removal of the exhaust products is possible in the power cylinder 26 as the cylinder can be designed with substantially no or minimal clearance volume, if desired. The fact that compression of the air and fuel mixture 34 occurs primarily within the mixing cylinder 22 allows there to be minimum clearance volume 100 within the power cylinder 26. In conventional engines, some clearance volume needs to exist to prevent the air and fuel from being compressed to extreme pressures, which can cause knocking in some scenarios, as was previously discussed.

In other embodiments of the invention, retaining some of the combustion products within the power cylinder 26 for admixing with the air and fuel mixture 34 of a subsequent cycle may be desired. Such strategies to recirculate exhaust gases can reduce NOx emissions of an engine. A portion of the combustion products may be retained in the power cylinder 26 either by including a clearance volume 100 in the power cylinder 26 or by timing the opening and closing of the second transfer valve 32 and the exhaust valve 40 of the power cylinder 26 accordingly.

In some embodiments of the engine, the end of the exhaust process may overlap with the beginning of the intake process. For instance, the embodiment of figure 4.8 has the transfer valve 32 into the power cylinder 26 open for approximately 10° while the exhaust valve 40 is open. This allows the incoming air and fuel mixture 34 to help purge the combustion products from the power cylinder 26. This particular embodiment also retains approximately a 20% portion of the combustion products for mixing with the incoming air and fuel mixture to help reduce NOx emissions. This also helps insure that the air and fuel mixture 34 is not allowed to escape through the exhaust port 96 and contribute to hydrocarbon emissions.

The power cylinder 26 comprises many conventional features that are typically used within a cylinder to support combustion therein. For instance, piston ring technology, cylinder surfacing technologies, cooling technologies, and other suitable features may be incorporated into the power cylinder design.

Alternative cycle embodiments

An entire engine operation cycle has been described according to an embodiment of the invention. However, other variations of the engine operation cycle may exist within the scope of the invention. For instance, while the motions of the mixing cylinder piston 22, intake valve 38, and mixing cylinder transfer valve 30 are similar, the power cylinder piston 24 moves out of phase with the mixing cylinder piston 20. The power cylinder transfer valve 32 and the exhaust valve 40 maintain a similar opening and closing relationship to the power cylinder piston 24. Many other cycles of the mixing cylinder and the power cylinder are possible, which are not shown here for brevity.

General engine construction

The various engine structures that may be used to provide the above-described cycles are now discussed. Figure 4.5(a) shows a cutaway schematic view of an embodiment of the invention. This particular embodiment is an inline, two-cylinder engine configuration where each cylinder has a swept volume of approximately 110 cm^3. However, other configurations, such as "V" configuration or "W" configuration engines, opposed cylinder engines, "H" engine configurations, or even Wankel-type engines, could employ features of the present invention to improve their emission characteristics. Furthermore, any number of mixing cylinders 22 and power cylinders 26 may be employed by a given engine.

While the engine operating cycle has been described with respect to a spark-ignition engine, and figure 4.5(a) depicts a spark-ignition engine, autoignition engines (e.g., diesel) may also benefit from many of the features of the present invention.

The apertures that provide fluid communication between the various portions of the engine, including the intake port 95, the mixing cylinder 22, the conduit 28, the power cylinder 26, and the exhaust port 96, may comprise any valving or porting means presently known or that may subsequently be developed. Such devices may include pressure-activated check valves or reed-type valves or ports that open fluid communication to a cylinder when a piston is located at a particular point as it reciprocates through the cylinder, as the invention is not limited in this respect. Additionally, solenoid-actuated valves may be used in the engine design. Solenoid-actuated valves can offer a wide range of flexibility as to when a given aperture is opened. These valves may also allow the valve opening time to be adjusted during the operation of the engine. Such opening and closing may be controlled by a programmable engine control module (ECM) that operates the engine for optimum performance.

Simulation results

Before the detailed features of the engine were designed, the effect of various parametric relationships and performance were simulated by the engine group at the Institute for Advanced Engineering in Korea.[11] Two of these results are shown in figures 4.10 and 4.11. Figure 4.10 shows the pressure change in the mixing chamber and the conduit. Figure 4.11 shows the pressure in the power cylinder.

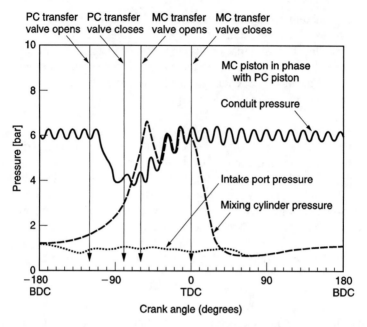

Figure 4.10 Pressure pumping loop for the mixing cylinder (0 phase angle).EV, exhaust valve of No. 1 mixing cylinder; IV, intake valve (**O, valve open; **C, valve closed); IV′, intake valve of No. 2 power cylinder (EV and IV′ are connected to the conduit). Conduit pressure is steadily increased from 4 to 6 bars when EV is open (EVO~EVC). First of all, the conduit is filled up for this duration, and in the next cycle, IV′ is open to push all mixture into the power cylinder so that the conduit pressure is going down from 6 to 4 bars.

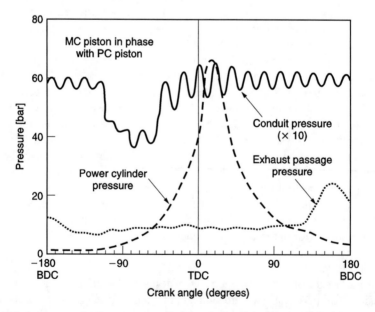

Figure 4.11 A pressure plot for a power cylinder and a conduit of an engine with a mixing cylinder piston moving in phase with a power cylinder piston.

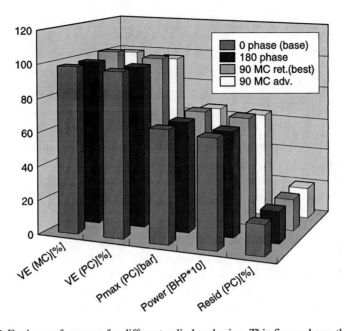

Figure 4.12 Engine performance for different cylinder phasing. This figure shows that engine performance does not vary much for different cylinder phasing. VE(MC), volumetric efficiency of No. 1 mixing cylinder; VE(PC), volumetric efficiency of No. 2 power cylinder; Pmax(PC), max. cylinder pressure of power cylinder; Power in BHP (brake horsepower) ×10; Resid(PC), residual gas remaining in power cylinder.

The simulation results were compared with the published data of the Honda engine, which was converted to make the proposed design. The comparison showed that the proposed engine should perform as well as the Honda engine (see figure 4.12).

4.5.3 Conclusion on the proposed decoupled engine

Recently the decoupled engine was built and tested, but the test was not conclusive because of many problems with the test setup and instruments. Further work needs to be done.

One of the major objectives of the proposed design is to reduce the emission of hydrocarbons, NOx, and CO without using catalytic converters. If we can achieve this goal, we will have reduced the complexity of conventional spark-ignition IC engines. The final verdict is not in yet. It is quite a technological challenge.

4.6   Elimination of Time-Independent Imaginary Complexity

In chapter 3, time-independent imaginary complexity was defined as a measure of uncertainty that is not real uncertainty but arises because of the lack of knowledge. Using a decoupled design with a triangular design matrix, it was shown that when the designer tries to satisfy the FRs without the benefit of the design matrix, the number of trials required to find the right combination is given by $n!$, where n is the number of FRs and DPs. The imaginary complexity was illustrated using the examples of the combination lock and the printing machine developed by a company.

Time-independent imaginary complexity may be responsible for the high cost of developing engineered systems. The number of unnecessary trial-and-error steps taken will increase the development cost. The minimum cost of developing a new product may be substantially less than the actual cost of developing the product, depending on the time and effort wasted on imaginary complexity problems. There are two anecdotal cases that support this conclusion: the creation of Acclaro software and the development of a chemical mechanical planarization (CMP) machine.

Acclaro[12] is a software system developed by a small company to assist designers in the practice of axiomatic design without fully understanding the theory. In developing the core of the software engine, which might have involved several hundred FRs and DPs, it took 4 human-months for the design of the software and 2 human-months of actual coding. Those who reviewed the software estimated that such a software system would normally take about 2 human-years, a factor of $4\times$ greater effort than when the software was developed without involving imaginary complexity. Acclaro has since been expanded to include various additional functions that the users of the software required for various specific tasks.

In the case of the CMP machine, a commercial-scale machine was developed at MIT by four master's degree candidates in two years (Suh, 2001). The machine, which is about 1 m wide, 2.4 m high, and 2.75 m long, has the mechanical part and the control/software part. The machine is more accurate than commercial CMP machines that sell for about $2 million. The machine performed to specification after it was assembled without much debugging. The remarkable thing is that these students did not have any prior experience in designing and developing products. If an industrial firm developed such a system using their normal development process, it would have taken about 60 human-years because of their trial-and-error process of developing new products, which would invariably involve much imaginary complexity as well as real complexity.

In addition to the random trial-and-error processes used, there are other sources of imaginary complexity. For instance, redundant designs can cause imaginary complexity problems when the fact that there are more DPs and FRs is overlooked in designing and implementing the system. Another cause of imaginary complexity may be related to defining a wrong set of FRs or solving a wrong problem, using a set of irrelevant DPs. This may occur when a protein that has only a secondary effect on a given FR of a biological system is chosen to study the behavior of the system. A large number of random tests done in many fields may constitute special cases of time-independent imaginary complexity.

4.7   Summary

This chapter presented a means of reducing time-independent real complexity and time-independent imaginary complexity. Real complexity exists when the system range of an

FR is outside the design range. To have no real complexity the system range must always be inside the design range.

Time-independent imaginary complexity may exist when we have a decoupled system. It is a result of not knowing that the design is a decoupled design and that the FRs must be changed in the sequence given by the design matrix. However, when the design equation is not written and the system is being developed through trial and error, imaginary complexity can be a source of major wasted effort and money. This happens very frequently in industry.

When a system is a decoupled or uncoupled design, any time-independent real complexity can be decreased by eliminating the bias and reducing the variance of FRs, which may be done systematically.

When the FRs of a system are coupled at one or more levels of the system design, time-independent real complexity can be reduced only through redesign to develop an uncoupled or decoupled design. In this chapter, the design of an uncoupled knob was used to illustrate how real complexity can be reduced through redesign. The conventional coupled design may have infinite real complexity.

Another example is presented that illustrates how a coupling can be removed from a coupled design—the spark-ignition internal-combustion engine. A new engine was designed to enable the satisfaction of all FRs independently. However, thus far, test results are insufficient to determine the impact on the real complexity. It may require further work to make the design more robust.

Notes

1. See section 2.6.

2. At MIT we have designed and built decoupled commercial software and precision machines at a fraction of the cost and effort that a typical industrial firm would have spent for similar development projects, but a thorough comparative study has not been made. These products worked from the beginning without much debugging.

3. This example was also presented in Suh (2001).

4. If the concentrated reaction force acts near the slot and the stress due to the interference fit is more or less uniformly distributed between the shaft and the knob, the grip will loosen when the force due to the torque is less than the total distributed stress due to the interference fit. Also continuous opening and closing of the slot may cause other problems, such as fatigue failure.

5. Trademark of Trexel, Inc., Woburn, MA, U.S.A.

6. For references to conventional IC engines, see Heywood (1988), Lumley (1999), Ferguson and Kirkpatrick (2000).

7. This engine is a property of Ecogin, Inc. Patent pending (2002), courtesy of Ecogin, Inc., which has been assigned to MIT.

8. This design was constrained by the fact that an existing commercial engine was modified. If a new engine were to be designed from scratch, the engine would be modified to make use of the freedom that the decoupled design offers.

9. The detailed design and construction of a prototype engine was done by Dr. N.H. Cho of the Institute of Advanced Engineering, Yong-In, Korea, under contract to Ecogin, Inc.

10. Although a direct fuel injector is shown, this engine will work equally well with other fuel-delivery systems, such as manifold port injection.

11. The simulation was done using "Wave" by Dr. Nam-Hyo Cho of the Institute for Advanced Engineering, Yong-In, Korea.

12. Trade name of Axiomatic Design Solutions, Inc., Boston, MA (www.axiomaticdesign.com).

References

Ferguson, C.R. and Kirkpatrick, A.T. 2000. *Internal Combustion Engines: Applied Thermodynamics*, Wiley, New York.

Heywood, J.B. 1988. *Internal Combustion Engine Fundamentals*, McGraw-Hill, New York.

Kishbaugh, L. 2004. Private communication from Trexel, Inc.

Lumley, J.L. 1999. *Engines: An Introduction*, Cambridge University Press, Cambridge.

Okamoto, K.T. 2003. *Microcellular Processing*, Hanser Publishers, Munich, Germany.

Suh, N.P. 1996. "Microcellular plastics," in Stevenson, J. (ed.), *Innovation in Polymer Processing*, SPE Books of Hanser Publishers, New York.

Suh, N.P. 2001. *Axiomatic Design: Advances and Applications*, Oxford University Press, New York.

Suh, N.P. and Cho, N.H. 2002. Patent application (pending).

Exercises

4.1. It was shown in chapter 3 that the time-independent real complexity is equal to the information content of axiomatic design. Consider the following two different designs that perform exactly the same functions but have a different physical feature, which are represented by the design equations:

$$\text{Design A} \begin{Bmatrix} FR_1 \pm \Delta FR_1 \\ FR_2 \pm \Delta FR_2 \\ FR_3 \pm \Delta FR_3 \end{Bmatrix} = \begin{bmatrix} 1 & 0 & 0 \\ 0 & 1 & 0 \\ 0 & 0 & 1 \end{bmatrix} \begin{Bmatrix} DP_1 \pm \Delta DP_1 \\ DP_2 \pm \Delta DP_2 \\ DP_3 \pm \Delta DP_3 \end{Bmatrix}$$

$$\text{Design B} \begin{Bmatrix} FR_1 \pm \Delta FR_1 \\ FR_2 \pm \Delta FR_2 \\ FR_3 \pm \Delta FR_3 \end{Bmatrix} = \begin{bmatrix} 1 & 0 & 0 \\ 0.6 & 1 & 0 \\ 0.5 & 0.6 & 1 \end{bmatrix} \begin{Bmatrix} DP_1 \pm \Delta DP_1 \\ DP_2 \pm \Delta DP_2 \\ DP_3 \pm \Delta DP_3 \end{Bmatrix}$$

Design ranges are given by

$$\Delta FR_1 = \Delta FR_2 = \Delta FR_3 = 0.05$$

When the products were actually made and their performance was measured, it was found that the actual design parameters (i.e., physical parameters) are given by the following:

$$\delta DP_1 = \delta DP_2 = \delta DP_3 = 0.2$$

Determine the difference in time-independent real complexity between these two products. Make suitable assumptions for pdf (probability density function) for the random variations in DPs.

4.2. The cost of developing a new product may be assumed to be linearly proportional to the number of trials made as an upper bound. PTS company designed a new machine, which may be characterized by the following equation:

$$\begin{Bmatrix} FR_1 \\ FR_2 \\ FR_3 \\ FR_4 \\ FR_5 \end{Bmatrix} \begin{bmatrix} X & 0 & 0 & 0 & 0 \\ X & X & 0 & 0 & 0 \\ 0 & X & X & 0 & 0 \\ 0 & 0 & X & X & 0 \\ 0 & 0 & 0 & X & X \end{bmatrix} \begin{Bmatrix} DP_1 \\ DP_2 \\ DP_3 \\ DP_4 \\ DP_5 \end{Bmatrix}$$

Unfortunately, the engineers in the company did not write down the above design equation. Instead they relied on a random set of tests varying the DPs in a random sequence. Therefore, they spent extra effort working on time-independent imaginary complexity problems. Determine the difference between the minimum cost and the actual cost of development.

4.3. Examine the goals and the structure of your organization. Provide a qualitative assessment of the real and imaginary complexity of your organization.

4.4. The suspension system for automobiles must provide two different FRs: precise steering and ride comfort. The existing designs of automobile suspension exhibit a high level of coupling between these two FRs, which leads to time-independent real complexity. Because of this real complexity involved in providing both comfortable ride and tight steering, some cars (e.g., German cars) are designed to have tight steering at the expense of ride quality and other cars (e.g., American cars) are designed to provide soft ride at the expense of tight steering. In both cases, the automobile companies fine-tune the system by trial and error.

Propose a design that can eliminate the real complexity by uncoupling the steering quality from the stiffness of the ride. The following background information is provided as an aid in solving this exercise.

Background information on existing suspension systems
All existing front-wheel independent suspension systems are variations of the four-bar mechanism. For instance, the parallel arm suspension, the short long arm (SLA) suspension and the McPherson strut suspension can be kinematically represented as shown in figure 4E.1 (See J. Reimpell, H. Stoll, and J.W. Betzler, *The Automotive Chassis*, Society of Automotive Engineers, Warrendale, PA, 1996; Donald Bastow, *Car Suspension and Handling*, Pentech, London, 1987.)

Orientation of the wheels and steering axes with respect to the vehicle frame and with respect to the terrain changes owing to suspension travel. Figure 4E.2 shows the wheel alignment parameters (WAP) that describe the orientation of the wheel and the wheel axis. Excess camber causes tire wear and camber spread causes directional instability. Caster spread causes directional instability. Toe change due to suspension travel causes *bump steer* and excess toe causes tire wear. Because of these factors, vehicles exhibit tire wear and directional instability due to suspension travel under conditions of overload, offset load, or road undulations.

The coupling in the suspension and steering systems manifests itself through the change in wheel alignment parameters due to suspension travel. This change in the WAP causes directional instability and tire wear. The approach of the industry to solve this problem has been twofold. The first approach has been optimization of suspension link lengths to reduce the change in WAP to zero. Since this is not possible with the existing architecture, the solution used is the optimization of the

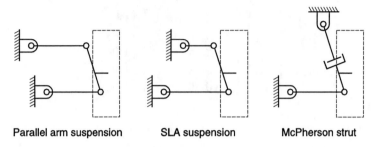

| Parallel arm suspension | SLA suspension | McPherson strut |

Figure 4E.1 Kinematic representation of existing front-wheel independent suspension systems (four-bar linkages).

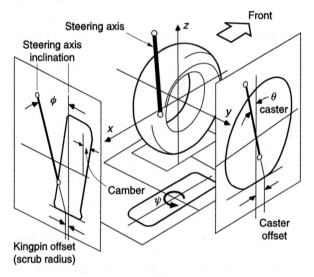

Figure 4E.2 Wheel alignment parameters. Toe angle, which is not shown, is the angle between the direction of travel of the vehicle and the plane of the tire.

spring stiffness K to get a compromise solution for comfort (which requires significant suspension travel and hence a soft spring) and directional stability (which demands the least possible change in wheel alignment parameters and hence a stiff spring).

The FRs that the four-bar linkage is expected to satisfy are given in the form of FR/DP decomposition in table 4E.1. Consider only the following three important FRs for simplifying the analysis in this exercise: permit relative Z motion, avoid track changes ($\Delta y = 0$), and avoid camber and caster changes ($\Delta \phi = 0$). Here Δy indicates tire scrub and $\Delta \phi$ indicates camber change.

Analysis of the parallel arm suspension shows that angle θ is capable of permitting relative Z motion as shown in figure 4E.3 and can maintain $\Delta \phi = 0$ as both joints of the steering axis have equal vertical motion during suspension travel. But the parallel arm suspension is incapable of satisfying FR_2: $\Delta y = 0$ during suspension travel. This causes tire scrub due to suspension travel as illustrated in figure 4E.3.

Table 4E.1 FR/DP decomposition of FR₁ (Maintain wheel alignment)

	Functional requirements	*Design parameters*
FR_{11}	Permit relative Z motion	Single degree of freedom system
FR_{12}	Avoid track changes ($\Delta y = 0$)	Effective swing axle radius
FR_{13}	Avoid camber and caster changes ($\Delta\phi = 0$)	Equal motion of steering axis joints

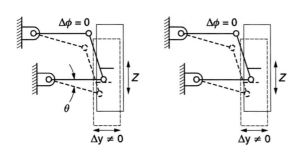

Figure 4E.3 Kinematic representation of parallel arm suspension showing tire scrub.

In the SLA suspension, we can achieve $\Delta y = 0$ (no tire scrub) through assignment of appropriate values to the link lengths, but this doesn't allow $\Delta\phi = 0$ during suspension travel. This causes camber change and caster change due to suspension travel. A compromise solution for Δy and $\Delta\phi$ can be obtained through optimization of the link lengths and joint positions, but we cannot satisfy all three FRs simultaneously using a four-bar linkage. Both Δy and $\Delta\phi$ can be reduced by increasing the link lengths, but this is limited by the constraints of cost, packaging, and unsprung weight of the vehicle.

The McPherson strut suspension, also a four-bar linkage with one prismatic joint, is also incapable of satisfying the three FRs simultaneously. It exhibits tire scrub as well as WAP changes due to suspension travel.

The foregoing considerations imply that in the existing designs, suspension travel affects the WAP. This coupling leads to several problems. The changes in camber angle and toe due to excess suspension travel under overload causes unnecessary tire wear. This could be a serious issue in trucks as the WAP could change significantly from unloaded to fully loaded condition. Under offset load, different suspension travel for the wheels could cause camber spread, caster spread, or toe spread leading to directional instability or *drift/pull* of the vehicle. Toe change due to suspension travel causes *bump steer* due to road undulations. Toe change due to suspension travel is also a possible source for the *nibble* problem, in which the high-frequency road noises are transmitted back to the steering wheel.

Very often in coupled designs, when one DP affects two or more FRs, these FRs require the DP to have different values. This leads to a tradeoff between the conflicting FRs and the designer has to resort to optimization of the DPs to achieve the best compromise solution. The coupling in the automobile suspension and steering system is manifested by the following tradeoffs to achieve compromise solutions:

1. Compromise between $\Delta \phi = 0$ and $\Delta y = 0$ through optimization of link lengths.
2. Compromise between control and comfort through optimization of spring stiffness K. Control demands a stiff suspension, whereas comfortable ride demands a soft suspension.

Since the existing designs cannot make the WAP independent of suspension travel, optimization of the spring stiffness has been the approach of the industry to get a compromise solution for FRs of comfort and control.

5

▬▬ Reduction of Time-Dependent Complexity through the Use of Functional Periodicity

5.1 Introduction

Time-dependent complexity plays an important role in engineered systems as well as in natural systems. In some cases, a system that initially behaves as a system with time-independent complexity degenerates into a system with time-dependent combinatorial complexity. Many systems fail when they are dominated by time-dependent combinatorial complexity. In such a system, the system range drifts away from the design range over its life cycle.

To reduce complexity and provide stability to a system, the time-dependent combinatorial complexity must be changed to a time-dependent periodic complexity by introducing a functional periodicity—if it can be done. If the functional periodicity can be designed in at the design stage, the system will be stable and last much longer than systems without it.

There are many different kinds of functional periodicity. However, the process used to introduce the functional periodicity is the same, although the physics that govern a specific functional periodicity may be entirely different. The purpose of this chapter is to show how time-dependent combinatorial complexity can be transformed into time-dependent periodic complexity.

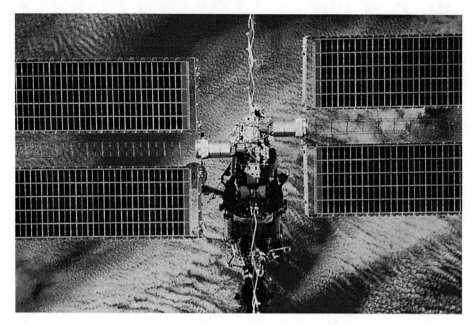

Figure 5.1 International Space Station (ISS) that orbits the Earth, showing the solar panel extending from the ISS through gimbals (shown in detail in figure 5.2). (Courtesy of NASA.)

An example of time-dependent combinatorial complexity

The International Space Station (ISS) shown in figure 5.1 is orbiting the Earth. It has a capacity for seven crew members, who conduct a variety of scientific and technological experiments.[1] The space shuttle and the Russian *Soyuz* have transported the crew and cargo to ISS. This space station is to be used for many decades to come. However, it appears that ISS may have a problem, which if not corrected soon may lead to a system failure in the future. This problem has developed after ISS has been in service in space, which is not the most convenient place to correct or repair the mechanism, notwithstanding the skills and dedication of astronauts.

ISS generates electricity using solar panels. The solar panels are attached to ISS through gimbals, shown in figure 5.2, which rotate once every 90 minutes to follow the sun. It appears that the FR (i.e., rotate the solar panel) may have a glitch. Once in a while, it requires a much higher current to turn the motor that rotates the solar panel because of the increased resistance to turning as indicated by the trace of the current going to the gimbals assembly (see figure 5.3). It is difficult to determine exactly what is causing the problem, but if nothing is done to deal with it, ISS may develop a serious problem of not generating sufficient electricity, which may lead to the entire system failure.

It is reasonable to assume that the problem is caused by a combinatorial process that somehow affects the FR of turning the solar panel. Therefore, unless this time-dependent combinatorial complexity is changed into a time-dependent periodic complexity, the solar panel may eventually fail to turn and lead to the loss of electric power in ISS. To prevent this possible failure mode, the complexity must be reduced by introducing a

Figure 5.2 The beta gimbals to which the solar panels shown in figure 5.1 are attached. (Courtesy of NASA.)

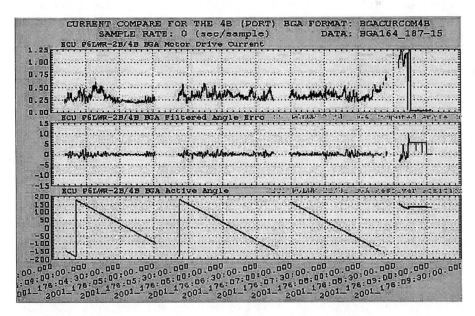

Figure 5.3 The current supplied to the gimbals assembly (shown in figure 5.2) that turns the solar panel shown in figure 5.1. Ordinate indicates the current and abscissa is a time scale. (Courtesy of NASA.)

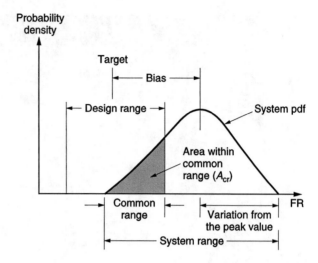

Figure 5.4 Probability
density function of one FR.

functional periodicity *now* so as to make ISS perform its mission for many decades to come.

What causes time-dependent combinatorial complexity?
When a system is newly designed, the minimum requirement is to have zero time-independent real complexity. That is, the system must have a zero information content, as stipulated by the Information Axiom. To have zero information content, the system ranges of all FRs of a system must be inside their respective design ranges at all times. However, if the system range is not inside the design range for one of the FRs because of poor design, as shown in figure 5.4, then there is a finite probability that the system will eventually fail.

To reduce the time-independent real complexity, we must create a system that satisfies the independence of FRs and reduces the information content by reducing the bias and the variance, as discussed in preceding chapters, or by introducing a new design as illustrated in chapter 4.

We have a more complicated system behavior when the system range changes as a function of time. Such a system has time-dependent complexity. One type of time-dependent complexity is time-dependent *combinatorial complexity*. Engineered systems can have time-dependent combinatorial complexity if the system range of one or more FRs moves away from the design range as a function of time, as shown in figure 5.5. Time-dependent combinatorial complexity exists when the future events of a system depend on prior decisions made or past events in an unpredictable way by having ever increasing and diverging possibilities.[2] The causes of time-dependent combinatorial complexity can be many: a physical process, a design parameter, noises, and so on, that make the system range of an FR change over time.

For example, when the FR is "create low friction surface," the generation and accumulation of wear particles may lead to increased friction and eventually to seizure and failure of a bearing. Similarly, when electric charges accumulate sufficiently at the LCD interface, it may cease to operate because of the relaxation behavior of liquid crystal polymers under a constant load applied by the electrical potential. Yet another example

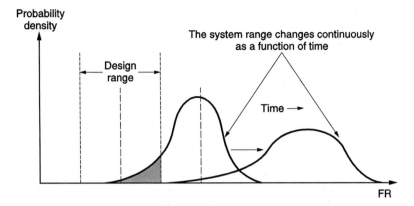

Figure 5.5 Time-dependent combinatorial complexity. This type of complexity may result in the system range moving away from the design range. When it moves completely out of the design range, we may have a chaotic state.

is the scheduling of robot transport in a manufacturing system, which can go into a chaotic state if the scheduling is done ad hoc using an expert system. Biological systems can go into a state of combinatorial complexity when they lose the ability to control their cyclic behavior. The functional requirement of such a system may go into a chaotic state and eventually fail.

To make a system with time-dependent combinatorial complexity stable, it must be converted into a system with time-dependent periodic complexity. In some cases, this may be accomplished by introducing functional periodicity, as discussed in chapter 3. In this chapter, the basic principles and methodology of introducing functional periods are presented.

There are different types of functional periodicity
As presented in chapter 3, there are many different types of functional periodicity. In the field of science and engineering, they are based on physical phenomena, information processing, biological processes, or temporal periods of certain functions, which are listed below:

1. Temporal periodicity.
2. Geometric periodicity.
3. Biological periodicity.
4. Manufacturing process periodicity.
5. Chemical periodicity.
6. Thermal periodicity.
7. Information process periodicity.
8. Electrical periodicity.
9. Circadian periodicity.
10. Material periodicity.

The types of functional periodicity listed above can be built into engineered systems during the design stage to make the system more stable and reliable. Such a system will also be safer since the FRs will be satisfied at all times through the renewal of functional performance on a periodic basis.

In socioeconomic–political fields, we have the following types of functional periodicity:

1. Organizational periodicity.
2. Economic functional periodicity.
3. Political functional periodicity.
4. Academic functional periodicity.

Although the basis of each functional periodicity is different from that of the others, they have common general characteristics. The purpose of this chapter is to highlight some of these common characteristics and develop methodologies for reducing the complexity of systems by transforming combinatorial complexity to periodic complexity.

Safety and reliability problems are caused by combinatorial complexity
If the system is so designed that the time-independent real complexity is zero and there is no imaginary complexity, then the only way the system can fail is if it has time-dependent combinatorial complexity, that is, if its system range drifts away from its design range. Therefore, lack of safety and reliability is a consequence of having time-dependent combinatorial complexity. To make a system with time-dependent combinatorial complexity safe and completely reliable, we must transform it into a system with time-dependent periodic complexity.

5.2   Conversion of Combinatorial Complexity to Periodic Complexity

The steps involved in transforming a system with combinatorial complexity into a system with periodic complexity are:

1. Determine a set of functions (FRs) that repeats on a periodic[3] (or cyclic) basis, in other words, has functional periodicity.
2. Identify the DP of a system that may make the system range of the FR undergo a combinatorial process.
3. Transform the combinatorial complexity to a periodic complexity by introducing functional periods,[4] that is,

$$T\{C_{\text{com}}(\text{FR}_a)|F_p\} = \{C_{\text{per}}\,(\text{FR}_a)\}$$

4. Set the beginning of the cycle of the set of FRs as $t = 0$.
5. Stop the process momentarily.
6. "Reinitialize" the system by establishing the state of each function at the instant of reinitialization. The initial state $t = 0$ is the instant at which the selected key FR function begins a new cycle.
7. Determine the best means of satisfying the FRs for the new period.
8. Allow the initiation of the next cycle.

This transformation can be attempted only if the system has a set of FRs that repeats on a periodic basis. The period is the functional period, which may be temporal, geometric, and so on. The functions could be related to performing a set of processes, inducing chemical reactions, controlling electric charge decay, forcing geometric contacts, and others. The functional periodicity is listed in section 4.1 and chapter 3.

The second step in the transformation process is the identification of FRs that are subject to combinatorial processes, such as scheduling of a robot motion or a wear process, that eventually lead to the failure of the system because the system range will move outside of the design range. Sometimes even if the FRs are not precisely identified, we can proceed with the transformation process by introducing functional periods.

The third step in this process is to determine how a design can be made uncoupled or decoupled. Once we have a design that satisfies the Independence Axiom, we can devise a means of reinitializing the FRs so as to create a functional period. If the system is coupled, it may not be possible to introduce a functional period because reinitialization of the FRs will be difficult.

The fourth step is to determine which FR will be used to establish the beginning of the cycle, that is, $t = 0$. In the case of scheduling a robot motion in a manufacturing system, it may coincide with when a robot picks up a part from the last station of a manufacturing cell. In the case of two sliding surfaces, the beginning of a new cycle is when the interface ceases to contact each other because of the undulated topography of the surface.

The fifth step is to stop the FRs momentarily (or instantaneously) at $t = 0$ so as to reinitialize the system. The need for reinitialization arises from the fact that at $t = 0$, the state of each FR is not always the same because of random variation of the system. In most cases, there may not be any physical or temporal disruptions other than determining the state of each component of the system in order to determine the remaining process times. Once the state of the system is reinitialized, we can determine the new values for the FRs or let the system renew its states without any external interventions. This step is then followed by the execution of the tasks of the system as defined by the FRs.

In some situations, it may be necessary to determine the best means of satisfying the FRs for the new period. For example, in scheduling the robot motion at time $t = 0$, the initial values of FRs can be used to determine the best robot schedule for the new period using a preestablished algorithm. In other situations, such as biological functional periodicity and geometric functional periodicity, the system may transit into the new period without the need to determine the best means of satisfying the FRs once all the FRs of the previous periods are satisfied.

These steps may involve physically different processes depending on the specific applications, as shown in the following chapters for various cases. However, they all follow the same process of the creation of functional periodicity and reinitialization.

EXAMPLE 5.1: POSSIBLE SOLUTION TO THE ISS COMBINATORIAL COMPLEXITY PROBLEM

The problem of the International Space Station (ISS) discussed in the introduction to this chapter is a time-dependent combinatorial complexity problem. The FR of rotating the solar panel once every 90 minutes is satisfied with difficulty from time to time, probably because of a tribological problem in the design of the gimbals or the motor.

When two surfaces slide over each other, wear particles are generated and often accumulate. When these wear particles collide and agglomerate under normal load, they form larger particles. Then the friction force goes up. When the agglomerated particles break up, the friction force decreases. If this friction and wear process continues in a geometrically confined space, eventually seizure can occur when the agglomerated wear particles are large enough to wedge in the interface.[5] Although it is difficult to state conclusively, it appears that the cause of the sudden jump in the current that flows into the gimbals of ISS may be an indication that it is in the early phase of this tribological process.

Solution

Since the International Space Station is in orbit, it is difficult to dismantle the system from Earth. We must try to correct the problem remotely. A suggested solution to NASA is as follows:

Instead of turning the solar panel one rotation every 90 minutes, rotate it once every 89 minutes. Then change the direction of rotation for 30 seconds at the same or slower speeds. Then reverse the direction again, returning to the original position in 30 seconds. After this one reversal of the rotating direction, resume the normal rotation for another 89 minutes for another complete cycle of rotation. Repeat this process periodically.

The rationale for the above suggestion is that reversing the rotational direction will disrupt the process of wear-particle generation and agglomeration by breaking up the particles already formed. This is equivalent to reinitializing the FRs once every 89 minutes, so that they can be satisfied as specified for another cycle. In other words, we are introducing a functional period, in this case in the form of a temporal period to "reinitialize" the interfacial friction condition by eliminating large agglomerated wear particles.

5.3   Identification of Combinatorial Processes and Functions for the C/P Transformation

To be able to implement the transformation of a combinatorial function to a periodic function, we must identify a set of functions (i.e., FRs) that repeat once a period. This period is not necessarily a temporal period, but rather a functional period. This task may be easy to implement functional periodicity if we are designing the system from scratch, since we have to state the FRs as part of the design process.

If we are to implement this functional transformation of the combinatorial-to-periodic process for an existing system, the task may be more complicated, but the first "tell-tale" sign may often be easy to identify. For example, if suddenly the coefficient of friction increases, it may indicate that there is a combinatorial process occurring. If a system suddenly fails after a prolonged service, we may suspect that there may be a combinatorial process in progress.

Many system failures have an exponential curve (solid line), as shown in figure 5.6. Systems that have these failure characteristics may undergo a combinatorial process, which results in the system behavior having time-dependent combinatorial complexity. If a functional period can be introduced to such a system during the initial design stage, we may be able to prolong the life of the system along the dotted line. In the

Failure

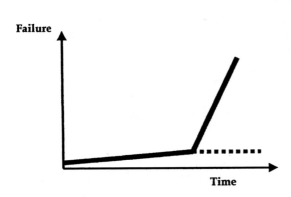

Time

Figure 5.6 Typical failure curve for a system (solid line). The system works well for a long time and suddenly fails. This indicates the likelihood of a combinatorial failure process at work. A combinatorial process depends on the prior history of the system. The life of the system may be extended (dotted line) by transforming this combinatorial process through the introduction of functional periodicity.

following chapters, some of these transformations will be described to illustrate the process.

What is the C/P transformation?
When we identify a system that has FRs with the failure characteristics shown in figure 5.6, it may be a good candidate for the combinatorial-to-periodic transformation. For brevity, this transformation will be identified as the C/P transformation.

If the combinatorial complexity cannot be identified at the highest-level FRs, we may have to decompose the FRs to identify the lower-level FR that may have the characteristics of time-dependent combinatorial complexity.

5.4   Reinitialization Methods

To perform the C/P transformation, we must "reinitialize" the FRs at the beginning of each period. The purpose of reinitialization is to establish the state of functions at the beginning of the new period and establish the best way of satisfying the FRs in the subsequent functional period. The purpose of reinitialization is not to establish the same initial state at the beginning of each cycle, but to establish the state of FRs. We can reinitialize the system either in the *functional domain* or the *physical domain,* depending on the function that repeats itself.

Reinitialization in the functional domain
The time-dependent combinatorial complexity problem exists because the system range of an FR does not lie inside its design range and changes as a function of time or subsequent events. Therefore, when a system is reinitialized, we need to determine the state of FRs at the instant of reinitialization so that we can determine the best means of satisfying the set of FRs in the subsequent period. An example of this case is the scheduling of the transport robot in a manufacturing system, where the period is set to be the sequence of robot motions to transport parts from machine to machine until the part is completely processed. In this case, we may have to establish the state of the FRs quantitatively at the beginning of each period.

Reinitialization in the physical domain

In some cases, reinitialization can be done more expediently in the physical domain by physical means. For example, the control of friction between two sliding surfaces so as to prevent the frictional behavior going combinatorial can be achieved by changing the surface geometry to introduce functional periods. In this case the functional periodicity can be introduced physically and/or qualitatively without the need to compute the initial state of FRs at the beginning of each period by geometric means. The physical conditions—the repeating geometry—determine the initial state of geometry for control of frictional behavior.

Reinitialization and modifying the set of FRs

In the foregoing discussion of functional periodicity, the need to identify a repeating set of FRs was emphasized. This would be the case if a system with a functional period were to renew itself automatically. However, in some cases, it may be necessary to modify the set of FRs by adding a new FR or subtracting an FR. This is equivalent to solving a new design problem at the beginning of each period for the newly modified FR set. This situation exists in dealing with socioeconomic–political systems. If the modification of the FR set is known a priori, a new design may be configured at the time of reinitialization using a predesigned algorithm.

EXAMPLE 5.2: LIQUID CRYSTAL DISPLAY

In a liquid crystal (LC) cell, light is modulated (turned on or off) by applying a voltage across the liquid crystal material.[6] The voltage results in reorienting the LC molecules, which in turn alter the polarization of the light as it goes through. The time scale of modulation is approximately 10 msec (video rate) up to several seconds or even longer for most applications. However, if the voltage were applied continuously for that long duration, it would result in permanent chemical change in the LC molecules, thereby destroying the cell. The reason is that charges can be transported through the molecules, and long application of voltage would lead to charge accumulation at the electrodes. This has two potentially deleterious effects: the screening field can depolarize the liquid crystal material, while the charges themselves can corrode the electrodes.

Solution

The solution is to transform the system with combinatorial complexity to a system with periodic complexity. How should we make the FR periodic?

Fortunately, the optical effect is the same if the polarity of the voltage is reversed. Thus, instead of applying a constant voltage, say V, throughout the duration that the cell needs to remain in the "on" state, the same effect is obtained by applying a much faster alternating voltage between +V and −V. Typically, the alternating voltage period is in the range of ~1 kHz. The response of nematic liquid crystals (which are the most commonly used in mass products) is in the range of 10 msec, so the alternating voltage actually does not result in switching, even though the undesired side-effects of DC voltage are removed. The alternate "off" state of the cell is accomplished with zero voltage, which has no effect on the cell.

EXAMPLE 5.3: THROMBOSIS DUE TO BLOOD CLOTTING

One of the major causes of death is thrombosis, which occurs when blood clots block capillary blood vessels.[7] This can occur, for example, when the artery of a patient is linked to an artificial heart. When the blood circulates through polymeric tubes in the artificial heart, proteins in blood can deposit on the surface of the tube and other artificial materials, forming platelets. Eventually it becomes a thick deposit. When a large deposit is dislodged from the surface, a solid particle large enough to block the flow of blood through capillaries may be flowing with the blood, causing thrombosis.

The dilemma is that we prefer to have some deposition of patient's proteins on these artificial surfaces to prevent the rejection of the human-made materials by the body, but if the platelets continue to grow, it can cause a serious problem. This represents a system with time-dependent combinatorial complexity. How would you solve this problem?

Solution

One of the key questions is: "What attracts the protein molecules to the surface of the artificial materials?" By introducing a functional periodicity, it may be possible to prevent thrombosis. The solution to this example is left to the reader as a homework problem.

EXAMPLE 5.4: REBOOTING OF COMPUTERS

All of us who use computers find it necessary to reboot them when they suddenly cease to execute the operation.[8] There may be many different causes for these failures. Some of them may be system failures that are caused by time-independent imaginary complexity because of the lack of rigor in formulating and specifying the FRs, DPs, and PVs. However, many of the software failures that occur when they are used for a long period of time may be due to time-dependent combinatorial complexity. How should we eliminate the failures caused by the time-dependent combinatorial complexity?

Solution

One of the causes of the time-dependent combinatorial complexity may be "memory leaks" or other forms of improper memory management. In modern computing systems, the operating system is responsible for allocating the computer's fixed resources, including system memory, to hosted applications. Once allocated to a hosted application, the operating system assumes, but minimally enforces, that the application manages the allocated resources properly. A "memory leak" occurs when the operating system allocates memory resource to an application, but fails to keep track of the resource and notify the operating system that the resource can be returned to the available pool when it is no longer required. In this instance, the system may continue to function within the design range for a long period of time, but will eventually exhaust all of its available memory unless the offending application is terminated, at which point the operating system automatically reclaims all resources allocated to the application.

A similar problem can occur when an application uses a resource that the operating system did not specifically allocate to the application. In this case, the offending application may not be adversely affected by its utilization of the resource. However, when the operating system allocates the unallocated resource to another application, which, in turn, modifies the resource without the knowledge of the original application, creating an undefined data state, the system range for either or both applications may suddenly lie outside their respective design ranges.

The common solution many end-users adopt to circumvent these issues is to terminate and restart either offending applications or the operating system itself (in the form of a reboot). Both approaches impose periodicity, by resetting the underlying system and resource state.

5.5 Summary

This chapter outlined the rationale and the procedure for transforming a system with time-dependent combinatorial complexity to a system with time-dependent periodic complexity.

The most important factor in insuring the safety and reliability of a system is to reduce the time-independent real and imaginary complexities to zero and make certain that time-dependent combinatorial complexity does not play a role in the performance of the system. Whenever possible, a system should be designed to have time-dependent periodic complexity by reinitializing the system behavior.

Notes

1. Currently there are only three astronauts in ISS because the Russian space capsule *Soyuz* can rescue only three astronauts in the case of emergency. The U.S. space shuttle, which is currently grounded after the *Columbia* accident, can transport up to seven astronauts.

2. Turbulent flow may be a system with time-dependent combinatorial complexity. Every fluid particle in the turbulent stream moves through a series of bifurcation points where the particle can move in two or more directions with equal ease from both the energetic and momentum points of view. The subsequent motion of the fluid particle as well as the overall fluid motion change dramatically depending on the specific "choice" made at each instant of its flow.

3. The period refers to "functional periods," which were given in the preceding section.

4. $T \langle \rangle$ is a symbol for transformation. C_{com} and C_{per} are combinatorial complexity and periodic complexity, respectively. F_p represents the selected key FR for the functional period.

5. This tribological process of friction and wear is further discussed in chapter 7.

6. This example was provided by Professor George Barbastathis of MIT, private communication (2003).

7. This example is a result of the author's discussion with Dr. Kurt Dasse of Pharos, Inc., Waltham, MA.

8. This example has been contributed by Christian Arangio of Axiomatic Software Design, Inc., Boston, MA.

Exercises

5.1. It is very important for an economy to be stable and to maintain a steady growth rate. Establish a set of repeating FRs of a national economic system and discuss how functional periodicity may be introduced to make the economic system more stable and robust.

5.2. Identify examples of engineered systems and natural systems that have time-dependent functional periodicity. Explain how the systems function.

5.3. Discuss the consequence of reinitializing the system before the completion of a functional period.

5.4. What determines the shortest functional period?

5.5. How should we introduce functional periodicity an ecological system? What would be the consequence of having an ecological system that has time-dependent combinatorial complexity?

5.6. A major problem in transmitting messages in telecommunication is the noise that disturbs the signal. Therefore, the rate of information transmitted is less than the maximum theoretical channel capacity. In other words, the noise introduces complexity in the system. For example, when we send a string of binary messages such as [01101], the receiving end may get a modified message such as [01110] because of the noise. One way of reducing the noise problem is to segment a message by introducing functional periodicity and send the same message several times to eliminate the effect of the noise. How would you introduce a functional periodicity to binary messages?

6

Reduction of Complexity in Manufacturing Systems

6.1 Introduction

Is a manufacturing system complex? Why?

Complexity is defined as the measure of uncertainty in achieving the FRs of a system due to poor design or to the lack of understanding and knowledge about the system. If the FRs can be satisfied at all times within the design range, the system is not complex according to this definition, although it may have many physical parts and must satisfy many functions. If the design range is small and if the system range cannot be inside the design range, the uncertainty associated with achieving the FRs will be large and therefore the task of achieving the FRs is complex.

The FR of a manufacturing system is to maximize its productivity, which is defined as (Total value added – Cost of manufacturing)/Total investment. Given this definition, which of the following is more complex: the operation of a large job shop[1] or an automated factory that manufactures automobile engine blocks using transfer lines?[2] How do we maximize the productivity of these manufacturing systems? To maximize the productivity of these manufacturing systems, we must reduce their complexities.

The complexity in manufacturing is caused by two factors: time-independent real complexity because of poor design of manufacturing processes and systems, and because of deterioration of the processes and systems as a function of time, which results in time-dependent combinatorial complexity. One of the goals in manufacturing is to reduce or eliminate these causes of complexity through design of rational products and processes.

The complexity of manufacturing systems depends on the specific FRs and the DPs that must be satisfied, specific manufacturing operations, the equipment available, the materials to be processed, and the skills of engineers and workers. Manufacturing systems must be designed so as to minimize their complexity. As the complexity is reduced, the reliability and the productivity of the manufacturing system should invariably improve.

Since the complexity is determined relative to our ability to satisfy the FRs of a given system, it depends on one's perspective. A fully automated factory may not appear to be complex to the operator of the factory because the FRs can be achieved at all times, whereas it may be a complex task to the designer of the fully automated factory because of the uncertainty involved in designing a factory that can achieve the FRs as specified at all times.

The purpose of this chapter is to show how the complexity of a manufacturing system can be decreased using a specific example.

Complexity and manufacturing productivity
To maximize the productivity of a manufacturing system, we must design the system so that the FRs of the production system may be easily satisfied at all times. Such a system will have zero complexity. When the complexity is finite, the productivity of the system will be less than the maximum and the quality of the products that are produced by the system may not be the best.

Ever since the Industrial Revolution, many approaches have been tried to maximize industrial productivity. The productivity of a manufacturing system is a function of many things, such as labor cost, the number of hours required to manufacture a specific product, the cost of capital investment, and the degree of automation. Depending on the relative importance of these factors in determining the manufacturing productivity, different approaches and measures have been used to increase productivity.

As socioeconomic conditions have changed over time, the driving force for pro-ductivity has changed. In the early twentieth century, labor productivity was the most important factor for determinig manufacturing productivity of many factories.[3] When the labor content of manufactured products was reduced through automated equipment, the cost of capital investment in equipment and factories constituted a large fraction of the manufacturing cost. Thus, much effort was devoted to reducing the capital cost. At the beginning of this new millennium, we find that the materials cost is the largest fraction of product cost. Furthermore, we now find that with supply of consumer goods exceed-ing demand, we cannot simply mass-produce products—we must satisfy customers' specific needs to increase market share. Therefore, the design of products and manu-facturing systems is becoming the most important factor in determining industrial com-petitiveness. Therefore, the complexity of manufacturing systems depends on the FRs that are relevant at a given instant in history.

Manufacturing systems are a subset of engineered systems
Manufacturing systems, such as a semiconductor manufacturing "fab"[4] and factories for automotive parts, are subsets of an engineered system. Some systems are large, func-tionally as well as physically. The supply-chain management system for large manufac-turers is an example of a large system. Some systems, such as a "cluster" of machines

that consist of an integrated lithography tool with a track tool in a semiconductor manufacturing fab, are a functionally large but physically small. Other systems may be physically large but functionally small.

Systems are often created through the integration of subsystems that have been acquired from many different vendors and original equipment manufacturers (OEMs) by "system integrators" to satisfy a given set of FRs of a system. In many industrial firms, system integration of two or more subsystems is based on prior experience using ad hoc approaches. During the process of system integration, many fundamental issues, such as mechanical interface, software interface, the scheduling of system operation, and transport of parts, must be addressed in order to satisfy a set of FRs such as throughput rate, quality of products, and reliability.

How do we decrease complexity?
When we are confronted with such a complex system as a manufacturing system, typical approaches that have been used in industrial operations have been either trial-and-error processes to improve the productivity or the optimization of manufacturing systems by trading off various FRs. These approaches have been used in part because the products to be manufactured change over the life of the manufacturing system. Therefore, to reduce capital investment and increase productivity, the flexibility of a manufacturing system has been an important factor in designing such a system. Nonetheless, there are many manufacturing systems that are dedicated to one product because of the special manufacturing operations involved and the high value-added nature of the products.

A manufacturing system must be designed correctly to maximize the productivity and quality of products by clearly stating the FRs of the manufacturing system as well as those of the products. The goal should be the reduction, if not total elimination, of the complexity associated with manufacturing systems through carefully planned and designed systems.

A manufacturing system may involve time-independent real and imaginary complexities as well as time-dependent combinatorial complexity. There will be real complexity if the system cannot make parts within the specified design ranges of the FRs, that is, if the tolerance cannot be satisfied by the manufacturing processes at all times. In this case, either the system must be changed or the product must be redesigned. Time-dependent complexity may be due to deterioration of manufacturing processes or poor design of the manufacturing system. It is also likely that empirical trial-and-error processes waste time and effort by creating imaginary complexity problems.

Clustering of machines to create a manufacturing system
Suppose that we created a manufacturing system by clustering two subsystems, each with many manufacturing modules (such as machine tools, furnaces, etc.), to manufacture identical parts. If one of the subsystems can process 100 parts an hour and the other subsystem 200 parts an hour, what should the cluster be able to produce per hour? One would expect to get the same throughput rate from the cluster that the slowest subsystem is capable of producing as a stand-alone subsystem, which in this case is 100 parts an hour. However, in most cases we get less than 100 parts an hour throughput rate for the cluster. Why?

System integration requires that several subsystems, often made by different manufacturers, be assembled together seamlessly so that the overall system satisfies the FRs of the system. The integration process can be quite challenging when the subsystems are not exactly compatible with each other and when the modules that make up each subsystem have random variation in process times. The integration must be done without compromising the independence of the FRs of the system. One of the FRs of a manufacturing system is the throughput rate of the integrated system.

This chapter deals with the scheduling problem of a system for which two or more subsystems that have many independent modules for different processes with different fluctuating cycle times must be integrated. Integration without understanding the fundamentals of system integration issues may lead to time-dependent combinatorial complexity or imaginary complexity. For example, changing DPs in a random sequence without regard to the best sequence outlined by the design matrix may lead to both time-independent imaginary complexity in the short term and time-dependent combinatorial complexity in the long term.

The performance of a system that is formed by assembling two or more subsystems can be improved by recognizing that the time-dependent combinatorial complexity problem can be transformed into a time-dependent periodic complexity problem.

How do we transform a system with time-dependent combinatorial complexity into a system with time-dependent periodic complexity?
Chapter 5 presented the basic methodology of transforming a system with time-dependent combinatorial complexity to a system with periodic complexity. This transformation of a combinatorial complexity problem to a periodic complexity problem can be done if there is a repeating (cyclic) set of FRs. A "period" begins when the same key FR of the FR set reinitiates the subsystem. The beginning of a new period coincides with the reappearance of a "key function"—triggered by an event either external or internal to the system. The key function is used to reinitialize the "initial" state of the system. A specific function must be chosen as the reference (key) function to signify the beginning of a new period. For reinitialization, the "initial" state of each function must be established by either measurements or calculations at the beginning of each functional period. The actual duration of the period may be either fixed or varied, depending on the variation of the process times. After reinitialization, the best sequence of satisfying the FRs must be determined.

This approach to scheduling can be used to integrate many different kinds of systems, such as factories with individual machines, manufacturing systems, supply-chain management systems for large retailers, and any system with two or more subsystems. However, when a variety of parts are made by a manufacturing system, the complexity of a time-dependent combinatorial problem can be reduced by using a "pull" system of process control where the parts are moved along a system when the part leaves the manufacturing system after undergoing the last operation on the last machine. Here the period is determined by the frequency of the finished part leaving the manufacturing system.

6.2   Complexity Associated with System Integration Task

Lessons learned
To illustrate the system integration process based on the transformation of a system with time-dependent combinatorial complexity into a system with time-dependent periodic complexity, the case of integrating two different manufacturing systems to form a

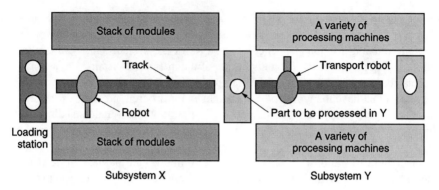

Figure 6.1 A manufacturing system formed by clustering two subsystems. The part is first processed in subsystem X by subjecting it to a variety of processes and then processed in subsystem Y using a variety of machine tools.

larger integrated cluster will be used (Suh and Lee, 2004). It will be shown that the highest productivity is obtained by making the system repeat the same process within a functional period even if the system can be speeded up in a given cycle rather than following the functional periodicity.

Example of a cluster of machines

Consider a simple system consisting of subsystem X and subsystem Y, which are physically connected to each other to make a product, as shown in figure 6.1. Each subsystem has many physical modules to process a part by subjecting it to a variety of different processes that require different process times. The process time in each machine is different and specific to the part to be made. In addition, although the nominal processing times are set for each subsystem, the throughput rate of each subsystem varies from the nominally set values because of the stochastic factors.

When the last operation in subsystem X is completed, the part must be taken by a robot of subsystem Y for the subsequent manufacturing operations in subsystem Y without delay, when the part processed in subsystem X must be immediately processed in subsystem Y before the parts age. The complexity is associated with determining the best sequence of the part transport to maximize the productivity of the cluster consisting of subsystems X and Y.

The modules in the subsystems will complete their processes at different times because the part arrives in each machine at different times and also because their process times are different. Furthermore, there will be inevitable random variation in the process times of individual modules in the subsystems. An additional complicating factor for scheduling is that some of these parts will be finished nearly at the same time, waiting to be picked up by the robot for the next operation. Depending on which part is picked up first by the robot, the subsequent pattern of part flow and the subsequent schedule of the robot motion will be different. Each one of the decision points is a bifurcation point and therefore the number of possible combinations for the flow paths for the product and for the robot motion will increase with the number of such decisions made. Such a system has time-dependent combinatorial complexity, that is, a system in which follow-on decisions are always affected by previous decisions made, a situation that increases the number of possible combinations. Reduction of system complexity through the transformation of the system to one of periodic complexity maximizes the productivity.

To create a system with time-dependent periodic complexity in the presence of random variation and other stochastic factors, several conditions must be met, as discussed in chapters 3 and 5. Functional periodicity requires:

1. Existence of a period in which all the functions are repeated.
2. At the beginning of each period, establishment of new initial conditions.
3. Determination of the shortest cycle time for the entire process from the newly established initial conditions.

The period begins when a key function—triggered either internally or externally—reinitializes the system at the beginning of a new period. The initial states—the values of FRs—are not the same at the time of reinitialization, since they may vary due to unexpected variations in process times as well as other factors.

6.3 General Formulation of the Scheduling Problem of a Cluster of Two Subsystems

If subsystem X consists of N processes with individual process times P_i and if the time to transport the part from module to module is constant and given by t_p, the total time required to process a part W is given by

$$W = \sum_{i=1}^{N}(P_i + q_i) + (N + 1) \cdot t_p \tag{6.1}$$

where q_i is the "buffer" time that may be inserted to prevent concurrent demand for the robot transport.

In steady state, the total number of the parts that can be processed in a unit time n is given by

$$n = 1/t_s \tag{6.2}$$

where t_s is the sending period, that is, the time between the feeding of the parts into the first module. Equation (6.2) represents the case where there is no delay in removing the finished part from the first module. The actual sending period t_s^a is given by

$$t_s^a = t_s + q_0 \tag{6.3}$$

where q_0 is the buffer time for the first module. Then t_s in equation (6.2) must be replaced with t_s^a in determining the throughput rate.

The minimum sending time $(t_s)_{min}$ is equal to the process time of the bottleneck process P_b plus associated transport time. If all q's are allowed to have nonzero values (as well as zero values), then we can always make the sending period be equal to $(t_s)_{min}$. This problem was solved by Oh (1999) and Oh and Lee (2000) to schedule the robots in a semiconductor-processing machine that had scheduling conflicts.

When all of the processes must be tightly controlled so that all q_i (i.e., $q_1, \ldots, q_N$) except q_0 must be equal to zero, then we need to determine the actual sending period that will yield the highest throughput rate. In this case, the sending period may be longer than $(t_s)_{\min}$. Then the maximum throughput rate can be obtained by determining a set of the sending times t_s^a, corresponding to a set of values of q_0, that can process the parts without creating any scheduling conflict.[5]

When the process times vary randomly for a variety of reasons, the processing time for a part is given by

$$W = \sum_{i=1}^{N}(P_i + q_i + \delta P_i) + (N+1) \cdot t_p \qquad (6.4)$$

where δP_i is the random variation of process time P_i. In the presence of random variation in process times, the beginning of each new period must coincide with the recurrence of a key function in the functional space rather than a key event in temporal space. The maximum throughput rate when such variations are present can be solved by means of "reinitialization" when a key reference function recurs.

6.4   Functional Periodicity in a Scheduling Problem

Given a process sequence and process times, a fixed set of FRs can be defined per period. For example,

$$\text{Process:}\quad LL_{IN} \Rightarrow P_1 \Rightarrow P_2 \Rightarrow \dots \Rightarrow P_N \Rightarrow LL_{OUT}$$

$$\text{Chamber:}\quad LL_{IN} \Rightarrow \{C_1\} \Rightarrow \{C_2\} \Rightarrow \dots \Rightarrow \{C_N\} \Rightarrow LL_{OUT}$$

where LL_{IN} and LL_{OUT} are input and output stages, respectively.

$S_i = \{C_i\}$ is called a stage. A stage may consist of one or more identical process modules to meet the target throughput requirement. Some short processes require only one chamber, for example, $S_{short} = \{C_{short}\}$, whereas long processes may need two or more chambers, for example, $S_{long} = \{C_{long,1}, C_{long,2}, C_{long,3}\}$.

Functional requirements for part-transfer are, then,

$$\{FR_0\} \quad \{\text{Transfer a part from } LL_{IN} \text{ to } S_1\}$$

$$\{FR_1\} = \{\text{Transfer a part from } S_1 \text{ to } S_2\}$$

$$\{\vdots\} \quad \{ \qquad \vdots \qquad \}$$

$$\{FR_N\} \quad \{\text{Transfer a part from } S_N \text{ to } LL_{OUT}\}$$

Define $u_i(t)$ as

$$u_i(t) = 0 \quad \text{if } t \le t^*$$

$$u_i(t) = 1 \quad \text{if } t \ge t^*$$

where t^* is the time FR_i is achieved.

Then, within one period, the function state **FR** should change from the initial state to the final (completion) state. That is, the function state vector, $\mathbf{FR}(t) = \{u_0(t), u_1(t), \dots, u_N(t)\}$, changes from $\mathbf{FR}(0)$ to $\mathbf{FR}(T)$ as

$$\mathbf{FR}(0) = \{0,0,\dots,0\}$$
$$\mathbf{FR}(T) = \{1,1,\dots,1\}$$

The number of total (feasible and infeasible) transition patterns is $(N+1)!$ A particular transition pattern corresponds to a particular schedule. Some of these patterns will represent feasible patterns, and others will not.[7]

Here, the period is defined by the time (or sequence of events in the discrete case) between $\mathbf{FR}(0)$ and $\mathbf{FR}(T)$, where T is the length of the period. T is not fixed, that is, it is not a constant, and its specific value is not important in our approach. Therefore, by taking a process time that varies significantly as T, we can reduce the impact of variation on the scheduling task.

Note that $\mathbf{FR}(T + \varepsilon) = \mathbf{FR}(0)$, where ε is a small positive number nearly equal to 0. This, leads to the following conclusion:

- Scheduling is defined as finding the best pattern of transition from $\mathbf{FR}(0)$ to $\mathbf{FR}(T)$ for the current period.
- After the completion of scheduling for one period, $\mathbf{FR}(t)$ returns to its initial state, $\mathbf{FR}(0)$.
- The scheduling task for the later period is to find another—not necessarily the same—transition pattern from $\mathbf{FR}(0)$ to $\mathbf{FR}(T)$.
- By conceiving a general scheduling procedure for one period, the subsequent scheduling task is done by repetition of the same procedure.

In a typical scheduling problem of a clustered system, the system continuously produces parts, each part being separated, on average, by the throughput time. This implies that each of the functions, FR_0, FR_1, ... , FR_N, appears, on average, once in a period. Therefore, a single, unique event, FR_k will initiate and terminate the functional period. In other words,

$$\mathbf{FR}(0) = \{u_0(0), u_1(0), \ldots, u_{k-1}(0), u_k(0), u_{k+1}(0), \ldots, u_N(0)\}$$
$$= \{0, 0, \ldots, 0, \mathbf{0}, 0, \ldots, 0\}$$

$$\vdots$$

$$\mathbf{FR}(T-\Delta) = \{1, 1, \ldots, 1, \mathbf{0}, 1, \ldots, 1\}$$
$$\mathbf{FR}(T) = \{1, 1, \ldots, 1, \mathbf{1}, 1, \ldots, 1\}$$
$$\mathbf{FR}(T+\varepsilon) = \{0, 0, \ldots, 0, \mathbf{0}, 0, \ldots, 0\} = FR(0)$$

FR_k is the "reinitializing" FR.

The design task is the scheduling (sequencing) of the transition $\mathbf{FR}(0) \Rightarrow \mathbf{FR}(T)$. To reinitialize, we have to know the exact state of each FR at the beginning of the cycle. This can be done by determining the remaining time for each process.

If we define R to be the remaining time for each process in a current period at the instant of reinitialization, $\mathbf{FR}(0)$ is mapped to R in each period as

$$R = \{R_0, R_1, \ldots, R_{k-1}, R_k, R_{k+1}, \ldots, R_N\}$$
$$= \{R_0, R_1, \ldots, R_{k-1}, T, R_{k+1}, \ldots, R_N\}$$
$$R_k = T$$

T is the functional period (in this case, the temporal functional period), and its exact value is unknown *a priori* because of potential variation.

Then the design task may be stated as follows:

$$\text{Find } \boldsymbol{R} + \boldsymbol{q} = \{R_0+q_0, R_1+q_1, \ldots, R_{k-1}+q_{k-1}, T, R_{k+1}+q_{k+1}, \ldots, R_N+q_N\}$$

subject to

- $|(R_i + q_i) - (R_j + q_j)| > \delta$
- additional constraints if necessary

where δ is the temporal clearance for successive transports (or minimum time for transport).

EXAMPLE 6.1: CLUSTER OF SUBSYSTEM X AND SUBSYSTEM Y

A manufacturing system for manufacture of LCD panels is to be created by integrating two subsystems X and Y.[6] The task is to create an algorithm that can maximize the productivity of the system consisting of subsystem X and subsystem Y. Subsystem X performs four different functions in a given sequence, *a, b, c,* and *d,* to produce LCD panels. Subsystem Y takes the semifinished panels from subsystem X and subjects them to subsequent processes. In figure E6.1a, processes *a* through *d* are performed in subsystem X by machine M_a through M_d, respectively. Each machine M processes only one part at a time. A robot located in subsystem X transports each part within subsystem X and passes it to subsystem Y.

The process time PT_a, PT_b, PT_c, and PT_d for each process *a, b, c,* and *d* is the time between the receipt of a part and the completion of all processing at the respective machine M_a, M_b, M_c, M_d. The cycle time CT_Y of subsystem Y is defined as the time between the receipt of a part at subsystem Y and the removal of the part from subsystem Y. We assume that the cycle time of subsystem Y varies significantly compared to the variation of process times of subsystem X. It should be noted that machine M_a is generally still occupied at a time PT_a after the receipt of a part, and machine M_a is not ready to take its next part until its current part is removed by the robot.

In this illustrative example, process *c* is assumed to be time-critical such that a part in machine M_c must be removed as soon as process *c* is completed. It is further assumed that considerations of economic efficiency render highly desirable the maximum utilization rate of subsystem Y.

Figure E6.1b depicts a physical configuration of the exemplary system with one of the possible robot travel paths in subsystem X. The configuration is characterized

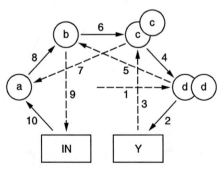

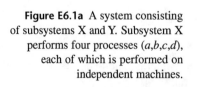

Figure E6.1a A system consisting of subsystems X and Y. Subsystem X performs four processes (*a,b,c,d*), each of which is performed on independent machines.

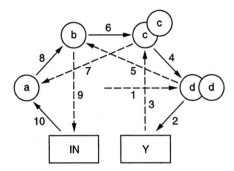

Figure E6.1b Robot motion in subsystem X. A robot travel without carrying a part is indicated by a dotted arrow, and a solid arrow represents a travel with a part.

by a transporter surrounded by multiple process machines M_a, M_b, M_c, M_d. Such a system is generally referred to as a cluster tool. The number of commonly labeled circles indicates the number of machines for the respective process, and is determined based on the process time and required throughput rate.

Perkinson et al. (1996) showed that the maximum steady-state throughput rate is determined by the nominal fundamental period, FP as[8]

Maximum throughput rate $= 1/FP$

$$FP = \max[(PT_i + 4t_{tr})/n_i] \qquad (i)^9$$

where t_{tr} is a transport time, PT_i is the process time of process i, and n_i is the number of machines for process i.

One of the assumptions made in deriving equation (i) is that the robot waits at its current position and begins moving toward a destination machine only after the process therein is complete. It is sometimes called *simple* scheduling.

Equation (i) can be modified to better model this example. In a particular setup of this example, the FP of the whole system depends on which of subsystem X and Y is slower. If the throughput rate of subsystem X is less than the throughput rate of subsystem Y and therefore determines the pace of the integrated system, the fundamental period FP is given by

$$FP = FP_X$$
$$= \max\left[\frac{PT_i + MvPk_i + MvPl_{i+1} + MvPk_{i-1} + MvPl_i}{n_j}\right], \qquad i = a, b, c, d \quad (ii)$$

where $MvPk_i$ is the time it takes for a robot to move to M_i and pick up a panel, $MvPl_j$ is the time it takes for a robot to move to M_j and place a panel. ($+1$) and (-1) in the subscripts indicate the next and previous machines M, respectively. For example, when $i = a$, ($i-1$) is IN and ($i+1$) is b.

If subsystem Y is slower than subsystem X, the fundamental period of the integrated system is given by

$$FP = FP_Y = CT_Y + Mvpk_{Y-1} + Mvpl_Y \qquad (iii)$$

where $MvPk_{Y-1}$ is the time for a robot to move to one of the last process machines M_d in subsystem X and pick up a part, and $MvPl_Y$ is the time for a robot to move the part to subsystem Y.

The fundamental period of the overall system FP is given by the larger of the fundamental period FP_X determined by subsystem X and the fundamental period FP_Y determined by subsystem Y, that is,

$$FP = \max[FP_X, FP_Y] \tag{iv}$$

Based on the process times and transport times, the number of machines M for each process is selected to achieve the required system throughput rate. For example, if the fundamental period FP_X of subsystem X is larger than the desired throughput time, and is determined by process P_i, adding more machine M_i to perform process P_i generally reduces the fundamental period FP_X. The new fundamental period FP_X is then generally determined by another process in subsystem X. This design progression is repeated until the desired fundamental period and thereby the throughput rate is achieved.

The following three cases illustrate the reinitialization of systems to achieve time-dependent periodic complexity for different subsystem throughput relationships:

Case 1: Subsystem X is slower than subsystem Y.

$$FP_X > FP_Y$$

Case 2: Subsystem X is faster than subsystem Y.

$$FP_X < FP_Y$$

Case 3: Both systems are about the same, with CT_Y fluctuating about its mean.

$$FP_X \approx FP_Y, \text{ i.e., } \min\{FP_Y\} < FP_X < \max\{FP_Y\}$$

In each case, the maximum productivity (throughput rate) is attained when the operations of the subsystems are subject to a repeated reinitialization implemented after the completion of the subsystem cycle. Reinitialization introduces periodicity and thus changes the scheduling problem from that of time-dependent combinatorial complexity to one of time-dependent periodic complexity. Each one of the situations will be considered separately.

Case 1: $FP_X > FP_Y$

Case 1 is directed to a system in which the fundamental period FP_X exceeds the fundamental period FP_Y. As a consequence of subsystem Y's being faster than subsystem X, subsystem Y has to wait until a next part finishes its processes in subsystem X. In other words, subsystem Y is operated in a starved mode. Therefore, it does not matter when subsystem Y finishes its processing and requests a new part as long as this relationship between FP_X and FP_Y exists. In short, the variation of CT_Y does not affect the operation of subsystem X. The maximum productivity is achieved simply when the throughput of the integrated system reaches that of subsystem X.

Table E6.1 shows the process times for processes a, b, c, d, the cycle time for subsystem Y, the number of machines for each process, and the associated transport times. According to equation (ii), the fundamental period FP_X is 80 sec. As stated above, CT_Y is assumed to vary within ± 5 sec, that is $CT_Y = 55$–65 sec. Equation (iii) yields the fundamental period FP_Y of 65–75 sec. Given FP_X and FP_Y, the fundamental period FP of the integrated system is 80 sec.

Table E6.1

Station	PT_i or CT_Y (sec)	No. of machines	$MvPk_i$ (sec)	$MvPl_i$ (sec)
IN	—	1	5	—
a	30	1	5	5
b	40	1	5	5
c	60	1	5	5
d	80	2	5	5
Y	60 ± 5	1	—	5

As mentioned above, subsystem Y operates in a starved mode since FP_X is larger than the maximum FP_Y. Therefore, variation in subsystem Y's cycle time, 55–65 sec, will not affect the system. In other words, even if subsystem Y requests a panel to process, no panel is available from subsystem X, and thus subsystem Y has to wait for a while. The variation in CT_Y needs not be considered in subsystem X's transport scheduling. Since there is no variation issue in scheduling, the steady-state scheduling can be directly used.

To solve this problem, it is very useful to visualize the situation using a timing diagram. Figure E6.1c is a part-flow timing diagram. The horizontal axis represents time and the vertical axis (row) represents different parts processed by the system. In particular, the first row represents the flow of the first part processed by the system, the second row represents the flow of the second part processed by the system, and so on. Since FP is constant at 80 sec, the incoming parts are assumed to be 80 sec apart from each other in steady state. In other words, the sending period—time from one panel entering the system to the next panel entering—is set to equal the FP of the system.

Figure E6.1c shows that there are two transport conflicts between transport tasks [1,2] and [5,6] and [3,4] and [9,10], and the conflicts occur repeatedly. It is clear that, depending on which panel is picked up first by the robot at the moment of the conflict, the subsequent pick-up schedule will be affected by the decision. Thus, the number of possible routes for the robot increases as additional decisions are made at the time of the conflicts. The system is therefore subject to time-dependent combinatorial complexity.

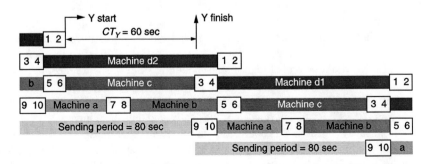

Figure E6.1c Part-flow diagram. Each row represents an individual panel processed by different machines. Transport tasks [3,4] and [9,10] are in conflict with [1,2] and [5,6].

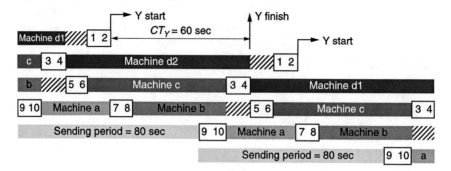

Figure E6.1d Part-flow diagram with postprocess waiting times. 10 sec of postprocessing waiting times at b and d resolve the conflicts.

To reduce the number of possible combinations and to make the robot schedule more deterministic, this time-dependent combinatorial complexity problem must be converted into a periodic complexity problem. If a fundamental period and waiting times at some of the machines are appropriately selected, the complexity of subsystem X can be converted into a periodic complexity.

From figure E6.1c, one can easily see by inspection, that 10 sec of postprocess waiting time in machine M_b and another 10 sec postprocess waiting time in M_d will simply resolve these conflicts without the need for real-time decision-making. In other words, those waiting times prevent conflicts from occurring in the first place. Figure E6.1d shows such a scheduling solution with the waiting time indicated by a hatched area. Note that the time subsystem Y starts processing is determined by subsystem X, which is not affected by the variation in CT_Y.

Case 1 is equivalent to a scheduling problem with no variation since subsystem X, which has no process time variation, determines the pace of the whole system. Therefore, steady-state scheduling such as that of Oh and Lee (2000) can be used directly to provide a scheduling solution as in figure E6.1d.

Case 2: $FP_X < FP_Y$

Table E6.2 shows the process times for processes a, b, c, d, the cycle times for subsystem Y, the number of machines for each process, and the associated transport times. Based on the equations (ii)–(iv), the fundamental period of the integrated system is determined by subsystem Y to be 85–95 sec. If CT_Y were constant at 80 sec, assigning appropriate waiting times with the system's sending period set to 90 sec would solve the scheduling problem simply, as in Case 1. The pattern shown in figure E6.2a would be one of legitimate scheduling solutions and repeat itself.

However, contrary to figure E6.1c, the pattern of timing diagram given the variation will not be the same for each period. Subsystem Y takes a part from subsystem X once every 85–95 sec, that is, nondeterministically, and thus the timing of the transport task from subsystem X to Y is not fixed with respect to other transport tasks. Thus the constant waiting time solution is not valid, and the schedule for robot motion must be recomputed each time subsystem Y picks up a semifinished part from subsystem X.

In this illustrative example, there are two constraints on subsystem X. First, the part just processed at machine M_c must be immediately picked up for transport to

Table E6.2

Station	PT_i or CT_Y (sec)	No. of machines	$MvPk_i$ (sec)	$MvPl_i$ (sec)
IN	—	1	5	—
a	30	1	5	5
b	40	1	5	5
c	50	1	5	5
d	80	2	5	5
Y	80 ± 5	1	—	5

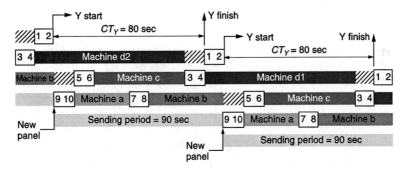

Figure E6.2a Part-flow diagram without variation in CT_Y—a simple case.

M_d. Second, a part and the robot must be available for subsystem Y when it is ready to take the part.

Depending on the variation in the cycle time CT_Y and the inherent conflict patterns, the task of scheduling can be significantly complicated. In particular, the difficulty in scheduling results from randomness in the transport conflict pattern. The variation in CT_Y along with decisions made at the "current" conflict can lead to combinatorial complexity. As previously described, when a time-dependent combinatorial complexity problem is converted into a periodic complexity problem, the design of the operations scheduling is simplified. The conversion requires that a (*functional*) *period* be imposed on the system. In such a period, the same set of tasks is performed cyclically in a predetermined way, and therefore a limited number of scheduling possibilities exists. The period is initiated by an internal or external key event, or reinitializing event, that needs to be appropriately defined.

In this example, a part-request from subsystem Y is chosen to be the reinitializing event for many reasons. First, subsystem Y limits the pace of the integrated system. Also, the pace of the integrated system has to be adjusted to accommodate the variations in the cycle time of subsystem Y. Third, a basic constraint on the system requires that delivery of a part to subsystem Y be completed as soon as possible. Because a part-request issued by subsystem Y is treated as the reinitializing event, the length of each period depends on CT_Y. Even though the length of each period is generally different, the same set of functions, that is, transport tasks, is performed by the robot in subsystem X within each period.

As described earlier, the central idea is to force the system to have a periodicity whose initiation is defined by the reinitializing event, which is a part-request from subsystem Y. At the time of reinitialization, the current state of subsystem X (i.e., which machines are available and the remaining process times of occupied machines) is determined. Appropriate waiting times are then calculated for each of the occupied machines, which determine a schedule for the current period. First, to ensure that the robot is always available during the time slot of the next reinitializing event, a no-transport-time is determined as

$$
\begin{aligned}
&\text{No-transport-time} = \\
&\{t \mid t \in [(MvPk_{Y-1} + MvPl_Y) + \min(CT_Y), \\
&\quad (MvPk_{Y-1} + MvPl_Y) \cdot 2 + \min(CT_Y)]\}
\end{aligned}
\tag{v}
$$

where $t = 0$ at the moment of the current reinitializing event. Second, assign all transport tasks that are determined at the instant of reinitialization (i.e., prefixed transport tasks). Prefixed tasks include transport [1,2] by the definition of the reinitializing event, and possibly transport [3,4] due to the need to move a part immediately from machine M_c. Remaining transport tasks are then allocated.

Figure E6.2b depicts a part-flow timing diagram for the steady-state operation,[10] showing a potential variation in cycle time CT_Y at the end. The vertical lines indicate the time when subsystem Y requests a part from subsystem X and therefore represent the moment of reinitialization. As shown in the last period, because of the variation in CT_Y, subsystem Y requests a semifinished part from subsystem X at some time at least 85 sec but no more than 95 sec after the reinitialization (75–85 sec cycle time plus 10 sec transport time for [1,2]). Therefore, the transport task is scheduled so that the robot is available for the period from 85 sec to 105 sec after the reinitialization.

After a part-request is issued by subsystem Y (vertical line), a renewal signal is generated to reinitialize the database of processes. First, the state of each machine M is identified as busy or idle and as empty or occupied. For example, at the onset of the second reinitialization (second vertical line), the following information is identified:

Empty	$\{\}$
Busy	$\{M_a, M_b, M_c, M_{d2}\}$
Occupied	$\{M_{d1}\}$

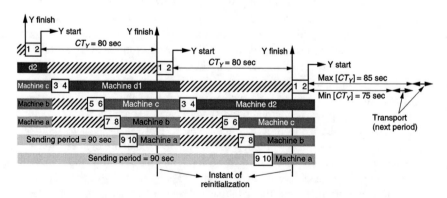

Figure E6.2b Part-flow diagram with variation in subsystem Y's cycle time.

t' = initialized t		0	5	10	15	20	25	30	35	40	45	50	55	60	65	70	75
CT_Y				0	5	10	15	20	25	30	35	40	45	50	55	60	65

	State	
Machine a	15	a
Machine b	15	b
Machine c	15	c
Machine d1	ocp	
Machine d2	15	d2

Figure E6.2c Information at the instant of reinitialization.

Once busy machines are identified, their remaining process times are monitored. Figure E6.2c depicts the remaining process times for the busy machines.

Based on this information, the transport schedule is constructed. In this example, no-transport-time is $\{t'|85 < t' < 105\}$. Transport task [1,2] and [3,4] are prefixed tasks. Transport task [1,2] occurs 0 to 10 sec after the moment of reinitialization. Another task [3,4] must occur from 15 to 25 sec after reinitialization because the part in machine M_c must be removed as soon as process c is complete. The allowable transport time slots are computed and the remaining transport tasks are assigned in the time slots.

One possible schedule is shown in figure E6.2d, in which the Xs signify the no-transport-time slot. Transport task [5,6] is delayed for a while because of the no-transport-time condition for the following task [3,4] in the next period. Transport tasks [7,8] and [9,10] simply follow task [5,6] at the earliest possible time according to the fundamental conditions for part transport (i.e., the current machine is finished, the next machine is empty, and the robot is available).

By applying the above-mentioned approach repeatedly, schedule for the next period can be determined relatively easily regardless of when subsystem Y picks up the part exiting subsystem X. Figure E6.2e depicts multiple intervals with different cycle times CT_Y for subsystem Y, and it is clear that the variation is well accommodated. Indeed, every interval is independent from the previous intervals except for the immediately preceding interval. In other words, the effect of the variation of the CT_Y does not propagate to the later periods.

Case 3: $FP_X \approx FP_Y$, i.e., $min[FP_Y] < FP_X < max[FP_Y]$

Case 3 is a hybrid version of Case 1 and Case 2 where the time taken by both systems is about the same, with CT_Y fluctuating about its mean. The cycle time FP_Y

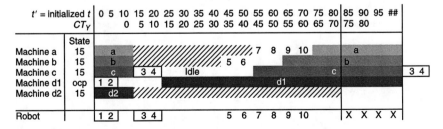

Figure E6.2d Part-flow diagram showing no-transport-time slots. Transport times [1,2] and [3,4] are fixed. X indicates no transport time.

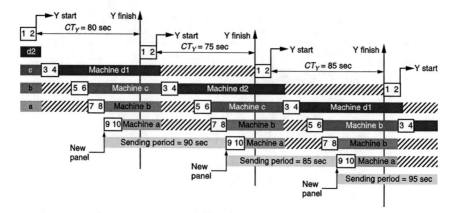

Figure E6.2e Resulting schedule for Case 2.

is sometimes less than FP_X and other times greater than FP_X. It is tempting to think that the effect of the period where subsystem Y is slower is canceled out by that of faster periods over the long run, and thus it is possible to operate subsystem X at subsystem Y's pace all the time. Unfortunately, the faster cycle with $\min[CT_Y]$ cannot be used to directly offset the slower cycle with $\max[CT_Y]$ because the duration of the actual cycle time CT_Y for the next period is not known *a priori*. In other words, since it is not known when subsystem Y will request its next part, subsystem X has to be ready to deliver a part at the *earliest* possible request time by subsystem Y if it is to keep pace. This is equivalent to operating subsystem X at a faster pace than its maximum capability, which cannot be achieved.

The process times, number of machines, and associated transport times are shown in table E6.3. According to equations (ii) and (iii), FP_X is 70 sec and FP_Y is 65–75 sec, that is, the pace of the integrated system is sometimes determined by subsystem X, and at other times by subsystem Y.

Figure E6.3a depicts one mode of steady-state operation of subsystem X with a sending period equal to the nominal FP of 70 sec. Limited to the illustrated instance, subsystem X appears capable of providing a part to subsystem Y even if

Table E6.3

Station	PT_i or CTY (sec)	No. of machines	$MvPk_i$ (sec)	$MvPl_i$ (sec)
IN	—	1	5	—
a	30	1	5	5
b	40	1	5	5
c	50	1	5	5
d	80	2	5	5
Y	60 ± 5	1	—	5

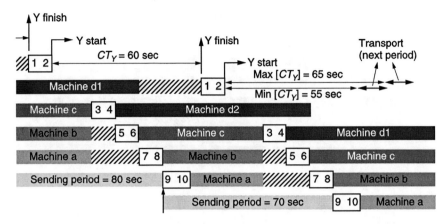

Figure E6.3a Steady-state operation with 70 sec sending period.

subsystem Y has a cycle time CT_Y of 55 sec. In particular, the fundamental conditions for part transport are satisfied because a part is ready at machine M_{d2}, the robot is available, and subsystem Y is ready to accept a part. Would it be acceptable as it appears? Would subsequent intervals with CT_Y of 65 sec compensate for this period?

In figure E6.3b, however, it can be shown that subsystem X cannot sustain a high system throughput over many intervals. In figure E6.3b(a), we begin with, a situation where CT_Y is first assumed to be 55 sec which is min[CT_Y], and afterwords is kept to 65 sec, which is max[CT_Y]. In other words, subsystem Y is *once* faster than subsystem X, and it becomes slower afterward. We used simple fundamental conditions of part transport with no-transport-time considered.

Figure E6.3b includes a series of part-flow timing diagrams arranged in chronological order. Figure E6.3b(b) immediately follows figure E6.3b(a) in time, figure E6.3b(c) immediately follows figure E6.3b(b) in time, and so on. Each row in the figures is numbered according to a specific part number and the horizontal axis represents increasing time. An interval is defined as the period of time between a "Y finish" and the immediately following "Y finish," for example, time from 0 to 1 in figure E6.3b(a).

In figure E6.3b(a), subsystem Y requests a part after it finishes its process with 55 sec CT_Y (see 1). Subsystem X is able to deliver a part for this early request because machine M_{d2} has completed its process and waits for part number 2 to be picked up (see 2). Thus, when subsystem Y completes its cycle, part number 2 is immediately provided. As a result, the throughput time for this period (time from 0 to 1) is 65 sec. In figure E6.3b(b), CT_Y is shown as 65 sec. Note that there are only four transport tasks, that is, [1,2], [3,4], [5,6], and [7,8], in the first interval of figure E6.3b(b). It is required that a no-transport-time duration of 20 sec (indicated by two vertical lines and the Xs) be available to handle variations in the cycle time CT_Y of subsystem Y. Consequently, transport task [9,10] cannot be performed during the first interval of figure E6.3b(b) and is instead delayed to the next interval (see 3).

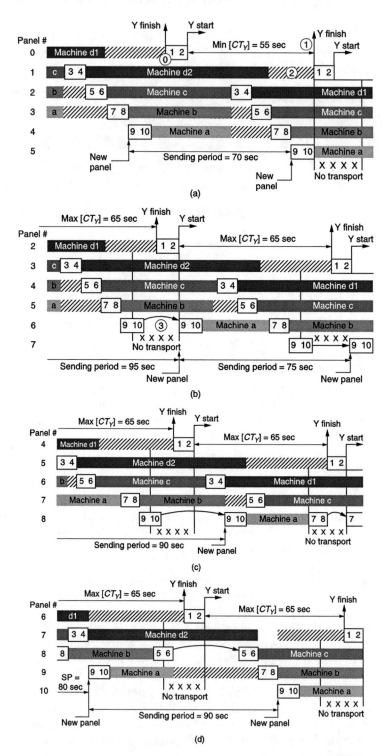

Figure E6.3b Chronological part-flow timing diagram. See text of explanation of parts (a)–(g).

158

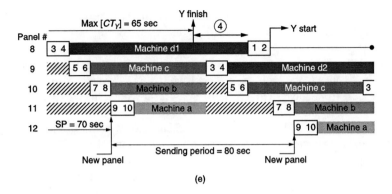

(e)

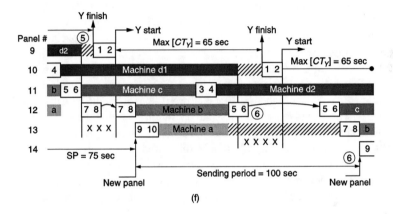

(f)

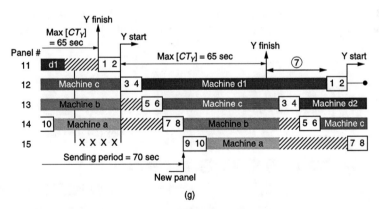

(g)

Figure E6.3b *(continued)*

The effect of the incomplete interval is first manifested in the elongation of subsequent sending periods beginning with the immediate increase shown in figure E6.3b(b). Some of the subsequent sending periods are longer than the 75 sec, which is required by the system's fundamental period of 75 sec based on the 65 sec cycle time CT_Y of subsystem Y. Unless there are a sufficient number of short sending periods to compensate for the effect of the long sending periods, the integrated system will not be able to produce parts per FP of 75 sec because of the shortage of parts introduced into the system. For the intervals up to and including figure E6.3b(d), subsystem X manages to follow the rate of request from subsystem Y. In figure E6.3b(e), however, there is no part in subsystem X ready to satisfy a part-request from subsystem Y (see 4). As a result, subsystem Y must wait for its next part to be available.

In the first interval shown in figure E6.3b(f), subsystem X regains its ability to immediately satisfy a part-request from subsystem Y (see 5). However, the no-transport-time condition in the next interval produces a long sending period of 100 sec (see 6), which returns the system back to the shortage state. Figure E6.3(g) illustrates another instance of delay in part delivery from subsystem X to subsystem Y (see 7). The pattern depicted in figure E6.3b(g) is the same as that of figure E6.3b(e); thus it is apparent that the system achieves a steady state and that an extension of the part-flow timing diagram of figure E6.3b is simply a repetition of figure E6.3b(e) to figure E6.3b(g).

In constructing figure E6.3b, it has been assumed that CT_Y is 65 sec, except for the first interval. Subsystem X is faster than subsystem Y when CT_Y is 65 sec because FP_Y is 75 sec, whereas FP_X is 70 sec. Thus it might be expected that subsystem X would evolve to a steady-state operation such that the fundamental period of the total system is 75 sec, and that subsystem X would always be able to feed a part to subsystem Y in time. Obviously, this is the case if the cycle time CT_Y is 65 sec without exception. Temporal location of the no-transport-time is fixed relative to the other robot moves; thus the same part-flow pattern is established in steady state as shown in figure E6.3c. Such a case is trivial since it is merely a transition from one steady state to another. Figure E6.3b demonstrates, however, that a single occurrence of a 55 sec CT_Y combined with an attempt to run subsystem X above its maximum speed—seen as a no-transport-time slot—results in a degradation of

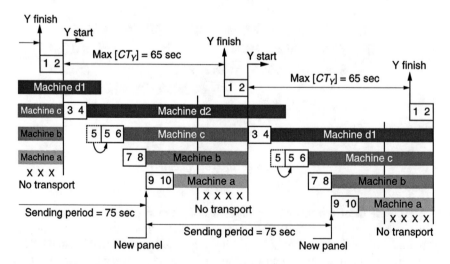

Figure E6.3c Steady-state operation with consecutive 65 sec cycle time of subsystem Y.

system performance. As a result, the system cannot even maintain a fundamental period of 75 sec at which it otherwise could operate. Every fourth interval, subsystem Y must wait for an additional 25 sec (see 4 and 7). Thus its average throughput time in steady state is 81.25 sec (i.e., $(75 + 75 + 75 + 100)/4$ sec), which is the same as the average sending period, as expected.

In the immediately preceding example, the longer interval of subsystem Y does not compensate for the shorter interval of subsystem Y. Even in the longer period, in order to cover the full range of *unknown* variation in cycle time CT_Y, subsystem X should complete its cycle (processes and transports) within the minimum cycle time CT_Y of subsystem Y, so that there is enough time to accommodate the no-transport-time slot. If subsystem X cannot complete a cycle CT_X within one interval of CT_Y, the scheduling problem becomes a time-dependent combinatorial complexity problem lacking periodicity. As a result, opportunities for nonoptimal scheduling decisions increase and the overall system performance can degrade.

The root cause of the problem—inability to achieve the desired productivity—is the attempt to run subsystem X beyond its maximum speed with a presumption that slower periods and faster periods would nullify the effect of each other. The attempt to overrun subsystem X caused an incomplete cycle. As a result, it yielded a low throughput rate, which was the opposite of what was intended. This implies that subsystem X can absorb subsystem Y's cycle-time variation only when subsystem X is sufficiently faster than subsystem Y. Under this condition, it can be reinitialized with appropriate intervals and thus, assure the *periodicity* of robot's transport functional requirements. In the present example, subsystem X must be reinitialized conditionally; if subsystem Y requests a part at a pace faster than subsystem X's maximum speed, then initialization must wait until it reaches FP_X, that is, subsystem X is ready. This, in turn, means that subsystem X must always be given enough time to complete its functional cycle before going into the next period. In other words,

$$t_{ini} = t_{request} \text{ if } t_{request} \geq 70 \text{ sec } (FP_X)$$

$$t_{ini} = 70 \text{ sec if } t_{request} < 70 \text{ sec} \qquad \qquad \text{(vi)}$$

Under this limitation, the scheduling procedures used in case 2 can be applied to the present example.

Figure E6.3d shows the remaining process times and state of each machine at the moment of initialization (see 0 in Figure E6.3b(a)). The no-transport-time is

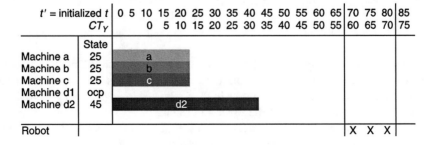

t' = initialized t		0	5	10	15	20	25	30	35	40	45	50	55	60	65	70	75	80	85
	CT_Y		0	5	10	15	20	25	30	35	40	45	50	55	60	65	70	75
	State																		
Machine a	25		a																
Machine b	25		b																
Machine c	25		c																
Machine d1	ocp																		
Machine d2	45			d2															
Robot																X	X	X	

Figure E6.3d Information at the instant of reinitialization.

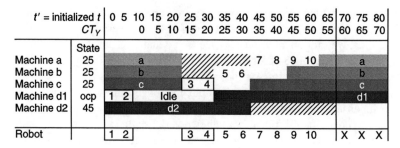

Figure E6.3e Result of scheduling for a period.

determined to be the duration between 70 sec and 85 sec after the moment of reinitialization. The time from 65 sec to 70 sec is excluded from the no-transport-time because, as previously described, the reinitialization cannot occur before the FP_X of 70 sec has expired. Prefixed transport tasks [1,2] and [3,4] are allocated outside of the no-transport-time slot. The other transport tasks are assigned based on the fundamental conditions of part transport. A finished part in machine M_{d2} does not leave until another part request is issued by subsystem Y. One possible schedule for a period is shown in figure E6.3e, and the same approach can be repeatedly applied to successive periods.

6.5 ## Summary

This chapter showed how the idea of reducing the complexity of system by transformation from time-dependent combinatorial complexity to periodic complexity can be applied to a manufacturing system. The manufacturing system considered was a cluster of two subsystems with many machines in each. We showed that the cluster can be a system with time-dependent combinatorial complexity because of the fact that the number of ways the parts can be moved from module to module increases rapidly as a function of time, eventually reaching a chaotic state. This increase in complexity will reduce the productivity of the cluster. Such a cluster cannot produce parts at the rate of the slowest subset of the system.

The complexity of this manufacturing system can be reduced since there is a finite set of FRs that repeats, that is, the functional periodicity was the set of robot motions that repeats cyclically. In this case, the system can be made into one with complexity by reinitializing the system when a key function repeats itself at the beginning of a new cycle.

The reinitialization step involved determining the state of each FR and then coming up with a repeatable strategy for moving the parts within the given period. By introducing such a period (i.e., rhythm), productivity can be maximized, in other words, can be equal to the throughput rate of the slowest machine.

When a system must be constructed by integrating various subsystems, various machine functions must be coordinated—subsystems with varying process or cycle times. To maximize the productivity and reliability of such a system, *periodicity* must be introduced by reinitializing the subsystems on a periodic functional interval and by eliminating scheduling conflict through the introduction of decouplers. This is equivalent to changing a combinatorial complexity problem into a periodic complexity problem, which increases predictability of system performance and productivity.

Notes

1. A large job shop manufactures parts that its customers bring in at random times in unpredictable quantities. A job shop handles the workload using general-purposed machine tools rather than dedicated special-purpose machines. The productivity of a job shop can be measured by the value added to the parts by its operations. In general, an attempt is made to fully utilize the existing machine tools.

2. A transfer line is made of an assemblage of special-purpose, dedicated machines and cutting tools. It processes identical parts through many different processes, where the parts are transferred from machine to machine using an automated transport. It involves a small team of operators.

3. Even nowadays we measure labor productivity, but it may not a true measure of the total productivity in many industries, since in many manufacturing operations the cost of labor is less than 10% of the total direct manufacturing cost.

4. "Fab" is the jargon used to mean the integrated factory that makes semiconductors.

5. Song et al. (1993) and Kats et al. (1999) showed that the values of q_0 can have multivalues. When there are two values for q_0, then one cycle (or a period) consists of two parts undergoing an identical set of processes, that is, a two-part cycle. Similarly, when q_0 has three values, the period of the system would correspond to a three-part cycle.

6. This example was done by Taesik Lee as part of his doctoral thesis at MIT (2003).

7. Exhaustive search for feasible transit patterns and selecting the best one through simulating all of them is not an economic solution to the problem, especially in the presence of significant process time variations. Therefore, we do not use the combinatorial search method. Some scheduling methods, for instance, Oh and Lee (2000), Jevtic (2001), and Rostami et al. (2000), involve a long-term planning (or fixed schedule). On the other hand, the commonly used reactive method involves no planning at all. Each has shortcomings: the fixed-scheduling is vulnerable to variation, and thus its applicability (or utility) is limited in a practical situation. The priority-based reactive schedule does not guarantee the performance.

8. The throughput rate of cluster tool is well studied by Perkinson et al. (1994) and Wood (1996). In this example, we follow Perkinson's notation and result.

9. This formula applies when the system is in process-limited mode of operation. Throughout the discussion of the example, the system is assumed to be process-limited.

10. There are many feasible steady states.

References

Jevtic, D. 2001. "Method and apparatus for automatically generating schedules for wafer processing within a multichamber semiconductor wafer processing tool," in *US Patent & Trademark Office*, Applied Materials, Inc., United States.

Kats, V., Levner, E., and Meyzin, L. 1999. "Multiple-part cyclic hoist scheduling using a sieve method," *IEEE Transactions on Robotics and Automation*, Vol. 15, pp. 704–713.

Lee, T. 2003. "Complexity theory in axiomatic design," Ph.D. dissertation, MIT, Cambridge, MA.

Oh, H.L. 1999. "Reducing complexity of wafer flow to improve quality and throughput in a single-wafer cluster tool," in *IEEE/CPMT International Electronics Manufacturing Technology Symposium*, Austin, TX, October 18–19, pp. 378–388.

Oh, H.L. and Lee, T. 2000. "A synchronous algorithm to reduce complexity in wafer flow," in *The First International Conference on Axiomatic Design*, Cambridge, MA.

Perkinson, T.L. et al. 1994. "Single-wafer cluster tool performance: an analysis of throughput," *IEEE Transactions on Semiconductor Manufacturing*, Vol. 7, pp. 369–373.

Perkinson, T.L., Gyurcsik, R.S., and McLary, P.K. 1996. "Single-wafer cluster tool performance: an analysis of the effects of redundant chambers and revisitation sequences on throughput," *IEEE Transactions on Semiconductor Manufacturing*, Vol. 9, pp. 384–400.

Rostami, S., Hamidzadeh, B., and Camporese, D. 2000. "An optimal scheduling technique for dual-arm robots in cluster tools with residency constraints," in *39th IEEE Conference on Decision and Control*, Sydney, Australia.

Song, W., Zabinsky, Z.B., and Storch, R.L. 1993. "An algorithm for scheduling a chemical processing tank line," *Prod. Plan. Contr.*, Vol. 4, pp. 323–332.

Suh, N.P. and Lee, T. 2004. "System integration based on time-dependent periodic complexity," U.S. Patent 6,701,205 B2.

Wood, S.C. 1996. "Simple performance models for integrated processing tools," *IEEE Transactions on Semiconductor Manufacturing*, Vol. 9, pp. 320–328.

Exercises

6.1. A "job shop" is a manufacturing facility that makes a random set of products when customers bring in the work. Job shops cannot normally anticipate the nature and volume of future work. Therefore, job shops have general-purpose machine tools and use the most appropriate machines to process the parts their customers bring in. Would it be possible to introduce a functional periodicity to job shop operations? How would you maximize the productivity of job shops?

6.2. Airlines use a "hub and spoke" system to maximize their revenue and minimize the operating cost and capital investment. For example, Northwest Airlines has three hubs: Detroit, Minneapolis-St. Paul, and Memphis. The airlines try to maximize the utilization of their aircraft, minimize the operating cost, and maximize the yield on passenger-miles. Show how an airline that has three hubs and twenty (20) airplanes can achieve a maximum profit using the concept of functional periodicity. State your assumptions clearly. Establish the functional periodicity for the schedule you designed.

6.3. Some universities use a "quarter" system rather than a semester system. The quarter system consists of four three-month quarters and thus universities with a quarter system can teach throughout the year, whereas the semester system consists of two semesters (each four and a half months long) with a three-month summer vacation. The perceived advantages of the quarter system are the maximum utilization of facilities and a greater flexibility of scheduling of courses. The functional periodicity for these academic systems is a temporal functional periodicity. Compare these two systems and identify the constraints that must be satisfied to maximize the productivity of these two systems.

6.4. In this chapter, the analysis done by Lee (2003) was presented. One issue not treated in the analysis is the question of the operating procedure when one of the modules in a subsystem breaks down. Assuming that the module b in subsystem X was down for 20 minutes, develop a strategy of maximizing the productivity of the manufacturing system.

6.5. In the past, various strategies for maximizing manufacturing productivity, including group technology and job shops, were tried. In group technology, parts with similar geometry were machined using the same set of machines. In job shops, machines that perform similar functions were grouped together, and parts were moved to different parts of the factory on carts. All of these operations entailed a large number of parts (i.e., work-in-progress) and inventory, since a batch of parts was moved form machine to machine and the required parts for assembly were not completed at the same time. To overcome the shortcomings of these manufacturing strategies, linked-cell manufacturing systems have been implemented in recent decades. In this system, the machines are arranged in a cellular structure and the worker moves in a loop opposite to the flow of the work piece. Does the linked-cell manufacturing system have a functional periodicity? When does the new functional period begin? Does the linked-cell manufacturing system have a higher productivity than the job shops that employ group technology?

6.6. One of the major issues in developing nanotechnology is the integration across length scales, since a nanostructure must be linked to microstructure and ultimately to the macrostructure. Indeed one of the major issues in the semiconductor industry is the packaging of the integrated chips so that they can be connected to other parts. Devise a means of introducing functional periodicity so as to facilitate the connection of nanoscale features to a microscale structure.

7

▓▓▓ Reduction of Complexity by Means of Geometric Functional Periodicity

7.1 Introduction

The subject of this chapter is the use of geometry as the basis for creating functional periodicity. We will use geometry to transform a system with time-dependent combinatorial complexity to a system with periodic complexity. Based on this basic concept, we can reduce the complexity of the following systems that depend on geometry:

(a) Sliding interfaces for low friction and wear.
(b) Seals for low wear and long life.
(c) Adhesive joints for strong bonding.
(d) Electrical contacts for low contact resistance and wear.
(e) Pin-joints.
(f) The interface "connection" between the nanosystem and the macrosystem.

The following case studies will be presented in this chapter: the use of undulated surfaces for reduction of friction, the Tribotek electric connectors, seals for long life, and a pin-joint.

In chapter 5, the process of transforming a system with time-dependent combinatorial complexity to one with periodic complexity was stated as:

1. Determine a set of functions that repeat on a periodic (or cyclic) basis.
2. Identify the design parameter of a system that may make the system range of the FR undergo a combinatorial process.
3. Transform the combinatorial complexity to a periodic complexity by introducing functional periods,[1] that is,

$$T\langle C_{com}(FR_a) | F_p \rangle = \langle C_{per}(FR_a) \rangle$$

4. Set the beginning of the cycle of the set of FRs as $t = 0$.
5. Stop the process momentarily.
6. "Reinitialize" the system by establishing the state of each function at the instant of reinitialization. The initial state $t = 0$ is the instant at which the selected key FR function begins a new cycle.
7. Determine the best means of satisfying the FRs for the new period.
8. Allow the initiation of the next cycle.

Tribology-related geometric functional periodicity
Geometric systems considered in this chapter are related to tribological applications. In designing a tribological surface, many FRs must be satisfied at the same time. The selected FRs must be satisfied by conceptualizing a design and by choosing an appropriate set of DPs. Both the FRs and the DPs constitute vectors, which are related to each other by a design matrix. To reduce complexity, design must be done so that the FRs are independent of each other and the independence of FRs must be maintained by selecting an acceptable set of DPs.[2] This axiom is satisfied if the design matrix that relates the FR vector to the DP vector is either diagonal (uncoupled design) or triangular (decoupled design). Many tribological surfaces designed by a trial-and-error process tend to be coupled designs. These coupled designs cannot be improved by simply changing some material properties or the geometry of devices arbitrarily. It is necessary to design it right, that is, satisfy the Independence Axiom, rather than try to correct problems generated by coupled designs.

Geometric functional periodicity is specific to a given set of FRs, which are in turn specific to a given task. For instance, low-friction surfaces that operate under constant normal load have a different set of FRs than the seals and pin-joints that operate under geometrically constrained conditions. When the interface of a geometrically constrained system cannot move perpendicular to the surface when a particle is present at the interface, therefore, the normal load increases, which results in much higher friction. When we design a tribological surface, we must consider these factors and introduce functional periodicity in a judicious way.

7.2 Review of Friction and Wear Mechanisms

Tribological products such as a pin-joint, low-friction surfaces, electrical connectors, or seals are systems with time-dependent combinatorial complexity. The time-dependent combinatorial complexity of these tribological products is a result of wear. As will be shown in this section, wear particles are gradually generated at the sliding interface with sliding, some of which agglomerate to form large particles. These agglomerated wear particles penetrate the interface deeper due to the larger load applied to fewer particles, increasing friction force and generating more wear particles by further plowing. This process continues until a steady state is reached.

To reduce the time-dependent combinatorial complexity of the current tribological products, we must understand the basics of tribology, otherwise it will be difficult to choose the right set of DPs to satisfy the FRs. Therefore, in this section, the basics of tribology that are responsible for friction and wear in typical tribological applications will be reviewed.

7.2.1 Review of friction mechanisms

Friction at the sliding interface of metals is caused by the following mechanisms (Suh, 1986):

(a) Plowing of the surface by wear debris and other particles.
(b) Removal of asperities by asperity interactions at the interface.
(c) Adhesion of the sliding interface.

The contribution of each of these mechanisms can be represented in the friction space shown in figure 7.1. The vertical axis is the friction coefficient, one of the horizontal axes ($w/2r$) represents the degree of plowing by wear particles, and the other horizontal axis represents the degree of adhesion, f. When $w/2r$ is equal to 1, half of the wear particle, which is assumed to be a sphere, is imbedded into the surface. When the asperities adhere completely and have the shear strength of the bulk, f is equal to 1. The slope of the asperity is represented by θ. The two surfaces corresponding to $\theta = 0$ (perfectly smooth surface) and $\theta = \theta^*$ (steady-state roughness) represent the upper and lower bounds of the friction space. Friction coefficients can be anywhere within the friction space. Therefore, friction is not an inherent material property, as it depends on the relative hardness, surface topography, and environmental conditions.

The largest contribution to friction in most engineered systems is due to the plowing component. At the sliding interface, wear particles are generated when asperities are

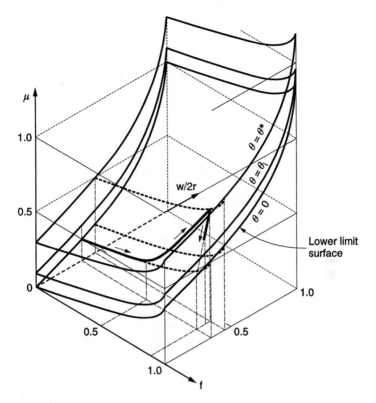

Figure 7.1 The friction space. The f-axis is for the adhesion component. When $f = 1$, all junctions are completely bonded. The $w/2r$ axis is for plowing, as shown in figure 7.2. θ is the slope of the asperity, that is, it indicates roughness of the surface.

removed by asperity interactions. They are also generated when one or both of the sliding surfaces are plowed by the wear particles or asperities that have been generated earlier. Under severe loading conditions, large delaminated wear particles are generated as a result of subsurface deformation and crack propagation. Figure 7.2 illustrates the plowing mechanism. It shows that to support the normal load applied at the interface, particles penetrate the surfaces. The depth of penetration is greatest when the two surfaces have the same hardness. These particles that have penetrated the surfaces plow the surface when the interfaces slide against each other under a normal load. Therefore, work must be done by an external agent to slide surfaces that are being plowed by wear particles. The work done to deform the surfaces plastically, which is the dominant energy-consuming mechanism between any two large-scale (macro) engineering surfaces, is much greater than the changes in surface energy or any interatomic forces. The force resisting the tangential motion is the frictional force, that is, the work done per unit distance slid.

   The tangential force required to shear the interface by plowing depends on the depth of penetration of a given size particle as shown in figure 7.3, which is a plot of the friction coefficient due to plowing as a function of the ratio of the penetration to the particle diameter. When friction is primarily due to plowing, the frictional force depends sensitively on the depth of penetration. The friction coefficient increases nonlinearly as

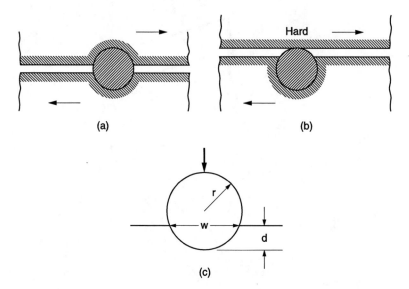

Figure 7.2 The plowing mechanism. A spherical particle entrapped at the sliding interface is indenting into (a) two metals of equal hardness, (b) soft metal when one metal is much harder than the other. The dimensions of the particles are shown in (c).

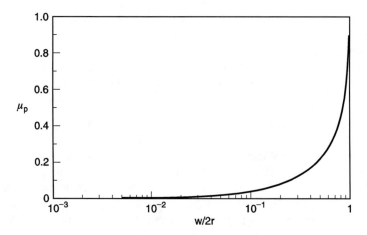

Figure 7.3 A plot of the plowing component of friction as a function of penetration by a sphere of radius r.

a function of the depth of penetration of the wear particle. Therefore, when two identical materials are sliding against each other, the coefficient of friction is higher than when dissimilar materials slide against each other, because while the particle penetrates both surfaces, the depth of penetration is the greatest when the hardness is the same.

With continuous sliding, the wear particles may agglomerate to form larger particles at the sliding interface, especially if the material can readily undergo plastic deformation, when there is sufficient pressure to deform the particles and agglomerate them

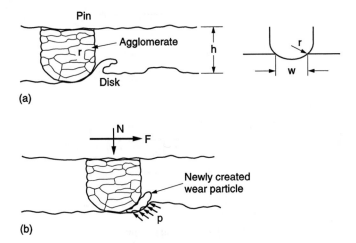

(a)

(b)

Figure 7.4 Wear particle agglomeration process at the sliding interface. (a) Wear particle generation by plowing of the surface by the agglomerated wear particle. (b) The normal load N and the friction force F are in equilibrium with the stress field around the plowed wear particle. (From Oktay and Suh, 1992.)

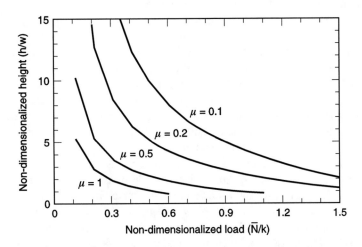

Figure 7.5 Height of agglomerated particles as a function of normal load and coefficient of friction. (From Oktay and Suh, 1992.)

as shown in figure 7.4 (Oktay and Suh, 1992). The particle may grow by agglomeration until it reaches a "steady-state size," as particles larger than a critical size may easily fracture into smaller particles. Particles agglomerate because they move at different velocities and collide. The collided particles deform together, leading to particle agglomeration. The maximum size of the agglomeration is a function of the normal load and the friction coefficient (see figure 7.5). As the particles agglomerate, the applied normal load is carried by a small number of large particles rather than by a large number of small particles. The agglomerated particles penetrate deeper into the surface than the smaller particles do, resulting in increased friction force (see figure 7.6).

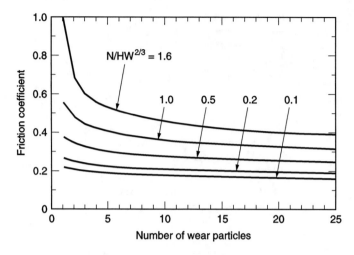

Figure 7.6 Friction coefficient by plowing of the surface by wear particles as a function of the number of agglomerated wear particles. (From Oktay and Suh, 1992.)

When the interface is lubricated with a boundary lubricant, the friction coefficient is lowered to about 0.1. Friction is still high even when there is a boundary lubricant at the interface because the metal-to-metal contact is still necessary to support the normal load. Careful examination of the lubricated surface after sliding shows that there are many fine plowing grooves, indicating that plowing occurs at the interface even when boundary lubricants are used (Shepard and Suh, 1982). These facts indicate that the friction force at the lubricated interface is caused by plowing. This conclusion is further supported by the analysis of the wear groove shape. From the measurement of the groove shape, the ratio of the penetration depth to the width can be determined, which is then used to predict the friction coefficient based on the plowing model. The predicted and the measured friction coefficients agree reasonably well.

Since the largest contribution to friction is made by the plowing mechanism, it is clear that one way of reducing friction is to eliminate the wear particles from the interface before they agglomerate and plow the surface. A means of removing particles from the sliding interface is to create "wear particle traps," into which wear particles can drop. This is demonstrated by creating undulated surfaces.

It may be noted that the idea of interrupting the combinatorial complexity process by creating undulated surfaces as a means of reducing friction was originally conceived to demonstrate that friction is not caused by the adhesion of asperities. The result shows that by eliminating plowing by wear particles, the friction coefficient can be reduced by a factor of 5 or 6 when identical metals are slid against each other.

7.2.2 Review of wear mechanisms in metals

In the preceding section, the mechanisms that cause friction between metal surfaces were discussed. It was stated that the plowing mechanism due to particles entrapped at the interface is primarily responsible for friction between typical engineered surfaces because of the plowing of the surface by the particles. A result of the plowing is the creation of additional particles, which agglomerate because of the sliding action. These agglomerated particles do greater damage to the surface because the normal load exerted on the interface is carried by a smaller number of large particles, leading to deeper penetration of the surface. To continue sliding by plastically deforming the surface, sufficient external force must be applied. The plastic deformation forms surface "furrows," which may ultimately become wear particles under repeated loading and deformation. However, the wear rate due to this plowing mechanism leads to slow wear of the surface, since the wear particles do not form readily if the normal load remains constant, such as when a pin slides on a surface under constant load.

When the applied load is high or when the interface is geometrically confined (e.g., the interface between the shaft and a bushing), the wear particles can cause severe damage. In a geometrically confined system, as the wear particles grow by agglomeration, the normal load on the particle increases continuously since the particle must be pushed into the surface (for geometrical compatibility in a geometrically confined space). When the normal load is high, it leads to higher friction forces. When high normal and tangential forces deform the surface plastically, the surface wears by delamination wear (Suh, 1973, 1986). The wear particle generated by delamination wear is flat and very large, which ultimately leads to higher friction forces and, in the geometrically confined case, to seizure.

The delamination wear mechanisms consist of the following steps:

1. Plastic deformation of the surface due to the surface traction, creating a large shear deformation gradient near the surface extending down to about 200 μm in typical cases. (The magnitude of the shear strain can be several orders of magnitude larger than typical strain encountered in a tensile specimen.)
2. Nucleation of subsurface cracks, typically around the second-phase particles.
3. Propagation of cracks due to the load applied at the surface that causes shear deformation.
4. These cracks typically propagate parallel to the surface, but eventually reach the surface. When they reach the surface, because of the stress field behind the trailing edge of the subsurface crack, they become loose wear particles. These particles are flat (typically 1 to 5 μm thick) and long (typically 20 to 100 μm) plates.

In typical engineering applications, the wear rate associated with delamination is much greater than that due to plowing of the surface, leading to rapid failure. The typical wear coefficient associated with sliding wear is about 10^{-4}.

Figure 7.7 shows crack nucleation around a hard particle due to displacement incompatibility when the matrix deforms as a result of the shear loading and the rigid particle remains spherical. Figure 7.8 is a micrograph of the cross-section of a worn iron specimen with tungsten particles, which shows crack nucleation and the early stages of crack extension and propagation. Figure 7.9 is the graphical representation of a subsurface crack. Its crack length extends each time cyclic loading is applied at the surface because of the sliding action of the asperity of the counter surface. Figure 7.10 shows a wear particle generated by delamination wear lifting up from the surface. The sliding direction was from left to right, that is, the delaminated wear particles often separate from the leading end of the subsurface crack tip.

Second phase particle

Crack nucleation

Figure 7.7 Crack nucleation around a rigid spherical particle when the matrix deforms under shear loading as a result of the displacement incompatibility.

Matrix before shear deformation

Matrix after shear deformation

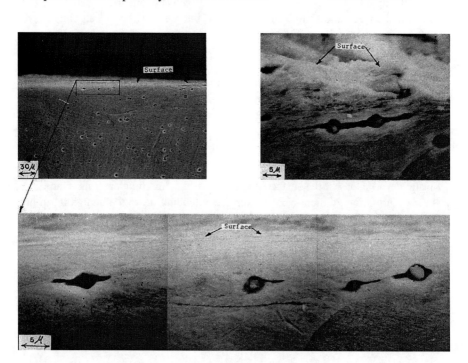

Figure 7.8 A micrograph of the cross-section below a wear track of worn annealed Fe–1.3%Mo. (From Jahanmir et al., 1974.)

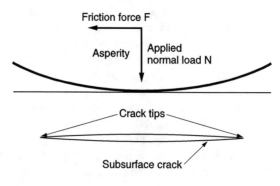

Friction force F

Asperity

Applied normal load N

Crack tips

Subsurface crack

Figure 7.9 Subsurface crack propagation due to the applied cyclic load. The crack is normally closed rather than open because of the compressive loading. The crack tip undergoes shear deformation that leads to crack propagation.

Figure 7.10 Micrograph of a delaminated wear sheet lifting off the surface of iron solid solution. The sliding direction is from left to right since only the crack tip behind the slider experiences a tensile stress field which propagates the crack in a direction 70° away from the crack direction. (From Jahanmir et al., 1974.)

When these delaminated wear particles form and then grow to form even larger wear particles that cannot be accommodated by the space available at the interface (such as between a shaft and a bushing), they become wedged in. Then the normal and tangential forces increase rapidly, mechanically interlocking the interface and leading to eventual seizure between two surfaces.

7.3 Geometric Functional Periodicity for Reduction of Friction

The time-dependent combinatorial complexity that occurs in sliding surfaces is caused by the wear particle generation and agglomeration process (Suh, 1986). We will introduce geometric functional periodicity by making use of the tribology knowledge presented in the preceding section to design a surface with low friction without the use of lubricants.

It was shown that the plowing by wear particles is primarily responsible for friction in many engineering surfaces. When these particles agglomerate,[3] the friction force increases significantly because of deeper penetration by a smaller number of larger particles. Other mechanisms that cause friction are the removal of asperities by the counterface and adhesion at the asperity contacts.

It was emphasized that friction between macroscale engineering surfaces is not an intrinsic material property like yield stress, modulus, and so on, Instead, friction depends on many relative quantities such as:

1. Ratio of the hardness of the two sliding surfaces.
2. Ratio of the apparent area to the actual contact area.
3. Ratio of the distance the particles must travel to the length of the contact area.
4. Ratio of contact temperature divided by the melting point of materials, that is, the homologous temperature.

Friction was represented in a friction space as a function of the plowing depth and the adhesion at asperity contacts. The friction coefficient can be anywhere in the friction space, bounded by two surfaces, one representing a smooth surface and the other the steady-state rough worn surface.

Various wear mechanisms that cause wear at the sliding surface were also presented. There are two kinds of wear processes, one dominated by the mechanical behavior of materials and the other dominated by the chemical behavior of materials. In normal sliding situations involving solid surfaces, many are dominated by the mechanical behavior of materials.

In many engineering applications, the wear of metal surfaces involves plastic deformation of the surface. When two surfaces are sliding relative to each other under a normal load, the original asperities deform and fracture, creating wear particles. When asperities or wear particles plow the surface, plowing grooves (i.e., furrows) are formed on the surface or small wear particles are generated. When the normal load is high, the wear is caused by delamination wear, especially in materials with second-phase particles. Delamination wear involves the plastic deformation of the surface and subsurface layer, nucleation of cracks at the subsurface, and propagation of these cracks under repeated sliding. When the cracks reach a critical size, they propagate to the surface, creating large wear sheets.

The wear particle generation and agglomeration process is a time-dependent combinatorial process, which can be made into a time-dependent periodic process.

7.3.1 Geometric functional periodicity for low friction

Design of low-friction sliding surfaces without lubricants
The system to be considered in this section is "engineered surfaces" for low friction without the use of lubricant. The goal is to design the topography of surfaces to achieve specific engineering goals based on our understanding of the fundamental mechanisms of friction and wear and of axiomatic design theory.

Consider the task of creating low-friction surfaces for sliding applications. The constraints are that metals must be used and that lubricants cannot be used because the sliders are to be used in applications where lubricants cannot be allowed. Based on a fundamental understanding of the friction mechanisms discussed in the preceding section, the FRs[4] of the interface for low friction may be stated as:

FR_1 = Support the normal load.
FR_2 = Prevent particle generation.

FR_3 = Prevent particle agglomeration.

FR_4 = Remove wear particles from the interface.

The constraint is that lubricants cannot be used.

The surface of the sliding interface is to be designed to satisfy the above set of FRs. The particle generation and agglomeration process is a time-dependent combinatorial process, which gives rise to time-dependent complexity. To disrupt this combinatorial process, one idea that has been tried is the creation of undulated surfaces that consist of pockets and pads. The pockets are created to trap wear particles and thus prevent the particle generation and agglomeration process. The pads carry the applied normal load (Suh, 1986).

One example of the undulated surface is the checkerboard pattern, where every other square is a pocket into which wear particles can fall, as shown in figure 7.11. Any particle generated and entrapped at the sliding interface is carried into the pocket by the sliding action of the surfaces, as illustrated schematically in figure 7.12. To disrupt the time-dependent combinatorial process (i.e., prevent plowing of the surface, particle generation, and particle agglomeration under the pressure and sliding action at the interface), the length of the pad must be small.

The DPs for the undulated surface that can satisfy the FRs are:

DP_1 = Total contact area of the pad, A.

DP_2 = Roughness of the planar surface of pads, R.

DP_3 = Length of the pad in the sliding direction, λ.

DP_4 = Volume and depth of the pocket for wear particles, V.

It may be noted that according to Theorem 4 of axiomatic design, an ideal design has the same number of FRs and DPs (Suh, 1990, 2001). In this proposed design, there are four FRs and four DPs. Figure 7.13 shows the design of the undulated surface with DPs.

What is geometric functional periodicity for this undulated surface?

The functional period is given by the combined length of the pad and the cavity. Each time the counterfaces start to make contact at the sliding interface after moving over the cavity for wear particles, the tribological process starts all over again, that is, the FRs are reinitialized.

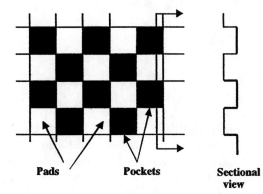

Pads   **Pockets**   **Sectional view**   **Figure 7.11** Undulated surface with a checkerboard pattern.

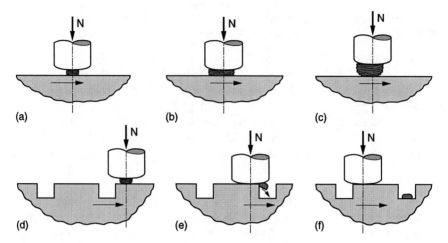

Figure 7.12 Schematic representation of wear particle agglomeration and wear particle removal by means of traps for wear particles. (a) Small wear particle is trapped at the interface. (b) and (c) show the particle growth. (d) The wear particle approaches the trap. (e) The particle falls into the pocket. (f) Wear particle in the trap.

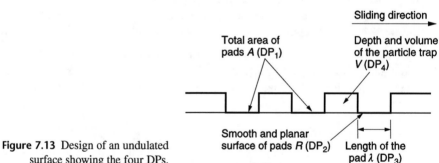

Figure 7.13 Design of an undulated surface showing the four DPs.

Determination of DPs

To a first approximation, the total area A of the pads is given by

$$A = \frac{L}{\sigma_y} = DP_1 \tag{7.1}$$

where L is the normal load and σ_y is the yield strength of the material. It may be necessary to decompose FR_1 and DP_1 to develop the design details further.

DP_2 requires that the surface be smooth without large asperities so as to prevent them from becoming wear particles. DP_3, the characteristic length λ of the pad, should be as small as possible to prevent particle agglomeration under the normal load at the interface. Initially we chose λ to be about 100 μm, which is a typical dimension of the asperity contact. DP_4, the depth and width of the pocket, should be large enough to accommodate the largest wear particle that is likely to be generated at the interface.

Reasonable dimensions for the pocket are 100 μm wide and 50 μm deep. The number and size of the wear particles are expected to be small, since plowing cannot continue generating more wear particles and since the wear particles cannot agglomerate because of the short sliding distance, that is, the length of the pad, λ. Therefore, the friction force should be small with a minimum wear rate at the interface since particle generation is limited and agglomeration is minimized.

The design matrix for the proposed design is given by

$$\begin{Bmatrix} FR_1 \\ FR_2 \\ FR_3 \\ FR_4 \end{Bmatrix} = \begin{bmatrix} X & 0 & 0 & 0 \\ 0 & X & x & 0 \\ 0 & 0 & X & 0 \\ 0 & 0 & 0 & X \end{bmatrix} \begin{Bmatrix} DP_1 \\ DP_2 \\ DP_3 \\ DP_4 \end{Bmatrix} = \begin{bmatrix} X & 0 & 0 & 0 \\ 0 & X & x & 0 \\ 0 & 0 & X & 0 \\ 0 & 0 & 0 & X \end{bmatrix} \begin{Bmatrix} A \\ R \\ \lambda \\ V \end{Bmatrix} \qquad (7.2)$$

The capital X signifies a strong relationship and the lower-case x signifies a weak relationship. The magnitude of X is much larger than that of x. The design matrix is triangular (almost diagonal) and therefore the design is a decoupled design, which satisfies the Independence Axiom of axiomatic design theory. Once we feel that the design is reasonable, we may model each FR/DP relationship based on physical principles to replace the X's and x's with exact physical parameters or constants. When $X \gg x$, the lower-case x can be neglected. Even if x is not negligible, the design is still acceptable since it is a decoupled design.

When the sliding speed is high, the heat generated by the rubbing action at the contact area must be removed rapidly, which may be treated as an additional FR. If we denote this FR as FR_5, DP_5 can be either the shape of the pad (slope of the sidewall to maximize thermal diffusion) or coolant if a coolant can be circulated around the pad.

7.3.2 Friction test results with dry undulated surfaces

Figure 7.14 shows the friction coefficients of copper pins sliding on copper as a function of the distance slid (Suh, 1986; Suh and Saka, 1987). It shows that the friction coefficient of flat copper specimens without undulation increases from about 0.2 to 0.7 with

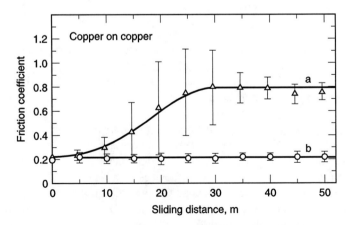

Figure 7.14 Friction coefficient of copper on copper with (b) and without (a) undulation.

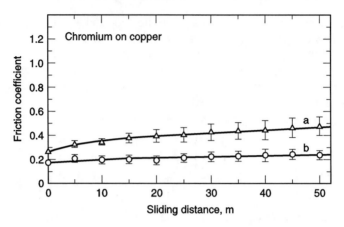

Figure 7.15 Friction coefficient of chromium on copper with (b) and without (a) undulation.

continuous sliding. However, the friction coefficient of undulated copper pins sliding on an undulated copper surface continues to have a low constant friction coefficient of 0.2. It should be noted that this is a low friction coefficient, because the friction of surfaces with boundary lubricants is about 0.1 to 0.2.

These results show that the undulated surface has transformed the time-dependent combinatorial complexity into time-dependent periodic complexity. Through this transformation, the combinatorial process that increases the friction coefficient has been disrupted and low friction has been maintained.

Similar results are obtained when the copper pin is replaced with a harder material. Figure 7.15 shows the case of chromium sliding on an undulated and a flat copper surface. The friction coefficient of the chromium pin sliding on copper increases from about 0.2 to 0.48. The friction coefficient on undulated copper remains at about 0.2. It should be noted that even the flat copper surface without undulation does not reach the same level of friction coefficient as the flat copper on copper. This difference is attributed to the difference in the hardness ratio rather than to any chemical effects. Because chromium is harder than copper, plowing of chromium by the wear particles imbedded in copper is less.

Geometric functional periodicity in tribological systems with lubricants
Similar results were obtained with lubricants. The introduction of geometric functional periodicity makes the lubricated surfaces perform better, even when the temperature exceeds 120 to 150°C, when most lubricants lose their effectiveness.

Geometric functional periodicity in microscale mechanical devices
Microscale mechanical devices (MEMS) are becoming increasingly important. As devices get smaller, friction and wear of small components can limit their performance. One of the problems encountered in these devices is the sticking of the surfaces and a large fluctuation of the friction coefficient. Cha and Kim (2000) have investigated the friction and wear of silicon wafers by creating undulated surfaces. Even in this case, the undulated surface shows much lower friction.

7.4 Geometric Functional Periodicity to Decrease Seal Wear

Seals are important mechanical elements used in machines, engines, and equipment with moving parts. Many of these seals have the characteristics of time-dependent combinatorial complexity, which can be transformed into a system with periodic complexity by introducing geometric functional periodicity.

7.4.1 Behavior of conventional elastomeric seals

Seals are used to shield bearings, engines, and other machine components from damage by abrasive particles. Figure 7.16 is a schematic drawing of a face seal that separates the lubricated section from the slurry section (Ayala et al., 1998). Many machines must be overhauled on a regular basis to replace worn seals.

Elastomeric seals wear when abrasive particles penetrate the interface between the seal and the surface of a machine element, which may be made of metal or ceramic. Often small clay particles (less than 2 μm in size) penetrate the seal/bushing interface. These particles agglomerate under the seal where the pressure is highest; the pressure increases from the edge and then drops down. The wear of these seals typically has a period of minimal wear followed by a rapid wear phase, resulting in a total failure of the seal, as shown in figure 7.17 (Ayala et al., 1998).

Initially, during the incubation period, abrasive particles do not penetrate the interface, and then after a period of low wear, suddenly the wear rate increases as well as the density of abrasive particles at the seal/metal interface. Furthermore, the abrasive

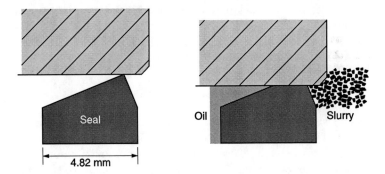

Figure 7.16 A schematic drawing of an elastomeric face seal separating slurry from the lubricated section of a machine. (From Ayala et al., 1998.)

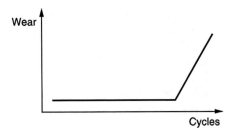

Figure 7.17 Wear of elastomeric seals in abrasive slurry as a function of the cycle of oscillation.

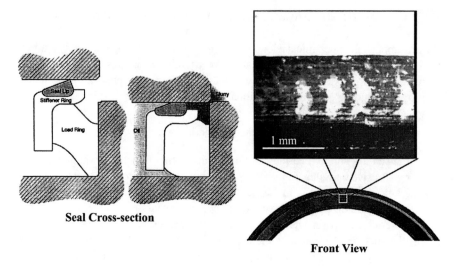

Seal Cross-section

Front View

Figure 7.18 An image sequence showing the motion of particles trapped in a seal contact band. These particles cluster into radially aligned tear-shaped particles. (From Ayala et al., 1998.)

particles accumulate and agglomerate forming radially elongated large particles, which look like raindrops, as shown in figure 7.18. The wear of elastomers is attributed to the abrasive action of the particles that have penetrated the interface between the seal and the moving surface. The particles may agglomerate if the pressure on the particles entrapped between the seal and the moving surface is sufficiently large. Hart and his associates at MIT (Ayala et al., 1998) demonstrated that the wear of these seals can be significantly reduced by introducing surface bumps.

7.4.2 Behavior of elastomeric seals with engineered surfaces

The face seal consists of a wedge-shaped seal lip, where the seal length is equal to the width of the seal.[5] The seal was in contact with the flat surface of an 84 mm diameter bushing that underwent a slow oscillatory rotational motion through a 30° arc. It oscillated at 60 cycles per minute. The seal was pushed against the surface. The average contact pressure was 8 MPa and the average sliding speed was 34 mm/sec. They also measured the lubricant thickness using a fluorescence technique. The seal was surrounded by abrasive particles, which were a mixture of fireclay and bank sand.

The normal face seal design was shown in figure 7.16. To the seal face, Hart and his group added three different features—pads, pads with holes, or just holes—as shown in figure 7.19. The wear rates plotted are the area worn as a function of the number of oscillatory cycles for a flat seal surface (figure 7.20) and for a textured surface (figure 7.21). The seal was rubbing against a glass surface. Figure 7.21 shows that the face seal with flat pad does not wear throughout the cycle tested. This shows that the pads are very effective. However, it should be recalled from the design discussion presented earlier that it is not simply the pads that make the difference, it is the entire design (i.e., pocket for wear particles, the seal lips, materials, and lubricant) that enhances the life of the seal.

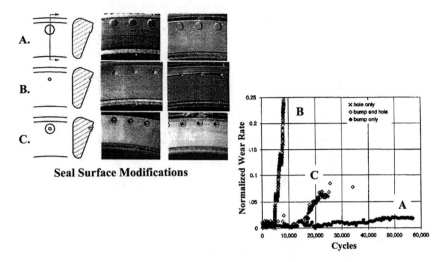

Seal Surface Modifications

Figure 7.19 Three face seal designs. The top figure (A) shows the surface of the seal with pads. The second figure (B) shows the seal face with holes. The last (C) shows the seal face with pads and holes. (From Ayala et al., 1998.)

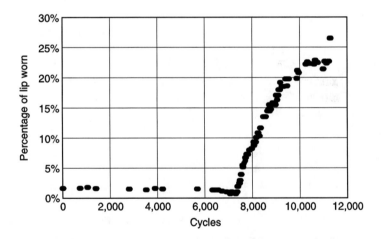

Figure 7.20 Percentage of lip worn as a function of the number of cycles for flat lip surface. (From Ayala et al., 1998.)

7.4.3 Geometric functional periodicity in seals with bumps

The FRs for the seal may be stated as:

FR_1 = Support the normal load.
FR_2 = Prevent particle migration across the seal.
FR_3 = Prevent agglomeration of particles in the seal/metal interface.
FR_4 = Provide lubricant to the interface.

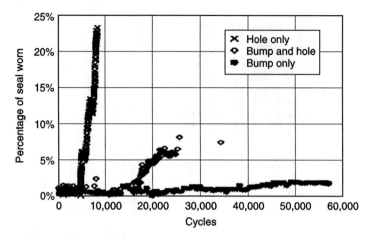

Figure 7.21 Percentage of lip worn as a function of the number of cycles for the three textured lip surfaces shown in figure 7.19. (From Ayala et al., 1998.)

FR_5 = Prevent the flow of the lubricant out of the sealed area.
FR_6 = Delay the initiation of the wear process.

The islands of circular pads collectively carry the applied normal load. Since the pressure at the pad surface is the highest, this is where the particle agglomeration can occur. The diameter of the pad is made small so as to prevent particle agglomeration and abrasive wear of the pad material by the particles. *The distance between the pads determines the functional periodicity.* The particles in the continuous pocket with lubricants do not agglomerate because there is virtually no compacting pressure, in addition to having the lubricants, which also break up particle agglomeration.

The DPs for the seal with bumps are:

DP_1 = Total area of the pad, A.
DP_2 = Seal lip.
DP_3 = Diameter of the pad, λ.
DP_4 = Lubricant.
DP_5 = Length of the seal lip, L.
DP_6 = Seal material.

The normal load exerted on the seal (FR_1) is carried by the total pad area (DP_1). This set of FR_1 and DP_1 can be further decomposed since the number of pads will depend on the total load exerted on the seal, the material properties, and the size of the pad (DP_3). The pressure on each pad should be as small as possible to minimize wear, but high enough to prevent the abrasive particles from penetrating into the interface between the pad and the surface. The seal lip (DP_2) is designed primarily to prevent the abrasive particles from penetrating (FR_2), which also affects the containment of the lubricant (FR_4). FR_2 will also be affected by the material properties of the seal (DP_6).

To prevent agglomeration of abrasive particles (FR_3), the pad size was chosen as DP_3. Agglomeration may also depend on the pressure at the pad surface, but it will be assumed that this pressure effect is negligible. Since the pad is supporting the normal load, it should provide good contact between the pad and the moving surface to prevent particle penetration into the interface. To provide lubricant to the seal surface (FR_4), we may put lubricant under pressure (DP_4), unless the surface tension can provide lubricants at all times. To minimize the leakage rate of the lubricant out of the sealed area (FR_5) at a given pressure and viscosity of the lubricant, the length of the seal lip can be selected as DP_5. Retarding the initial wear rate (FR_6) is very desirable as it determines the onset of the rapid wear process. Material can be chosen as DP_6 to satisfy FR_6.

The design equation and the design matrix for the seal may be represented as

$$
\begin{Bmatrix} FR_1 \\ FR_2 \\ FR_3 \\ FR_4 \\ FR_5 \\ FR_6 \end{Bmatrix} = \begin{bmatrix} X & 0 & 0 & 0 & 0 & X \\ 0 & X & 0 & 0 & 0 & x \\ 0 & 0 & X & 0 & 0 & x \\ 0 & 0 & 0 & X & 0 & 0 \\ 0 & x & 0 & 0 & X & x \\ 0 & 0 & x & 0 & 0 & X \end{bmatrix} \begin{Bmatrix} DP_1 \\ DP_2 \\ DP_3 \\ DP_4 \\ DP_5 \\ DP_6 \end{Bmatrix} \tag{7.3}
$$

The design matrix shows that the material property affects many FRs. The design is a coupled design, but coupling can be eliminated if we can make the lower-case x for FR_6 (the x of the last row of the design matrix) negligible. This can be done if the length of the pad is made small enough to prevent particle agglomeration. Then the design becomes decoupled. To finalize the design, it is necessary to model the design and generate quantitative dimensions of the seal.

The functional periodicity provided by the distance between bumps and the size of the bumps has provided the long life of the designs shown in figure 7.19. Further improvements may be made if we can provide another periodicity in the seal lip by providing grooves that extend radially toward the outer edge but not reaching the edge itself.

7.5 Case Study: Reduction of the Combinatorial Complexity of Pin-Joints

In this section, we will show how time-independent combinatorial complexity can be reduced by developing a decoupled design to replace a coupled design; this also transforms a design with time-dependent combinatorial complexity to one with time-dependent periodic complexity. To achieve this goal, we will take a conventional pin-joint used in industry, which is a coupled design, and develop a decoupled pin-joint design based on axiomatic design. It will be shown that the real complexity of the new design is less than the complexity of the coupled design.

7.5.1 Introduction

Pin-joints are used in many different kinds of mechanical devices and equipment, ranging from door hinges, to military tanks, to earthmoving equipment. Pin-joints are made of pins and bushings to join two mechanical parts that are rotating with respect to each other. A pin-joint allows rotational motion about its axis and carries shear force and bending moments. They are used in mechanical equipment and in many devices, either with or without lubricants. Often they are used in dirty environments where abrasive particles are present, for example, in excavators to hold the bucket to the load-carrying support frame and in undercarriages of bulldozers to link the treads of the track. They are also used in semiconductor processing tools where organic materials and particles cannot be allowed and also in medical devices where lubricants cannot be used.

To design a pin-joint, the FRs that the tribological system must satisfy should be defined first. Then a system solution must be sought to satisfy all of the FRs by conceptualizing a solution and choosing an acceptable set of DPs that can satisfy the FRs and the Independence Axiom. The conceptualization must be based on a thorough understanding of basic friction and wear mechanisms and the mechanical behavior of materials. The designer must also know the specific operating conditions under which the proposed tribological system will be used.

All pin-joints are tribological systems that must satisfy the following set of FRs and constraints at the same time: a pin-joint must join two separate mechanical parts, rotate about its axis, withstand large tangential and bending loads, have low friction and low wear rates in the presence of abrasives, and operate with and without lubricants in corrosive environments and at extreme temperatures. One of the major FRs is that it should never seize, which will be satisfied if the friction force is always low.

To satisfy a set of FRs, we must conceptualize a design that also has an equal number of DPs and that maintains the independence of FRs (see Suh, 2001). However, earlier pin-joints were designed empirically, for example, by simply increasing the hardness of pins and bushings, rather than treating pin-joint design as a systems-design task that must satisfy many different requirements. As a result, these pin-joints fail in service, which is an important industrial problem that requires advanced technological solutions.

This case study describes the design and the test results of innovative pin-joints that can satisfy a set of FRs.[6] All of the facts and knowledge of tribology as well as the FRs and constraints, together with axiomatic design theory,[7] have been used to design the pin-joint.

Complexity of conventional pin-joints
The FRs of an unlubricated pin-joint may be stated as:

FR_1 = Carry the load.
FR_2 = Allow rotational motion.
FR_3 = Minimize frictional force.
FR_4 = Minimize wear.

A simple conventional pin-joint consists of a shaft (normally made of hardened steel) and a bushing (normally made of a softer alloy than the shaft material). The DPs of this design are:

DP_1 = Shaft/bushing dimensions.
DP_2 = Clearance between the shaft and the bushing.
DP_3 = Clearance between the shaft and the bushing.
DP_4 = Material properties.

There are three DPs and four FRs. Therefore, it is a coupled design. Furthermore, FR_3 and FR_4 are coupled by DP_2 and DP_4.

Since an unlubricated conventional pin-joint is a coupled design, the combinatorial complexity of the conventional pin-joints approaches infinity since the pin eventually seizes in the bushing. Therefore, the reliability and the life of these pin-joints are not good.

7.5.2 Design of pin-joints with geometric functional periodicity

Many machines have pin-joints to allow a rotational motion of two rigid bodies connected together through a pin and a bushing. Some of these joints are lubricated and some are used without any lubrication to satisfy other FRs. Even lubricated pin-joints do not rotate at high enough speeds to create hydrodynamic lubrication. Therefore, these pin-joints are lubricated with boundary lubricants or grease to prevent metal-to-metal seizure. Eventually, as a result of both the generation of wear particles and the penetration of abrasive particles, these pin-joints wear out and ultimately the pin seizes in the bushing because of mechanical locking of the pin and bushing.[8]

The pin/bushing system that we wish to design must allow easy rotation, minimize wear, and prevent seizure. The FRs may be stated as:

FR_1 = Carry the load.
FR_2 = Allow rotational motion.
FR_3 = Minimize frictional force.
FR_4 = Minimize wear.

The DPs for the pin-joints are chosen to be:

DP_1 = Dimensions of the pin and the bushing structure.
DP_2 = Cylindrical pin/bushing joint.
DP_3 = Low-friction system.
DP_4 = Wear-minimization system.

The design equation for this highest-level design may be written as

$$\begin{Bmatrix} FR_1 \\ FR_2 \\ FR_3 \\ FR_4 \end{Bmatrix} = \begin{bmatrix} X & 0 & 0 & 0 \\ 0 & X & x & x \\ 0 & 0 & X & x \\ 0 & 0 & 0 & X \end{bmatrix} \begin{Bmatrix} DP_1 \\ DP_2 \\ DP_3 \\ DP_4 \end{Bmatrix} \tag{7.4}$$

The design represented by equation (7.4) is a decoupled design. Note that capital X signifies a strong relationship between the DP and the FR and lower-case x signifies a weak relationship. Note also that we are assuming no interaction between the low-friction system (DP_3) and FR_4 (minimize wear), which should be checked after the design is completed. Since, at this stage of design, the detailed design has not been done yet, equation (7.4) represents a conceptual design with certain design intentions.

FR_3 may be decomposed as

FR_{31} = Remove wear particles from the interface.
FR_{32} = Apply lubricant.

The DPs chosen to satisfy FR_{31} and FR_{32} are:

DP_{31} = Particle removal mechanism.
DP_{32} = Externally supplied grease.

The design matrix for FR_{3x}'s and DP_{3x}'s is a triangular matrix, that is, a decoupled design, as shown below:

$$\begin{Bmatrix} FR_{31} \\ FR_{32} \end{Bmatrix} = \begin{bmatrix} X & 0 \\ x & X \end{bmatrix} \begin{Bmatrix} DP_{31} \\ DP_{32} \end{Bmatrix} \tag{7.5}$$

It should be noted that the friction is less even without the application of the grease when DP_{31} is implemented. However, grease will reduce the friction further.

FR_{31} may be further decomposed as:

FR_{311} = Prevent agglomeration of the particles that are generated by plowing of the surface.

FR_{312} = Do not allow the increase in the normal force due to the presence of particles at the interface.

FR_{313} = Trap wear particles.

To satisfy these FR_{31x}'s, we will make use of an undulated surface pattern on the shaft and a slightly "deformable liner" between the pin and the bushing as shown schematically in figure 7.22. The exact shape of the "deformable liner" can be different from the one shown. Then the corresponding DPs may be stated as:

DP_{311} = The length of the pad on the liner that actually comes in contact with the pin.

DP_{312} = Flexible ring (elliptical or with many sinusoidal wave forms).

DP_{313} = The indented part of the undulated surface (i.e., the space between the pin and the flexible ring).

For the purpose of illustration, the flexible ring is shown as a stiff elastically deformable ring. To increase the number of contact points between the pin and the flexible ring, the circumference of the ring may have a sinusoidal shape. The flexible ring has "ears" that lock into the rigid bushing to make certain that the flexible ring does not slide (i.e., rotate) within the bushing. Other shapes that satisfy FR_{311}, FR_{312}, and FR_{313} are equally acceptable.

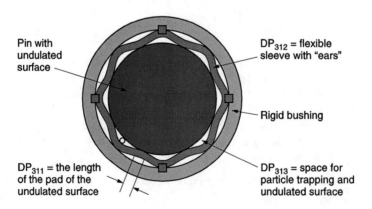

Pin with undulated surface

DP_{312} = flexible sleeve with "ears"

DP_{311} = the length of the pad of the undulated surface

Rigid bushing

DP_{313} = space for particle trapping and undulated surface

Figure 7.22 A schematic diagram of the "eternal" pin-joint. The flexible ring may have a sinusoidal corrugation so as to provide many contact points with the pin.

We may now write down the design equation as

$$\begin{Bmatrix} FR_{311} \\ FR_{312} \\ FR_{313} \end{Bmatrix} \begin{bmatrix} X & 0 & 0 \\ x & X & 0 \\ 0 & 0 & X \end{bmatrix} \begin{Bmatrix} DP_{311} \\ DP_{312} \\ DP_{313} \end{Bmatrix} \qquad (7.6)$$

FR_4 (minimize wear) and DP_4 (wear-minimization system) should also be decomposed, since DP_4 cannot be implemented without further detailed design. They may be decomposed as follows:

FR_{41} = Prevent particle penetration into both surfaces.
FR_{42} = Minimize the shear deformation of the pin and the flexible liner.

DP_{41} = Hardness ratio of the two materials.
DP_{42} = Coating of the pin with a thin layer of low shear-strength material.

If the hardness of the pin and that of the flexible liner are sufficiently different (DP_{41}), the particles will penetrate only into the softer material and plowing will be minimized. We will make the pin much harder than the flexible liner. By coating the pin with a material softer than the flexible liner (DP_{42}), the shear deformation will occur in the coated layer and thus prevent the wear of the flexible liner. In an abrasive environment, the soft coating may not last for a long time, but it should prolong the life by retarding the initial wear process.
   The design equation is:

$$\begin{Bmatrix} FR_{41} \\ FR_{42} \end{Bmatrix} = \begin{bmatrix} X & 0 \\ 0 & X \end{bmatrix} \begin{Bmatrix} DP_{41} \\ DP_{42} \end{Bmatrix} \qquad (7.7)$$

A master design matrix should be constructed to be certain that coupling has not been introduced inadvertently during the decomposition process. The master design matrix may be constructed as follows:

$$\begin{Bmatrix} FR_1 \\ FR_2 \\ FR_{311} \\ FR_{312} \\ FR_{313} \\ FR_{32} \\ FR_{41} \\ FR_{42} \end{Bmatrix} = \begin{bmatrix} X & 0 & 0 & 0 & 0 & 0 & 0 & 0 \\ 0 & X & x & x & x & X & 0 & 0 \\ 0 & 0 & X & 0 & 0 & 0 & 0 & 0 \\ 0 & 0 & x & X & 0 & 0 & 0 & 0 \\ 0 & 0 & 0 & 0 & X & 0 & 0 & 0 \\ 0 & 0 & 0 & 0 & x & X & 0 & 0 \\ 0 & 0 & 0 & 0 & 0 & 0 & X & 0 \\ 0 & 0 & 0 & 0 & 0 & X & 0 & X \end{bmatrix} \begin{Bmatrix} DP_1 \\ DP_2 \\ DP_{311} \\ DP_{312} \\ DP_{313} \\ DP_{32} \\ DP_{41} \\ DP_{42} \end{Bmatrix} \qquad (7.8)$$

   Equation (7.8) shows that, contrary to the argument presented earlier, DP_{31} (particle removal mechanism) has an effect on FR_{32} (apply lubricant), making it a decoupled design, because the grease supply may be affected by the topography of the undulated surface and the open space between the pin and the flexible ring. The tribological system represented by equation (7.8) is a decoupled system, since the design matrix is a triangular matrix (which can be seen by rearranging the matrix—by putting FR_2 and DP_2 as the last row and column).

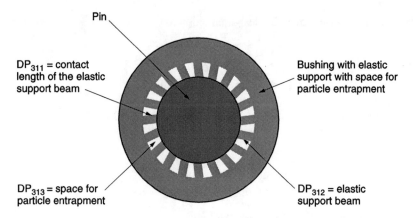

Pin

DP$_{311}$ = contact
length of the elastic
support beam

Bushing with elastic
support with space for
particle entrapment

DP$_{313}$ = space for
particle entrapment

DP$_{312}$ = elastic
support beam

Figure 7.23 Alternative embodiment of low-friction pin-joint.

To obtain the final physical dimensions, X and x of equation (7.8) can be replaced by numbers or equations by modeling the relationships between FRs and DPs. This will yield the final dimensions of various parts of the tribological system. It should be noted that in addition to these DPs, other dimensions must be determined, based on either constraints or the availability of materials, and so on.

There can be many physical embodiments that can satisfy equation (7.8). Figure 7.23 shows another embodiment. In both designs, the particles generated or entrapped at the sliding interface do not affect the interfacial conditions since the particles are accommodated by local compliance. These designs also have functional periodicity, which eliminates the likelihood of the problem transforming into a case of time-dependent combinatorial complexity.

7.5.3 Testing apparatus

A test apparatus[9] was constructed to compare the torque, friction, and wear performance of unlubricated conventional round bushings with the performance of low-friction, low-wear bushings that were designed based on the FRs stated in the preceding section. A diagram of the test apparatus is shown in figure 7.24.

The test apparatus was designed to apply a variable known load to a test pin-joint that was built around a 1 inch diameter test pin. The test bushing was placed inside a support housing and prevented from turning using a standard square key and keyway. The test pin is connected to an adjustable slip clutch and electric motor using a flexible coupling. The pin is supported on each end by a pillow block bearing. The test section consists of a pair of arm-mounted flanged bearings and a bushing block. The bearing arms and bushing block are connected through an end-plate assembly using a threaded rod and belville washer stack. The belville washers are loaded against the end-plate assembly across a load cell by turning a nut on the threaded rod. This allows the applied radial load on the bushing to be adjusted and measured. The total torque exerted on the test section by the test pin is measured using another load cell mounted under the end-plate assembly. The torque measured is the sum of the bushing torque and the torque from the flange bearings. The apparatus is designed for comparison testing, and the

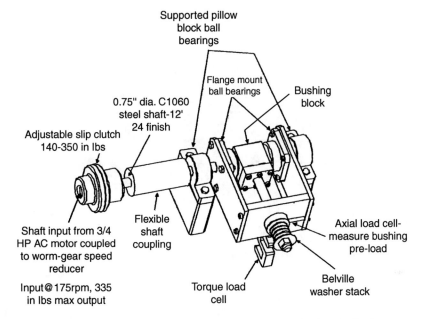

Figure 7.24 Schematic of test apparatus. (Courtesy of Tribotek, Inc.)

torque from the flange bearings can be shown to be constant from one test to another, so can be neglected for the purpose of bushing comparison.

7.5.4 Test results with pin-joints

The prototype test bushing was designed along the lines of the alternative embodiment shown in figure 7.23. A detailed side and end view of the prototype bushing are shown in figure 7.25. The nominal load for testing was 500 N. Both the conventional round and prototype bushing were manufactured from aluminum–bronze alloy 954, and a new C-1060 steel pin, hardened to BHC 60–65, was used on each test. The commercial bearing is shown schematically in figure 7.26. The results from multiple tests are shown in figure 7.27.

As can be seen, for the same radial load, the prototype low-friction bushings produce less total torque than the conventional round bushings. The round bushings start at a higher friction (higher torque) and increase steadily until the bushing seizes to the test pin. The low-friction bushings continue to run with no increase in friction or torque and will continue to run indefinitely, never seizing. Figure 7.28 shows the tests results when a sand mixture consisting of particles less than 0.4 mm in diameter was added to the prototype bushing. While the running torque was greater than for the same bushing without any sand, the same amount of sand caused instant seizure on the conventional round bushing, so no test data were available.

7.5.5 Determination of the time-dependent combinatorial complexity

The real complexity of the conventional pin-joint design is large because the design is a coupled design and the FRs cannot be satisfied within the design range at all times.

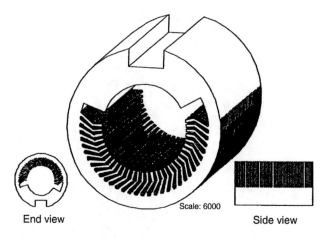

End view   Scale: 6000   Side view

Figure 7.25 Detailed view of the test bearing. The load was applied top-down and thus only the load-bearing surface was machined. (Courtesy of Tribotek, Inc.)

Figure 7.26 A schematic drawing of a commerical bearing made of multiple-stacked layers of stamped sheets with internal support only. (Courtesy of Tribotek, Inc.)

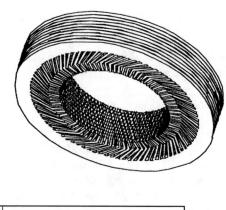

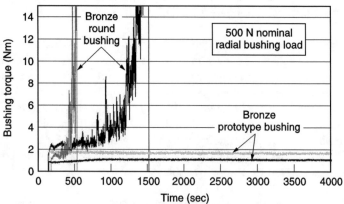

Figure 7.27 Prototype test results showing the torque applied to turn the shaft in the bushing. (Courtesy of Tribotek, Inc.)

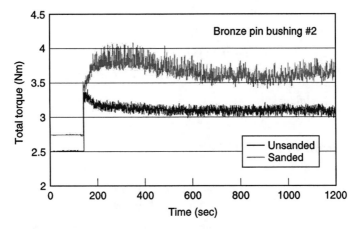

Figure 7.28 Test comparison with heavily sanded prototype bushing. (Courtesy of Tribotek, Inc.)

If there are any preexisting particles, the friction will be high from the beginning. Even if there are no preexisting particles, wear particles will be generated, which eventually brings about high friction; ultimately the pin-joint seizes and does not turn. This seizure is a result of the design transforming into a time-dependent combinatorial complexity problem.

We were successful in reducing the real complexity by making the system range always be in the design range through the introduction of locally compliant surfaces. We also transformed the time-dependent combinatorial complexity into time-dependent periodic complexity by introducing a geometric functional periodicity that enabled the reinitiation of sliding conditions by means of geometry.

To determine the time-independent real complexity of the new pin-joint design, we need to know the design ranges of the FRs of the pin-joint. Normally it would be a good idea to define the design range before we actually design the product, but in this case it was not done explicitly. We will evaluate real complexity by using ideal performance as the design range.

7.5.6 Conclusions on pin-joints

1. The new design proposed is a decoupled design and has negligible real complexity. We have transformed the pin-joint system with time-dependent combinatorial complexity to a system with time-dependent periodic complexity.
2. The time-independent real complexity of a conventional unlubricated pin-joint is shown to be very large, because it is a coupled design. Furthermore these pin-joints fail in service by seizure and wear owing to their large magnitude of time-dependent combinatorial complexity. To overcome these shortcomings, a new design of pin-joints that consists of pins and bushings and satisfies the following FRs using the following DPs was created:

FR_1 = Carry the load.
FR_2 = Allow rotational motion.

FR_3 = Minimize frictional force.
FR_4 = Minimize wear.

DP_1 = Dimensions of the pin and the bushing structure.
DP_2 = Cylindrical pin/bushing joint.
DP_3 = Low-friction system.
DP_4 = Wear-minimization system.

These highest-level FRs and DPs were decomposed as:

FR_{31} = Remove wear particles from the interface.
FR_{32} = Apply lubricant.

FR_{41} = Prevent particle penetration into both surfaces.
FR_{42} = Minimize the shear deformation of the pin and the flexible liner.

DP_{31} = Particle removal mechanism, such as open crevices to trap particles.
DP_{32} = Externally supplied grease.

DP_{41} = Hardness ratio of the two materials.
DP_{42} = Coating of the pin with a thin layer of low shear-strength material.

These second-level FRs and DPs were again decomposed to the leaf-level FRs and DPs as follows:

FR_{311} = Prevent agglomeration of the particles that are generated by plowing of the surface.
FR_{312} = Do not allow the increase in the normal force due to the presence of particles at the interface.
FR_{313} = Trap wear particles.

DP_{311} = The length of the pad on the liner that actually comes in contact with the pin.
DP_{312} = Flexible ring (elliptical or with many sinusoidal wave forms) or elastic support system.
DP_{313} = The indented part of the undulated surface (i.e., the space between the pin and the flexible ring).

3. This design is a pin-joint with bushing that allows the removal of particles—both wear particles and externally introduced particles—by providing the space for powder to fall in, and elastic load-supporting fingers that support the applied load and that prevent the indefinite increase of the contact pressure between the shaft and the bushing. This design reduced the time-independent real complexity and also transformed time-dependent combinatorial complexity into time-dependent periodic complexity.

7.6   Geometric Functional Periodicity for Electrical Connectors

7.6.1 Introduction

Electrical connectors are ubiquitous in the computer and electrical industry.[10] Computer components, printed circuit boards, and backplanes must be connected to other systems, boards, and components. With the advances in chip-making technology,

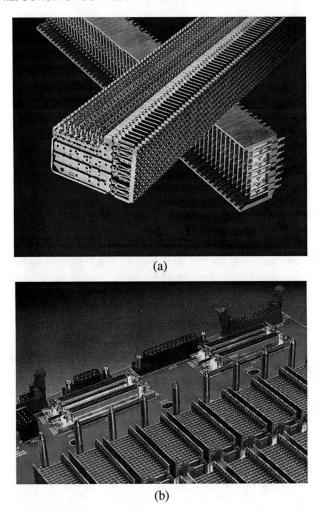

Figure 7.29 (a) Electrical connectors (socket and pins). (b) Electrical connector socket for backplanes. (Courtesy of Teradyne, Inc.—Connection Systems Division.)

electric/electronic circuits and components have become increasingly smaller and more powerful, but the technology for connectors that connect them together has not similarly evolved. Each individual connector is still comparatively large in physical size to accommodate more lines of interconnections.

There are many different types of electric connectors, which may be divided into two groups: data signal connectors and power connectors. Modern data connectors must be able to transmit high-speed data without degrading the quality of the electrical signal (e.g., no cross-talk), whereas modern power connectors must transmit high electrical current density to meet increasing system power requirements at low operating voltages. All these connectors must be made smaller because of the limited space available in computers and other equipment.

A conventional electrical data connector is shown in figure 7.29(a). It consists of a female "socket" and pins. The socket is made by injection molding engineering plastics and inserting a large number of stamped gold-plated brass strips. Large connectors have a large number of pins. The socket and pins have tight dimensional tolerances to ensure that all the pins develop good electric contacts. One of the design criteria for these conventional connectors is that the electric contacts must have a "wiping" action under large load to have a good electric contact. As a consequence, a large insertion force is required to insert the pins into the socket. A large connector requires 500 lb of insertion force, which requires mechanical cams for insertion. Figure 7.29(b) shows backplane printed circuit board connections.

7.6.2 Complexity of the conventional connectors

Electric contacts in a conventional electrode are made when a pin is inserted into a hole with two metal strips, as shown in figure 7.30. The strips exert normal and frictional forces on the pin. The number of electric contacts along the strip increases with the normal load. Under an ideal condition, the number of electric contacts should be as large as possible under a given normal load. Normally there are many pins and strips in a given electrical connector. The position of the strips in the electrical connector and the spacing between a set of two strips are determined by the accuracy of the injection molding that holds the strips together in a given electric connector. When a large number of connections must be made, the worst contact in the series must have an acceptable electrical resistance, which may sometimes overload other connections.

The conventional connectors are coupled designs, and therefore the time-independent real complexity is large. The normal force exerted by other neighboring pins affects the electric contact at each pin. Furthermore, when a large particle is entrapped at a point at the sliding interface, all other contacts of the electrode may also lose contact because the contacting points along a strip of contact are affected by the separation of the contacting surface at any point along the electrode. Furthermore, the metallic strip inside the female socket must have a high yield strength so that the strip can exert a large normal load on the pin without undergoing permanent deformation. Therefore the strip is made of a copper alloy rather than pure copper. The use of an alloy increases the

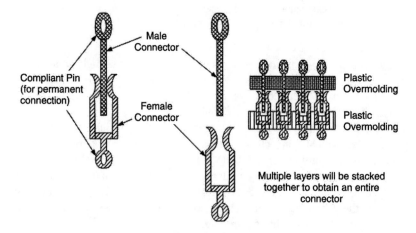

Figure 7.30 A schematic diagram of electric contacts made in a typical conventional electric connector. (Courtesy of Tribotek, Inc.)

electrical resistance, compromising the FRs of high electric conductivity and a low temperature rise at the connector. This is a classical coupled design. These coupled feature of the design reduce the allowable tolerance. Therefore, the time-independent real complexity is large. The real complexity increases with the number of pins because there is a greater probability of creating coupling as the number of pins increase.

These connectors also have time-dependent combinatorial complexity because of the wear of the contacts during repeated insertion and the relaxation of the injection-molded plastic as a function of time, especially at high operating temperatures. When the dimensional changes exceed the compliance of the contact, the number of contacts will decrease as a function of time, increasing the contact resistance continuously. Because of the time-dependent combinatorial complexity, the connector will eventually cease to satisfy the functional requirements.

To eliminate the real complexity and the combinatorial complexity, completely new connectors have been invented, which are described in this section. These connectors allow a large number of interconnects (i.e., greater circuit density) and a greater power density with greater reliability. They are designed to overcome the shortcomings—size, contact resistance, limited pin density, temperature rise, cross-talk, and so on—of the conventional connectors used in industry today. These new connectors are manufactured by entirely different manufacturing processes at lower costs than are the conventional connectors.[11]

7.6.3 Functional requirements (FRs) and constraints (Cs) of electrical connectors

Electrical connectors must satisfy several FRs to provide a long and reliable service life for a variety of applications. The highest-level FRs for data connectors may be stated as:

FR_1 = Mechanically connect and disconnect electrical terminals.
FR_2 = Control contact resistance (should be less than 20 mΩ).
FR_3 = Prevent cross-talk (i.e., interference) between the connections.

The FRs of the power connector are:

FR_1 = Mechanically connect and disconnect electrical terminals.
FR_2 = Control contact resistance (should be less than 20 mΩ).
FR_3 = Control the current density.

The constraints (Cs) for both data connectors and power connectors are:

C_1 = Low cost.
C_2 = Ease of use.
C_3 = Long life (> 1 million cycles).
C_4 = Maximum temperature rise of 30°C.
C_5 = Low insertion force.

The specific required amperage depends on applications. The detailed specifications are given by manufacturers (see, for example, http://www.teradyne.com/prods/tcs/).

7.6.4 Physical embodiment

The invention disclosed herein is for an electric connector that can satisfy the FRs stated in the preceding section. It consists of a "woven tube" and a rod with multiple electric

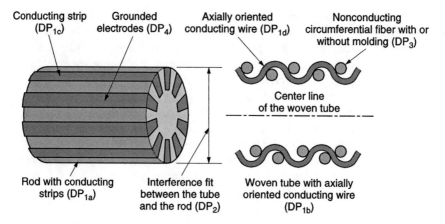

Figure 7.31 Schematic diagram of the "rod with conducting strips on the surface" and the "woven tube with axially located conducting fibers and circumferentially wound nonconducting high-strength, high-modulus, elastic fibers."

contacts. Electric connection is made when the rod is inserted into the tube. A physical embodiment of the data connector is illustrated schematically in figure 7.31. The "woven tube" is made of a woven mesh of conducting wires and nonconducting fibers. Conducting and insulating fibers in the warp are woven with nonconducting fibers in the weft to form the first half of the two-piece connector. A multitude of redundant contact points are formed from the sinusoidal pattern of the warp. The weft is tensioned to apply normal force to maintain reliable surface contact. The electrically conducting wires are separated from each other, creating discrete circuits, by the nonconducting fibers that are weaved into the warp. Conducting shields can likewise be included in the warp. The electric current may be passed through either all the conducting wires—when cross-talk is not an issue—or every other wire, if the shielding of electric signals is necessary to prevent cross-talk by grounding every other wire. When the shielding is necessary to prevent cross-talk, every other wire will be grounded. Conducting fibers are terminated to compliant pins.

In the case of power connectors, the insulating fibers in the warp are replaced with conducting fibers. Unlike conventional electrical connectors that require the use of beryllium copper for stiffness, these woven connectors use pure copper, which minimizes the power (i^2R) loss and the temperature rise. This design uncouples the FR for stiffness from the FR for high conductivity.

The pin for the data connector is made of conducting metal strips mounted on the surface of a nonconducting rod, whereas the pin for the power connector is a solid metal. Electric contacts are made when the conductor on the rod comes into contact with the wires in the warp of the "woven tube." Since the conductors on the tube and the pin must always be in contact under pressure, the pin diameter should be slightly larger than the tube diameter so as to have interference fit between the rod and the tube. To exert pressure, the nonconducting fibers in the weft will be made of a medium[12] modulus material that has a large maximum elastic elongation with no permanent set under tensile loading. The idea is to have a large local compliance using these nonconducting fibers so as to provide good electrical contact at the contacting point of the individual conducting wire that is in the form of a sinusoidal wave.

Therefore, the particles must be eliminated from the electric contacts by providing space into which the particles can fall and be trapped. Because of the nature of the weave, there is enough space between the contact points for this process to occur naturally. This will also eliminate further wear of conductors, which, if they are not removed, can further reduce the reliability of electric contacts.

The largest component of friction between two sliding surfaces is due to the plowing of the two surfaces by particles. The presence of particles (i.e., wear particles and other particles introduced externally) at the electric contact will increase the frictional force during insertion of pins as well as increasing the electric contact resistance. This component of friction and wear can be reduced in the woven connector design by providing space for particles to fall into, as stated above, but can also be reduced because of the localized compliance that is another benefit of the woven connector design. With localized compliance, when a wear particle is generated between two contacting points, the two contacting points are able to separate, which prevents the wear particle from plowing either surface. The wear particle is pulled from between the contacting surfaces and allowed to fall into one of the particle traps. By having multiple contact points for each electrical connection, good electrical contact can be maintained with much lower forces between mating parts. This results in lower insertion forces and less wear on the contacting surfaces.

7.6.5 Design parameters (DPs) of the electric power connector

The FRs considered for the data connector and the power connector should now be stated. However, the design is rather similar for both set of FRs, therefore only the design of the power connector will be presented in this section.

The DPs corresponding to the FRs for the power connector may be selected to be:

DP_1 = Cylindrical assembly of the woven tube and the pin.
DP_2 = Locally compliant electric contact.
DP_3 = Number of conducting wires.[13]

DP_2 states that when the compliance of the nonconducting tension fiber of the weft is large, good electrical contacts will be made everywhere since all the contacts will exert a contact force of similar magnitude. The force exerted at a given contact is nearly independent of the forces applied at other contacts. Furthermore, since the normal load at the contact point is just enough to provide the electrical contact, and since we do not overload the contacts to ensure that the worst contact point has the required maximum resistance, the friction and the insertion force will be at a minimum.

The constraint on the temperature rise will be taken care of if we use pure copper wire as the conducting wire. It should be noted that in conventional connectors, pure copper cannot be used because of their low yield stress and stiffness.

All these FRs and DPs may be decomposed further to be able to implement the design concept. For example, FR_1 (Mechanically connect and disconnect electrical terminals) and DP_1 (Cylindrical assembly of the woven tube and the pin) may be decomposed as:

FR_{11} = Align the rod axially inside the tube.
FR_{12} = Locate the axial position of the rod in the tube.
FR_{13} = Guide the pin.

DP_{11} = Long aspect ratio of the rod and the tube.
DP_{12} = Snap fit.
DP_{13} = Tapered tip of the pin.

Similarly, FR_2 (Control contact resistance to be less than 20 mΩ) and DP_2 (Locally compliant electric contact) may be decomposed as:

FR_{21} = Prevent oxidation of the conductor.
FR_{22} = Remove wear particles.
FR_{23} = Control line tension/deflection of the non-conducting fiber.

DP_{21} = Gold-plated metal surface.
DP_{22} = Space created in the crevices between fibers.
DP_{23} = Spring.

DP_{22} will also affect the insertion force (FR_3).

FR_3 (Control the current density) and DP_3 (Number of conducting wires) do not need to be decomposed as they can be implemented without further detailed design at this time.

7.6.6 Design equation and design matrix

The design equation for the highest-level design of the data connector shown in the figure above is

$$\begin{Bmatrix} FR_1 \\ FR_2 \\ FR_3 \end{Bmatrix} = \begin{bmatrix} X & 0 & 0 \\ X & X & 0 \\ 0 & X & X \end{bmatrix} \begin{Bmatrix} DP_1 \\ DP_2 \\ DP_3 \end{Bmatrix} \qquad (7.9)$$

The design is a decoupled design.

The decomposition of FR_1 and FR_2 yielded an uncoupled design. Therefore, the highest-level design matrix given by equation (7.9) is satisfied by the lower-level FRs and DPs.

7.6.7 Commercial connector

Figure 7.32 shows the commercial power connector designed by Tribotek, Inc., based on the design concept described in the preceding paragraphs. The circumferentially oriented set of fibers are nonconducting fibers that are attached to springs to apply a constant load on the pure copper wire that is woven around them. The copper wire, which conducts electrical current, is in contact with the conducting metal pin at a number of points where the nonconducting fibers exert force on the copper wire. It is a highly compliant design that makes it a robust design. Any wear particle, if it is present at all, cannot remain at the pin/fiber contact and generate more particle, since it will fall into the open space between the fibers.

Figure 7.33 shows a data connector that has many electrical contacts, which is similar in construction to the power connector shown in figure 7.32. Each strip of the gold-plated conductor on the pin makes a contact with two woven pure copper wires. The insulating fibers separate the copper wires. All of these axially oriented fibers are woven around tensioning fibers (circumferentially oriented) that exert the contact force

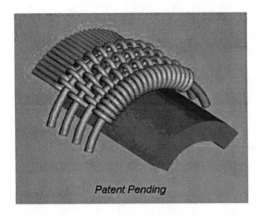

Patent Pending

Figure 7.32 Schematic drawing of a Tribotek power connector.

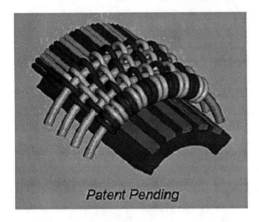

Patent Pending

Figure 7.33 Schematic drawing of a Tribotek data connector.

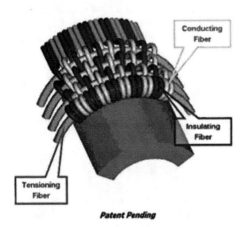

Conducting Fiber

Insulating Fiber

Tensioning Fiber

Patent Pending

Figure 7.34 Schematic drawing of a Tribotek data connector showing the location of fibers.

between the conductor strip on the pin and the copper wires. Figure 7.34 shows the location of these fibers. Because of the unique design of the woven connector, each connector can transmit many electrical signals concurrently.

Tribotek, Inc., has developed mass production machines that make these woven connectors economically. The Tribotek power connector has many advantages over conventional power connectors (Moran et al., 2004). Its current density is twice as much as that of the conventional power connector, the insertion force is as low as 5% of the conventional connector, and the electrical contact resistance is less than 5 mΩ. As a result, the insertion force required for Tribotek connectors is a small fraction of the force required for the conventional connectors. Since these connectors do not wear, they can have a long useful life. These connectors are robust and reliable, because they are uncoupled designs. More information can be obtained at http://www.tribotek.com.

7.7 Summary

1. Geometric functional periodicity has been used to disrupt the time-dependent combinatorial process to lower friction force between sliding surfaces and reduce wear.
2. The wear life of seals can also be increased through the application of geometric functional periodicity.
3. The reduction of friction, wear resistance, and durability of pin-joints were achieved by decoupling pin-joints by means of locally compliant design and by introduction of geometric functional periodicity.
4. The life of electric contacts for both power transmission and signal transmission has been improved by means of geometric functional periodicity.

Notes

1. $T\langle\ \rangle$ is a symbol for transformation. C_{com} and C_{per} are time-dependent combinatorial complexity and time-dependent periodic complexity, respectively. FP is the functional period.

2. Note that the Independence Axiom states that the FRs—not the DPs—must be independent from each other.

3. The particle agglomeration was observed experimentally with soft materials such as zinc. It is assumed that a similar phenomenon occurs with harder materials, although it is difficult to observe the effect directly.

4. It should be noted that we always state FRs starting with verbs and DPs with nouns. This makes it easier to distinguish between FRs and DPs.

5. Elastomeric seals with surface pads, which showed a dramatic improvement in seal life, were created by D.P. Hart and his research group at MIT (Ayala et al., 1998).

6. Patent pending, Tribotek, Inc., Cambridge, MA.

7. In the past, many ad hoc empirical approaches—such as improving the material properties or improving wear resistance by coating the surface with hard materials or incorporating lubricants—have been tried to improve the life of pin-joints. However, empirical approaches are inefficient and require exhaustive tests. Such an empirical approach was a result of poor understanding of friction and wear mechanisms and the lack of theoretical knowledge for system design. To develop tribological systems based on theoretical predictions, correct understanding of the underlying friction and wear mechanisms is necessary. When wrong tribology theories are used in the design of systems, the performance of the system will deviate from the predicted behavior. Consequently, engineers in industry have relied on test results rather than theoretical predictions in developing pin-joints. For example, designers who believe in the adhesion theory

of friction and wear may not be able to develop a rational tribological system that exhibits the low friction and wear behavior described here, because the adhesion theory will lead the designer to a wrong reasoning path.

8. This conventional pin-joint is a case of a design with time-independent real complexity due to coupling, which, if nothing is done, eventually transforms into a problem with time-dependent combinatorial complexity. This ultimately results in failure due to seizure.

9. Designed by Dr. Matt Sweetland, Tribotek, Inc., 2002.

10. The worldwide connector market in 2003 is about $34 billion, with ten connector companies controlling about 54% of the market.

11. Patents for these connectors are owned by Tribotek, Inc., Burlington, MA.

12. For robust design, the modulus should be low to allow the pressure to be in the right range, although there may be large variations in the interference fit. If the modulus is too high, the pressure will be a sensitive function of the variation in the interference fit. Also if the modulus is too low, it will be difficult to obtain a sufficient contact pressure without a very large interference fit. Therefore, we will use the term "medium" to denote the modulus in the right range.

13. DP_3 for the data connector is the insulation between the conducting wires of the tube and the insulating material between the conducting stripes on the pin.

References

Ayala, H.M., Hart, D.P., Yeh, O., and Boyce, M.C. 1998. "Wear of elastomeric seals in abrasive slurries," *Wear,* Vol. 220, pp. 9–24.

Cha, K.H. and Kim, D.E. 2000. "Tribological properties of micro-undulated silicon under light load," presented at International Tribology Conference 2000, Nagasaki, Japan.

Jahanmir, S., Suh, N.P., and Abrahamson, E. 1974. "Microscopic observation of the wear sheet formation by delamination," *Wear,* Vol. 28, pp. 235–249.

Moran, J., Sweetland, M., and Suh, N.P. 2004. "Low friction and wear on non-lubricated contact surfaces," IEEE Holm Conference on Electric Contacts, Seattle, WA.

Oktay, S.T. and Suh, N.P. 1992 "Wear debris formation and agglomeration," *Journal of Tribology, Transactions of the ASME,* Vol. 114, pp. 379–393.

Shepard, S.R. and Suh, N.P. 1982. "The effect of ion implantation on friction and wear of metals," *Journal of Lubrication Technology, Transactions of the ASME,* Vol. 104, pp. 29–38.

Suh, N.P. 1973. "The delamination theory of wear," *Wear,* Vol. 25, pp. 111–124.

Suh, N.P. 1986. *Tribophysics,* Prentice-Hall, Englewood Cliffs, NJ.

Suh, N.P. 1990. *The Principles of Design,* Oxford University Press, New York.

Suh, N.P. 2001. *Axiomatic Design: Advances and Applications,* Oxford University Press, New York.

Suh, N.P. and Saka, N. 1987. "Surface engineering," *CIRP Annals,* Vol. 36, pp. 403–408.

Exercises

7.1. We want to introduce a geometric functional periodicity to the inner surface of a polymeric tube to transport blood without causing the deposition of protein molecules on the surface and blood coagulation. The idea is to limit the contact area between the blood and the surface by introducing an undulated topography on the surface and by balancing the pressure of the fluid with the surface tension forces. (*Note*: The wetting angle is defined as follows: When the contact angle between a solid surface and a liquid droplet is 0°, the surface is completely wet. When the angle is 180°, the droplet does not wet the surface.) When the blood is not in contact with a surface, it is assumed that the electrostatic forces would not attract the protein molecules to the surface.

Define the FRs of the tube and propose a design of the tube surface that will prevent the protein deposition. The flow rate of the blood is 10 ml/min. The diameter and the maximum length of the tube are 2 mm and 2 m, respectively.

7.2. Examine natural systems around you and identify a natural system that has time-dependent geometric functional periodicity. State the FRs of the natural system and the role of geometric functional periodicity.

7.3. Engineering artifacts are getting smaller from macroscale to microscale and to nanoscale. To create macroscale engineering systems that are created from nanoscale structures, we must integrate across several length scales from nanoscale to macroscale. However, the current technology is limited to integration across the length scale of the order of 10^4 in typical manufacturing operations and 10^6 in semiconductor manufacturing and microelectronic processing because of the achievable tolerances involved. For example, in machining a 1 cm bar, typical accuracy that can be achieved using conventional machine tools is 10^{-4} cm. In making integrated circuit chips with a critical dimension (CD refers to the width of the circuit line) of 100 nm, the accuracy that can be achieved with a lithography machine is about 10 nm and the largest image that can be printed with the same machine is about 10^4 μm. Therefore, we need to combine two or more different levels of geometric functional periodicity to create macroscale engineering systems. Describe how biological systems deal with this issue by investigating the structure of biological cells. How would you apply your findings to fuel cells so that we can create macroscale fuel cells starting from microscale fuel cells?

7.4. There is no sharp demarcation line between different kinds of functional periodicity. For example, geometric functional periodicity overlaps with material functional periodicity, as in rope and woven fabric, in terms of the FR that prevents crack propagation, the FR that maintains strength, and the FR that allows bending. Explain the role of geometric functional periodicity in reinitializing the FRs of the streets and avenues of New York City. Can the city planning be improved?

7.5. When certain polymeric parts are made by injection molding, especially foamed plastics, the surface may have some "optical" flaws, showing the "flow marks" of the polymer. Develop a means of introducing geometric functional periodicity to hide the optical flaws.

8

▓▓ Reduction of Complexity in Materials through Functional Periodicity

▓8.1▓ Introduction

In chapter 3, it was stated that functional periodicity is a requirement for long-term stability of systems that are not at an equilibrium state and that many materials found in nature have functional periodicity as shown in table 3.1. Atomic elements follow a periodic structure as given by the periodic table of the chemical elements, crystalline materials have ordered structures, even some biological molecules have repeating structures, and biological cells divide on a periodic basis. The functional periodicity makes these natural materials stable. However, at a macrocontinuum scale, these materials have defects that may lead to unstable behavior and ultimate failure. This is an example of a system with time-dependent combinatorial complexity.

    The purpose of this chapter is to show how macroscale materials with time-dependent combinatorial complexity can be transformed into a system with time-dependent functional periodicity. As was briefly discussed, it will be shown that we can introduce one-dimensional (1-D) functional periodicity, two-dimensional (2-D) functional periodicity, and three-dimensional (3-D) functional periodicity to improve the material properties. Examples of materials with 1-D, 2-D, and 3-D functional periodicity are fibers, fabrics, and structural composites with continuous fiber reinforcement, respectively.

Some materials fail during service because certain physical properties undergo processes that involve time-dependent combinatorial complexity. This type of failure occurs because the system range drifts away from the design range as a function of time. Some examples of physical processes that cause time-dependent combinatorial complexity are: crack propagation in materials under cyclic loading that leads to fatigue fracture, work-hardening of metals due to dislocation generation and interactions during large plastic deformation that limits their ductility, delamination wear of materials under sliding conditions, and permanent morphological changes in liquid crystals when they are subjected to electrical potential.

These time-dependent combinatorial processes ultimately lead to uncertainty in satisfying certain FRs of materials and may result in unacceptable failures. Over decades or centuries, many of these failures have been addressed in ingenious ways, often empirically. Most of these solutions relied heavily on extensive testing and empirically acquired knowledge. In essence, these solutions involved the introduction of functional periodicity, which transforms the time-dependent combinatorial complexity that causes failure into time-dependent periodic complexity.

The use of functional periodicity in the materials field has a long history. Steel cables have long been used to hold wine kegs together, to hold cable cars, and to suspend bridges. Wire drawing uses functional periodicity as well. Plywood is tough and strong because of functional periodicity. More recently, with the emergence of the microelectronics industry, functional periodicity has been intentionally introduced to alter material properties by creating superlattices. An example of widely used technology is the alternating voltage field we use to prevent the permanent setting of the molecular orientation in LCD, which was discussed in example 5.2.

Even nature depends on material functional periodicity. Skeletal muscle derives its unique properties from its functional "periodic" structure.[1] Why does the periodicity exist? One possibility is that it is difficult for the body to build muscle units that are greater in size than the myocytes (the muscle cells that make them). After actin–myosin units are formed, hundreds are joined together to make a long fiber. Another possibility may be that, like cables, periodicity confers strength and resistance to injury.[2] Another reason may be related to the need to impart stability to skeletal muscle.

In this chapter, some materials technologies that exploit functional periodicity are reviewed. In many materials, their desired properties are enhanced through the use of time-dependent periodic complexity.

8.2   Functional Periodicity to Prevent Unstable Crack Growth

Many materials have to withstand cyclic loading during their use. The FR is to provide long fatigue life. In materials with limited ductility, once a crack nucleates, it can continue to grow under cyclic loading until it becomes an unstable crack, leading to fracture of materials. This type of failure of materials due to crack propagation is an example of time-dependent combinatorial complexity. To prevent crack propagation

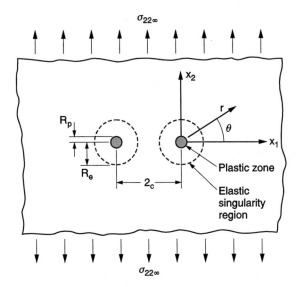

Figure 8.1 Schematic of the state of stress and strain around the tip of an initially sharp crack. The stress distribution inside a radius R_e is essentially given by the crack tip singularity solution. Plastic deformation has occurred in a region of R_p, which is much smaller than R_e.

and fracture of materials, functional periodicity can be introduced to these materials to transform a system with time-dependent combinatorial complexity into a system with time-dependent periodic complexity.

A crack in an infinite plate is shown in figure 8.1. This type of loading, called Mode I loading, is perpendicular to the crack position, and therefore the crack tip opens up under tension. When the loading is parallel to the crack, it is called Mode II. Mode III loading refers to the case when the shear stress is applied along the x_3-axis. When the entire crack is surrounded by a plastically deformed zone, the material deforms plastically and cracks do not propagate. However, when only a small zone at the crack tip undergoes plastic deformation and is surrounded by an elastically loaded region, cracks will extend and propagate. The crack propagation rate is controlled by the stress intensity factor, k_1, which is defined as

$$k_1 = \sigma_{22\infty}\sqrt{c} \qquad (8.1)$$

where $\sigma_{22\infty}$ is the stress away from the crack and c is half of the crack length.

When the material shown in figure 8.1 is subjected to a cyclic loading (repeated tensile and compressive loading), the state of deformation and stress changes at the crack tip. Therefore, the stress intensity factor undergoes cyclic change. Under compression, the stress intensity factor is zero and when the applied load is tensile, the crack tip opens up and the crack propagates. Cracks grow if the cyclic stress intensity factor is greater than the threshold stress intensity factor.[3] Initially, the crack propagation rate under constant cyclic loading is small, but the rate of growth under tensile loading increases in proportion to the square root of the crack length, that is, the cracks grow faster as they get longer. Eventually, when the crack reaches a critical length that corresponds to the critical stress intensity factor, the crack propagates unstably, resulting in fracture of the material. This is shown in figure 8.2. The crack propagation rate may be expressed as a function of the stress intensity factor as

$$\frac{dc}{dN} = A\left(\frac{\Delta k_1}{\sigma_Y}\right)^n \qquad (8.2)$$

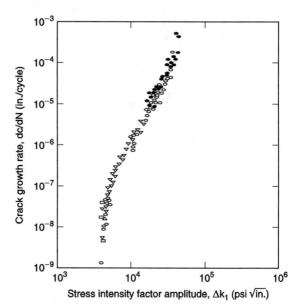

Figure 8.2 Correlation of crack growth rates with the stress intensity factor amplitude of the load cycles for A533-B steel. Different symbols are for different tests under different conditions. (From Paris et al., 1972.)

where N is the number of cycles and exponent n can vary from 2 to 5 depending on the material. σ_Y is the yield strength of the material.

The failure of materials due to crack propagation is "combinatorial" because cracks continue to grow somewhat unpredictably, depending on prior loading history, crack length, material properties, and the specific stresses applied to the crack.[4] A number of techniques have been developed to reduce the complexity associated with this process by transforming the combinatorial complexity to a periodic complexity. For instance, the size of the solid may be reduced to prevent the combinatorial process. When the size of the body shown in figure 8.1 is of the same order of magnitude as the plastic zone size in front of the crack, crack propagation cannot occur and the material will simply yield. Therefore, one way of transforming the time-dependent combinatorial complexity is to replace a large solid by layers of thin parts, either by laminating them or by mechanically holding them together. Another is to replace a continuum with fibers, which is an example of 1-D functional periodicity.

8.2.1 Cable rope and wire rope: 1-D functional periodicity

Well-known materials technologies that exploit functional periodicity are textile yarns made of bundle of fibers, and the metallic cable rope made of wires. In these structures, the failure of one fiber or wire does not result in the failure of the entire yarn or the cable rope. This introduction of 1-D functional periodicity prevents catastrophic failure of yarns and rope.

The idea of using cable rope is quite old, going back to mid-nineteenth-century Europe. Wire rope is used in elevators, hoisting, hawsers, guys, power transmission, oil-well drilling, in addition to suspension bridges and cable cars. Cable cars and suspension bridges do not use a large-diameter rod to support their weight. These thick rods, in addition to being stiff to bending, can fracture spontaneously when a crack in the rod reaches a critical length because the rod is subjected to cyclic loading, as discussed in the previous section.

For these and other applications, a cable rope made of strands of thin steel wires is used, each of which carries primarily tensile loads. Steel with 0.4–0.9% carbon is heat-treated (called patenting) to make a wire with a fine grain structure, with iron carbide distributed through the ferrite in a finely divided form. Wire rope is made of strands, each of which is made by twisting a group of individual wires helically with a uniform pitch. A number of these strands are then similarly grouped to form a wire rope. By introducing functional periodicity in the form of individual "thin" wires in a cable rope in place of a thick rod, even if a crack nucleates in one wire, it can propagate across only that wire or simply deform plastically if the plastic zone size at the crack tip reaches the outer boundary of the wire. Each time a crack grows and cuts across one of the wires, the process has to start all over again. Furthermore, unstable crack propagation may not occur in wires of very small diameter. Therefore, a cable can withstand repeated cyclic loading over a long period of time, since cracks cannot propagate across the entire strand of wires.

The use of a cable rope rather than a thick rod transforms the time-dependent combinatorial complexity (represented by crack propagation in a thick rod, resulting in failure) to a time-dependent periodic complexity. This reduction of time-dependent complexity removes the uncertainty associated with the prediction of the long-term behavior of cables.

8.2.2 Fabrics: 2-D functional periodicity

Textile technology is one of the oldest technologies, perhaps going back to 5000 BC, when Egyptians produced plain linen textiles. Textiles replaced animal skins as the material for clothing.

Textiles are woven fabrics with 2-D functional periodicity. Before the invention of polyamide (nylon) in 1938, all fabrics were manufactured by weaving natural fibers, which were made by spinning cotton, wool, and silk. Now many kinds of man-made fibers—polyamides, polyester, liquid crystal polymers (e.g., Kevlar), polypropylene, and so on—are used to make fabrics.

Nonwoven fabrics are also used, which are made by layering chopped fibers with different orientations to make a mat. Nonwoven fabrics are not as strong as woven fabrics, which have continuous fibers, but they are easy to form into shapes and cheaper to manufacture. This type of fabric is used in furniture, diapers, and certain kinds of artificial leather.

The individual woven fibers impart a two-dimensional (2-D) functional periodicity to the fabric, which makes it relatively tougher and stronger than a continuous sheet made of the same material (in the case of synthetic fibers). In a continuous sheet that does not have this functional periodicity, once a crack is nucleated, it will continue to run until the sheet is torn into two or more pieces. The functional periodicity in fabrics is also responsible for many other properties of textiles: light weight, flexibility, toughness, moisture permeability, strength, and durability. Furthermore, fabrics can be made to conform easily to human physique, making it the material of choice for styling and comfort. The use of functional periodicity in woven fabrics removes the uncertainty in satisfying the many FRs of fabrics.

8.2.3 Fiber-reinforced composites: 3-D functional periodicity

One of the modern structural materials is a composite, which consists of a reinforcing phase typically made of continuous fiber reinforcement and a matrix phase, which is typically a polymeric resin that bonds the reinforcing fibers together to form a continuous three-dimensional structure. There are many different kinds of composite structures and many different fabrication techniques. Figure 8.3 shows different kinds of composite structures. In advanced composites that are used in airplanes, sporting goods, and machine tools, the reinforcing fibers are carbon (or graphite) fibers, which have high specific stiffness and specific strength. Glass fibers are also used in manufacturing composites for automotive, chemical, and marine applications. These composites with continuous fibers represents a class of materials that have 3-D functional periodicity. In these composites, unstable cracks cannot propagate in the direction perpendicular to the fibers. These composites can be designed to satisfy a specific set of FRs, which depends on applications. They are used in aircraft, machine tools, automobiles, ships, and many other applications. The design of these composite structures based on axiomatic design is given in Lee and Suh (2004).

One of the special applications of composites is in making tribological components. Composites are used to arrest the subsurface crack propagation that is responsible for delamination wear,[5] which was discussed in chapter 7 (Suh, 1986). Functional periodicity is introduced to the material to reduce wear rate by embedding reinforcing fibers perpendicular to the surface to stop the crack propagation process. When the fibers are placed perpendicular to the sliding direction, the crack propagation rate decreases because of the reduced crack tip sliding displacement. When cracks reach the fibers, cracks cannot propagate any further. This reduces the wear rate, and thus the time-dependent complexity.

Figure 8.3 Various composite structures. (a) Unidirectional lamina with continuous fiber, (b) unidirectional lamina with discontinuous fiber, (c) random orientation in a plane, (d) 3-D random orientation short fibers, (e) laminate, (f) woven fabric composite, (g) hybrid-fiber composite, (h) 3-D woven/stitched/braided composite. (From Lee and Suh, 2004.)

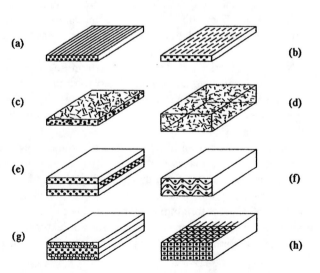

8.3   Functional Periodicity for Control of Material Properties

Ideal metals have a periodicity
Metals have a crystalline structure where atoms are stacked in a periodic structure. Typical structures are body-centered cubic (b.c.c.), face-centered cubic (f.c.c.), and hexagonal close packed (h.c.p.). These crystals have preferred planes on which one part of the crystal slips with respect to the other. These planes are called *slip planes*. In f.c.c. and h.c.p. crystals, the slip planes are close-packed planes. In b.c.c. metals, which do not have close-packed planes, the slip planes vary from metal to metal. The ideal shear strength of metal is about $G/6$, where G is the shear modulus of the metal. However, the yield strength of typical metals is orders of magnitude smaller than the ideal strength because of the presence of defects.

Defects create a time-dependent combinatorial complexity
The primary defect that is responsible for the reduction of shear strength of metals is called a *dislocation*. Dislocations are a line defect in the crystalline structure shown in figure 8.4.[6] When the line defect is an extra plane of atoms on one side of the *glide plane*,[7] as shown in figure 8.4(a), it is called edge dislocation. In this case, the dislocation is perpendicular to the relative slip across the glide plane. When the dislocation moves across the entire crystal, it has deformed plastically by a distance called the Burgers vector. When the Burgers vector does not lie on the glide plane, dislocations cannot move. These dislocations are called sessile dislocations. When the defect line is a screw type—the defect line between the perfect crystal and the partially displaced crystal is parallel to the direction of crystal displacement—as shown in figure 8.4(b),

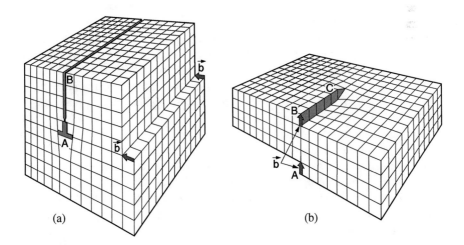

(a)   (b)

Figure 8.4 Partially slipped crystals containing line defects that separate the slipped portion from the unslipped portion. These defects are called dislocations. (a) An edge dislocation is at A. An extra half plane of atoms ABC has been created by the slip. $\vec{b}$ is the slip created by the slip vector, called the Burgers vector. (b) A screw dislocation is at C. $\vec{b}$, is the slip vector. Note that the slip vector of an edge dislocation is perpendicular to the dislocation line, and that the slip vector of a screw dislocation is parallel to the dislocation line.

Figure 8.5 Stress–strain curve of a typical metal under uniaxial loading. After the elastic–plastic transition at the yield point, the metal strain-hardens when the number of dislocations increases, forming sessile dislocations and a dislocation network. Plastic instability occurs when the rate of strain-hardening is less than the increase of local stress due to the decrease in cross-sectional area. The metal fractures when the local stress exceeds the fracture strength of the metal.

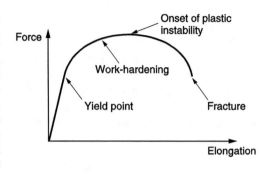

it is called screw dislocation. For the screw dislocation, the glide plane is not unique, since the dislocation vector and the Burgers vector are in the same direction. Plastic deformation of the crystal results when these dislocations move through the crystalline structure. When the glide plane is the same as the slip plane, it takes much less force to move dislocations through the crystalline structure than moving the entire layer of crystals at the same time, because this localization of plastic deformation near the dislocation at a given instant lowers the shear strength of metals.

The dislocation density in annealed commercial metals is of the order of 10^6 dislocations per cm^2. During plastic deformation, new dislocations are generated. The dislocation density continues to rise with further plastic deformation, and eventually may reach a density of 10^{12} dislocations per cm^2. As the number of dislocations increases, they interact with each other, forming sessile dislocations and a dislocation network that cannot slide on the glide plane. Sessile dislocations are dislocations that do not lie on the glide plane and therefore cannot move. The stress field generated by other dislocations, sessile dislocations,[8] and the dislocation network impedes the motion of mobile dislocations. Therefore, as the dislocation density increases with plastic deformation, the flow strength of the metal increases, thus requiring greater stress to continue plastic deformation of the metal. This process is called "strain-hardening" or "work-hardening" or "cold-working." When the rate of work-hardening is less than the rate of the stress increase due to the decrease in the cross-sectional area, plastic instability (sometimes called necking) takes place. Eventually the stress required to plastically deform the metal exceeds the fracture strength of the metal, and the metal fractures. This is shown in figure 8.5 for uniaxial elongation.[9] This process of strain-hardening and local plastic instability is a combinatorial process. Therefore, the elongation and work-hardening of metal may be considered to have time-dependent combinatorial complexity.

The combinatorial complexity can be stopped by "annealing" metals
Metals can be annealed at high temperatures, which reduce the dislocation density. At high temperatures, the polycrystalline metal deforms as a result of grain boundary sliding, as well as the dislocation motion. At high temperatures, even sessile dislocations can "climb" perpendicular to the glide plane when atoms diffuse to the dislocation site,

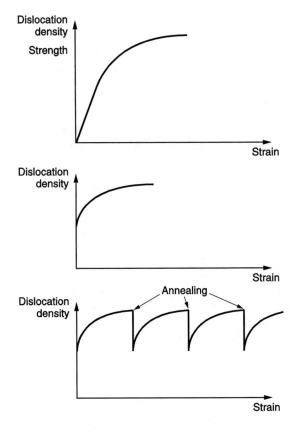

Figure 8.6 Annealing as a means to "reinitialize" the dislocation density to attain periodicity.

which then may react with other dislocations. They annihilate each other when dislocations with opposite orientation interact, thereby decreasing the total number of dislocations. Also at high temperatures, dislocations glide at lower stress and may be eliminated at the grain boundaries. Therefore, the effect of strain-hardening is "annealed" out at high temperatures. By annealing, we "reinitialize" the metallic system (see figure 8.6).

It is difficult to make defect-free bulk materials because of various combinatorial processes
When we make bulk materials, it is extremely difficult to make them defect-free. Commercial materials have dislocations and grain boundaries. They also have other defects, such as point defects (e.g., vacancies) and secondary-phase particles in the crystalline structure. Small amounts of impurities that are unintentionally dissolved in the crystal may also form secondary-phase particles that may adversely affect electronic and optical properties. We have to grow single crystals under a carefully controlled thermal environment. For example, to make electronic materials such as silicon wafers, single-crystal silicon is grown by slowly pulling the crystal from an impurity-free molten silicon bath without inducing thermal stress. When semiconductor wafers made from these ingots are processed to make integrated circuit chips, care must be taken not to generate dislocations by inducing mechanical and thermal stress.

Thin films provide functional periodicity

Since it is extremely difficult to make highly pure bulk materials with no line or point defects, we may circumvent the problem by depositing thin films by chemical vapor deposition (CVD) or physical vapor deposition (PVD). Layers of thin films provide functional periodicity, which is a means of controlling "defect-free" structures within thin layers. Many electronic devices, including integrated circuits (IC), utilize thin film deposition and etching technologies. Thin films are also used because the physics of transport phenomena in thin films, including heat transfer and electromagnetic wave propagation, are quite different from those in bulk materials, which form the basis for creating unique devices.

8.3.1 Functional periodicity in wire drawing

To make a thin copper wire, a copper rod of a larger diameter is drawn down through a series of "conical" dies to reduce its diameter in a number of steps. By *pulling* the wire (typically referred to as "wire drawing") through the die (rather than pushing it through a die, which is called "extrusion"), the diameter of the wire is reduced. During this drawing process, the metal work-hardens and may fracture before it is drawn to the final diameter. We must prevent fracture of the wire before it is drawn to the desired final diameter. To be certain that this large reduction of the diameter is achieved without fracture, we introduce a functional periodicity through annealing, which reinitiates the material microstructure.

The maximum reduction ratio of copper wire is limited by the fracture strength of the copper, the resistance to bulk plastic deformation, and the frictional force between copper and die.[10] The reduction of the cross-sectional area that can be attained in a single drawing is a function of material and temperature, normally less than 40% reduction. Therefore, to make a fine wire, we must use a cascade of dies (as many as 22) to reduce the diameter of the wire in several steps. To prevent fracture of the wire between the dies and enable further drawing, the wire must be annealed periodically by inductive heating of the wire between the dies. Figure 8.7 shows such an arrangement schematically.

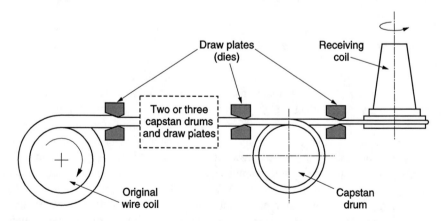

Figure 8.7 Schematic arrangement of wire drawing. The wire is annealed inductively between the dies when the reduction of the area is large. (From El Wakil, 1989.)

When the temperature of the wire is elevated, the dislocations glide more easily and are eliminated through annihilation of dislocations due to dislocation climb and dislocation reaction that occur at high temperatures. Dislocations may also be eliminated at internal surfaces such as grain boundaries at high temperatures. Without annealing, the wire may break before it can be reduced to the final diameter.

This example shows how the transformation of time-dependent combinatorial complexity (due to dislocation generation and interactions) to time-dependent periodic complexity reduces complexity and removes uncertainty in achieving the FR, which is to draw down the wire by inducing large plastic deformation.

8.3.2 Microcellular plastics: functional periodicity through the creation of crack arrestors

One way of preventing unstable crack growth is through the introduction of tiny bubbles in the plastic to blunt the crack tip and arrest the unstable crack propagation, which was discussed in section 4.4 in relation to time-independent real complexity. Functional periodicity can be introduced in the material if tiny bubbles that are uniformly distributed throughout the material can be created.

At MIT, microcellular plastics technology—a means of introducing a plethora of micron-size bubbles into plastics[11]—was developed and was later commercialized by a licensee of MIT. These bubbles can be as small as 0.1 µm to as large as tens of microns. The cell density is in the range of 10^8 to 10^{15} cells per cm^3. In a given plastic, the bubble size and density are fairly uniform and can be controlled. The density reduction can be from a few percent to tens of percent. This technology was originally developed to save materials without sacrificing their toughness, while maintaining part geometry. Microcellular plastics can be made of almost any thermoplastics and thermosetting plastics. Figure 8.8 shows the microstructure of microcellular foam of extruded polypropylene.

Microcellular plastic parts can be extruded or injection-molded. Unlike regular injection-molded parts, injection-molded microcellular plastic parts have minimal residual stress as discussed in chapter 4. Therefore, these injection-molded parts do not distort and

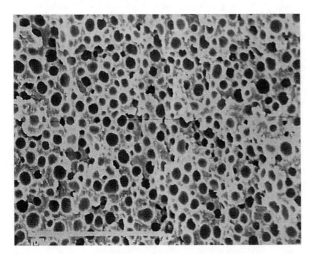

Figure 8.8 Micrograph of an extruded microcellular polypropylene sheet. (Courtesy of Trexel, Inc.)

are dimensionally accurate. Microcellular plastics technology enables the decoupling of the total mass of the plastic injected into the mold from the final geometry of the part made by the mold because the parts have the same geometric shape as the mold cavity regardless of the exact mass of the material injected into the mold. The parts are geometrically accurate because of low residual stress. The parts do not distort even after prolonged use.

Also, because of the large quantity of gas (either CO_2 or N_2) dissolved in the plastics under high pressure in a supercritical state, the solution of polymer/gas has a lower viscosity and a lower melting point than the polymer without the dissolved gas. Consequently, the plastic can be injection-molded at low temperatures, which shortens the cycle time of the molding process and reduces the molding pressure at a given injection rate, respectively. When making engineering parts, expensive engineering plastics with fibers have been used in order to retain geometric shapes during the product's life cycle. Using microcellular plastics technology, these expensive materials can be replaced by lower-priced polymers without any reinforcing fibers. Two of the MuCell parts[12] made by injection-molding microcellular plastics are shown in figure 8.9.

The functional periodicity in microcellular plastics is provided by the microbubbles that are more or less uniformly distributed throughout the plastic part. Because of these bubbles, unstable cracks cannot propagate, since the stress at the crack tip is relieved when the crack reaches the bubbles, that is, the bubbles blunt the crack tip. Therefore, fracture toughness of microcellular plastics is enhanced by the functional periodicity provided by the tiny bubbles when they are uniform and smaller than a critical size. In addition, since these bubbles expand in microscale regions, the local expansion of these bubbles eliminates any local residual stress that could otherwise develop because of the differential shrinkage that can occur during cooling. The injection-molded microcellular plastics replicate the shape of the mold cavity with limited warpage.

8.3.3 Superlattices

In microelectronics, layers of thin films are deposited to control the thermal and electronic properties of devices.[13] The thickness of these films is controlled to atomic accuracy. These periodic structures of thin films are called *superlattices*. They were first proposed in the 1970s by L. Esaki (Noble Laureate) and have since become the cornerstone of quantum structures and are widely used in photonic and semiconductor structures. A wide range of superlattice structures, including Si/Ge superlattices and quantum dots superlattices $CoSb_3/IrSb_3$ and Bi/Te, have been studied. A much lower thermal conductivity than what was achieved in bulk materials has been achieved by adding phonon rattlers in $CoSb_3/IrSb_3$ superlattices.

Superlattices have unique properties. Heat conduction in superlattices is quite different from that in bulk materials and involves complicated quantum and classical size effects. Potential applications for superlattices are thermoelectric cooling and power generation. They are also important in optoelectronic devices such as semiconductor lasers. In superlattices, phonons bounce off at the interface between films, thus reducing the thermal conductivity perpendicular to the film plane, while electronic transport is controlled both in the plane of the film and perpendicular to the film. We can control the thermal conductivity of superlattices to be smaller than the minimum conductivity of bulk materials.

Superlattices may exploit the idea of time-dependent periodic complexity to create new materials. One requirement in making superlattices is the lattice constant match.

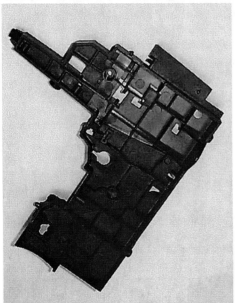

Figure 8.9 Precision microcellular
plastic parts for a printer made
by injection-molding. (Courtesy
of Trexel, Inc.)

For example, GaAs and AlAs have closely matched lattice constants, and thus people
are able to grow high-quality superlattices. On the other hand, Si and Ge have 5% lattice constant mismatch and growing Si/Ge superlattices was very challenging. People
have been successful in growing Si/Ge superlattices only with artificial substrate such
that Si and Ge layers are symmetrically strained, that is, one goes through compression

and the other is tensile. The reason for such difficulties is the very large stress created by the mismatch of the lattice constants.

To facilitate the growth of defect-free films, we may introduce a functional periodicity. To reduce the time-dependent periodic complexity in making thin films of Si on Ge, which have a large difference in lattice constants, we may deposit Si and Ge layers in a periodic manner so that tensile and compression strains are compensated. In this manner, we can grow thick superlattices with relatively low dislocation densities. The introduction of functional periodicity enables us to control desired properties that cannot be achieved in the bulk material. The functional periodicity in the form of periodic thin film layers reduces time-dependent complexity.

8.4   Summary

1. Material properties can be enhanced through the introduction of functional periodicity, which reinitializes the microstructure. In materials, 1-D, 2-D, and 3-D functional periodicity can be introduced to control properties in one, two, or three directions.
2. Defect-free superlattices may be created by periodically alternating two different materials to reduce the interfacial stress at the boundary.
3. The effect of strain-hardening can be eliminated by periodic annealing of the cold-worked metals to introduce functional periodicity.
4. Introducing functional periodicity in the form of thin wires and wire ropes can eliminate the catastrophic fracture of structures that occurs when unstable cracks propagate across the structure.
5. Introducing fibers that are periodically placed in the bulk material perpendicular to the sliding surface can reduce wear of materials by delamination.
6. Introduction of functional periodicity is shown to be an important tool to enhance the behavior of materials and to control material properties.

Notes

1. There are two main types of muscles, skeletal and smooth muscles. Skeletal muscles are highly structured, comprising bundles of fibers contained in a long syncytium of multiple cells. Skeletal muscles perform our voluntary movements, such as raising a hand to ask a question. Smooth muscles are not organized like skeletal muscle; they comprise sheets of muscle cells and are decidedly less periodic in structure than the skeletal muscles. Smooth muscles are present, for example, in the bladder. Smooth muscles can generate a very strong force, but typically the contractions are slow and sustained. Smooth muscles are responsible for moving food through the gut or for modulating blood flow by changing the diameter of a blood vessel. (Private communication from Jeffrey D. Thomas.)

2. Muscle injuries rarely tear a muscle in two; instead the damage is diffused across the width, depth, and length of an area of the muscle. Because of their relative inelasticity, the tendons that attach the muscle to bone are much more commonly injured than is the muscle itself. (Private communication from Jeffrey D. Thomas.)

3. There are many textbooks that cover fracture mechanics. See chapter 8 of Suh and Turner (1975).

4. The mechanics of crack propagation is given in many books such as Suh and Turner (1975).

5. This process of wear due to subsurface plastic deformation, crack nucleation, and propagation is called delamination wear (Suh, 1973, 1986). This delamination wear is a time-dependent combinatorial process. Initially the crack propagation rate is small, but increases with the crack length. The crack propagation depends on the prior crack propagation history, the microstructure of materials, and the state of loading. The wear process accelerates as these wear particles agglomerate at the sliding interface, where they penetrate deeper into the sliding interface and increase the surface traction. The increased friction force accelerates the delamination process.

6. More details can be found in many textbooks, for example, chapter 5 of Suh and Turner (1975).

7. The slip plane is the plane on which the crystal slides, resulting in a permanent plastic deformation.

8. Sessile dislocations can form when a sliding screw dislocation is cut by a dislocation, which forms a "jog" that is perpendicular to the direction of dislocation motion.

9. The stress–strain curve shown in figure 8.5 is for a polycrystalline metal. When the metal is a single crystal, the deformation can be highly anisotropic, especially if the crystal has a limited number of slip planes, for example, h.c.p. metals.

10. Many standard textbooks on metalworking and mechanics provide analysis of wire drawing, for example, Suh and Turner (1975).

11. The rationale for microcellular plastics can be found in Suh (2001), and more details are given in Suh (1996), which provides references to the contributions made by the author's students. See also www.trexel.com.

12. These are parts for printers made by Averplast of Singapore. MuCell is a trade name of Trexel, Inc.

13. The author acknowledges the helpful discussion he had with Professor Gang Chen of MIT.

References

El Wakil, S.D. 1989. *Processes and Design for Manufacturing*, Prentice-Hall, Englewood Cliffs, NJ.

Lee, D.G. and Suh, N.P. 2004. *Axiomatic Design and Fabrication of Composite Structures*, Oxford University Press, New York.

Paris, P.C., Bucci, R.J., Wessel, W.G., Clark, W.G., and Mager, T.R. 1972. "Extensive study of low fatigue crack growth rates in A533 and A508 steels," *Stress Analysis and Growth of Cracks*, ASTM Special Technical Publication No. 513, p. 143.

Suh, N.P. 1973. "The delamination theory of wear," *Wear*, Vol. 25, pp.111–124.

Suh, N.P. 1986. *Tribophysics*, Prentice-Hall, Englewood Cliffs, NJ.

Suh, N.P. 1996. "Microcellular plastics," in *Innovation in Polymer Processing: Molding* (ed. J.F. Stevenson), Hanser/Gardner Publications, Inc., New York, pp. 93–149.

Suh, N.P. 2001. *Axiomatic Design: Advances and Applications,* Oxford University Press, New York.

Suh, N.P. and Turner, A.P.L. 1975. *Elements of the Mechanical Behavior of Solids,* McGraw-Hill, New York.

Exercises

8.1. Does the existence of the periodic table of chemical elements indicate the existence of functional periodicity in atomic structure?

8.2. Turbine blades used in the hot section of jet engines must have creep resistance at high temperature, since the thermodynamic efficiency of a gas turbine increases with the temperature of the gas. Creep refers to the phenomenon of materials undergoing continuous strain (i.e., the time rate of change of strain is positive) under constant stress. In most metals, this occurs when the temperature is greater than one-third of the melting temperature. Because of creep, turbine blades eventually fail when the turbine blades elongate and hit the shroud (i.e., the wall of the chamber) or fail by creep-rupture. Such a failure cannot be stopped. This constitutes a case of time-dependent combinatorial complexity.

To minimize creep failure, some turbine blades are made of single crystals that do not have any grain boundaries, since grain boundaries and dislocations contribute to the mechanisms that cause creep. (*Note*: The causes of creep are known to be the following: (a) climb of dislocations when atoms diffuse along the dislocation, enabling dislocations to get around hard particles that intersect slip planes, (b) annihilation of dislocations, (c) grain boundary sliding, and (d) diffusion of atoms across grain boundaries.)

Is there a way of introducing a material functional periodicity to prevent the creep of turbine blades under uniaxial loading?

8.3. In fatigue failure of metals, cracks nucleate and propagate under cyclic loading. The crack propagation rate increases with the crack size and therefore cracks propagate faster and faster as they grow, which is a time-dependent combinatorial complexity problem. Suggest a means of transforming this combinatorial complexity problem to a periodic complexity problem by introducing a material functional periodicity to increase the fatigue life of metals.

8.4. In chapter 7, it was shown that the geometric functional periodicity in the form of an undulated surface decreases the friction and wear between two sliding surfaces. Propose a means of reducing friction and wear by introducing a material functional periodicity.

8.5. When a sharp crack develops in a welded butt joint of steel plates, unstable cracks may propagate along the seam when the stress intensity factor at the crack tip exceeds the critical stress intensity factor, i.e., a system with combinatorial complexity. A possible means of preventing the fracture is to blunt the crack tip or reduce the stress concentration at the crack tip. Introduce either geometric functional periodicity or material functional periodicity to prevent the fracture along the seam.

8.6. Glass is an amorphous material. It does not have a crystalline structure nor any material or geometric functional periodicity. Yet this type of amorphous material seems to be stable. The stability argument presented in chapter 3 stated that the stability requires either equilibrium states or a functional periodicity. Is there a contradiction between this argument and the behavior of glass?

8.7. Composite materials date back perhaps to prehistoric days. The walls of mud houses had straw in them as reinforcement so as to prevent their fracture. Windows of old historical buildings have glass panels set in sashes rather than a large single glass panel. These types of material functional periodicity have been invented to overcome the limitations of materials or manufacturing processes. Identify three other historical composite materials that have material functional periodicity.

9

▓▓ Complexity of Biological Systems

▓9.1▓ Introduction

As shown in table 3.1, many natural systems possess functional periodicity. Based on this observation and the behavior of engineered systems, a hypothesis was presented in chapter 3, which stated that

> For long-term stability of natural and engineered systems, the system must be either in an equilibrium state with its surrounding or have a functional periodicity. Systems without stability because of the lack of functional periodicity have a transitory existence or are in a chaotic state and eventually disappear or mutate into another system or matter which is either in equilibrium or is periodic.

Cells have functional periodicity, many living beings follow the circadian cycle, and the life–reproduction–death cycle supports the view that biological systems have functional periodicity.

In this chapter, we examine the complexity of biological systems. We will apply complexity theory and axiomatic design theory to the field of systems biology to relate system-level physiological functions to the behavior of proteins, DNA, cells, and so on. One of the issues of particular interest is the role of functional periodicity in reducing the complexity and providing the robustness of biological systems.

What are the systems-level issues in biology?
During the past half-century, biological sciences have been changing the world, scientifically, technologically, and socially. Yet the field is still evolving. One of these subfields of biology may be called the field of *systems biology,* a study of the systems aspect of biology. From the systems point of view, some of the major issues may be grouped as:

1. *Systems issues.* Molecular biology is yet to be related to the higher-level functions of cells, tissues, and physiology. How do we explain the physiological behavior of biological systems in terms of molecular-level interactions? Will the construction of the master design matrix for biological systems explain the behavior of biological systems better? Why are biological systems so robust? Is it because of biological functional periodicity?
2. *Complexity associated with the behavior of biological systems.* Biological systems are complex because there is a great deal of uncertainty in determining causality. For example, when a patient complains about chest pain, the diagnosis of the illness is not obvious. The doctor may treat the illness as coronary artery disease, whereas the real cause may be esophageal spasm. Why is the diagnosis of biological systems so complex? What causes uncertainty in diagnosis and thus complexity? Are these a time-independent real complexity? Is the time-independent imaginary complexity making the fields of medicine and biology more complicated than they really are?
3. *Complexity associated with our ability to predict.* We cannot predict—with 100% certainty—the behavior of organs and cells from first principles when changes are introduced to the system. Why does the Western diet cause more colon cancer? Why does the Asian diet cause more stomach cancer?[1] Will the creation of the master design matrix for biological systems provide a more effective diagnostic tool?

Systems issues
One approach to systems issues advocated recently in the literature is the systems control perspective: develop various models for biological interactions and then apply modern control theory to understand the dynamic behavior of the system through simulation. Such an effort may enhance our understanding of the interactions between proteins and DNA molecules and increase our understanding of the system behavior, but it may not lead to complete understanding of the relationship between the *functions* of higher-level systems and molecular behavior. Furthermore, the feedback mechanism used in control theory may not be applicable in systems that are made stable by the presence of functional periodicity. A stable system with functional periodicity may go into the next period when all the functions are completed rather than be triggered by feedback signals.

The problem in biology is analogous to understanding gears and cams and their interrelationships in an automobile transmission, but not being able to describe the functions of the total system (i.e., transmit torque, match mechanical impedance, change the input/output speed ratio) that are made of these gears, cams, and so on. These functions of higher-level systems—biological or mechanical—cannot be discerned by simply refining the understanding of the lowest-level physical entities. The typical reductionism of science, which is a powerful tool in understanding molecular behavior, may not yield a system-level understanding.

We may not be able to identify and characterize system-level functions without using a different kind of thinking process from that of the reductionist approach. Instead, we need to synthesize the aggregated behavior of systems based on an understanding of molecular-level behavior. To achieve this goal, we will need a new scientific framework that will guide the synthesis process. When a set of FRs must be satisfied at the same time, the coupling effect of certain DPs on some of these FRs must be understood to be sure that the biological system can function as a robust system.

Complexity of the behavior of biological systems
One of the remarkable things about biological systems is that living beings are robust and survive for relatively long periods of time, even when the external conditions continuously vary. The fact that they are robust implies that the effect of time-dependent combinatorial complexity is suppressed by a variety of functions that have functional periodicity and undergo time-dependent periodic complexity. As discussed in chapter 3, the complexity of a system is reduced when a system with time-dependent combinatorial complexity is transformed into a system with time-dependent periodic complexity.

To shed insight on the systems view of biology, the implications of complexity theory on biology will be explored in this chapter. The ultimate goal, which cannot be attained at this time, is to link molecular biology to the system-level behavior and support the conjecture that biological systems will not survive when they lose functional periodicity to sustain their biological functions.[2]

Complexity associated with our ability to predict the behavior
of biological systems
Our lack of ability to predict system behavior from molecular behavior of biological systems creates complexity. One of the big components of complexity is likely to be time-independent imaginary complexity, since biologists and engineers may be spending time and effort on imaginary problems that arise in studying decoupled systems. This problem will not be solved until we are able to write the design equation that relates the functions at all levels to the molecular interactions. Also our lack of understanding of the functions that can control functional periodicity may prevent us from understanding biological systems. Furthermore, the inability to distinguish time-dependent combinatorial complexity from time-dependent periodic complexity may make the system seem more complex than it really is.

In biology, the complexity of the molecular interactions implicated in cell regulatory networks is thought to be much more complex than engineered systems. The known interactions of the cell cycle regulatory network are represented in the form of a diagram, map, and/or database without the recognition that system design can be understood only by relating the functional and the physical domains.

This issue is exacerbated by genomic research, where biological entities are described without any relationship to functions. In contrast, traditional biochemistry research deduces the existence of bioentities (e.g., enzymes) based on the observation of functions.

A diagram convention is also used, which is designed to represent networks containing multiprotein complexes, protein modifications, and enzymes that are substrates of other enzymes (Kohn, 1999). This approach does not explicitly include functions nor does it differentiate the hierarchical nature of system functions and physical parameters.

Why do we think that biological systems follow the same set of axioms and complexity theory as engineered systems?
A biological system may be amenable to the same principles used in designing engineered systems to create software, hardware, and other engineered systems. There are several reasons for this belief:

1. According to axiomatic design theory, a coupled design cannot survive in an environment where the physical parameters change randomly and rapidly (Lipson et al., 2002). Such a system will have large time-independent real complexity. However, living systems appear to be robust, indicating that the system is well "designed" (i.e., not a coupled system). Many biological systems appear to be decoupled systems.
2. Natural system behaviors conform to many of the conclusions of complexity theory and axiomatic design theory. Biological systems seem to be consistent with the requirement that a robust design must have a functional periodicity (Suh, 1998, 1999, 2001; Suh and Lee, 2002). Cells have a periodicity, for example, cell cycles. Even cells that have very long mitotic cycles have a functional periodicity. For example, depolarization in neurons provides these cells with a periodicity.
3. Axiomatic design theory states that for a decoupled design, the physical parameters must be changed in a given sequence to fulfill FRs. Cell cycles, for example, behave as a decoupled system.
4. The pathway used to describe communication in biological systems is similar to the flow diagram for engineered systems in that there is a clear pathway for interactions of the lowest-level entities to achieve higher-level functions.

9.1.1 Unresolved questions

There are many unresolved questions that we hope to answer eventually as we pursue our systems biology research. Some of the questions are the following:

1. *Implications of uncoupled, decoupled, and coupled designs in biological systems.* Coupled systems are less robust than uncoupled or decoupled systems, especially when random changes occur in the environment. Biological systems are robust. Therefore, many biological systems must be either uncoupled or decoupled systems. Is this hypothesis true? Do biological systems act as either uncoupled or decoupled systems?
2. *How can we relate the higher-level functions of biological systems to molecular-level interactions?* As stated earlier in this chapter, one of the goals of systems biology is to relate the functions of biological systems (e.g., the lung) to molecular biology. How can we achieve this goal? A "V-model" will be presented in the next section to demonstrate how it might be done. We will use the lung as an example. Can we provide similar decompositions of other organs so as to relate their functions to their molecular behavior?

3. *Do biological systems have functional periodicity?* One of the conclusions of complexity theory is that when a system undergoes continuous and random changes throughout its life, it must have a functional periodicity to prevent the system from assuming time-dependent combinatorial complexity. Do all biological systems have functional periodicity? How is it regulated?

9.2 Biological Systems

9.2.1 What is a biological system?

A typical biological system satisfies a set of FRs at a given level of the system by aggregating a number of physical elements and biochemical processes within a boundary. There are also constraints such as temperature and pressure that bound the ability of biological systems.[3]

9.2.2 V-model for construction of biological functions from molecular behavior

To understand biological functions based on an understanding of molecular interactions, a V-model[4] shown in figure 9.1 is proposed for the necessary conceptual framework. The left leg of the V-model represents the decomposition of a system from the highest-level FRs of the system to the leaf-level DPs. The right leg of the V-model represents the integration process that leads to physical assembly of biological systems.

The decomposition process is to generate the leaf-level details (in this case, it could be molecules and molecular-level interactions) from the highest-level FRs of the biological system. Once we obtain all the DPs at the leaf level, we have to integrate (or assemble) them to create the system by going up the right leg of the V-model. Without

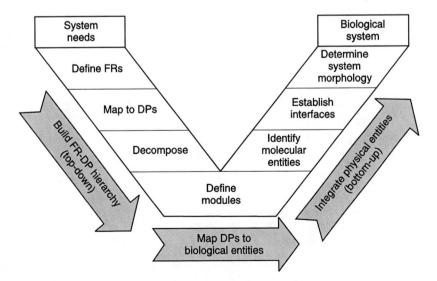

Figure 9.1 V-model for biological systems to relate biological functions to molecular interactions.

going up the right leg of the V-model, we cannot determine how the DPs must be put together to create a system (e.g., the lung).

9.2.3 Decomposition of FRs and DPs of biological systems: the left leg of the V-model

Biological systems must be decomposed to understand the interrelationships between the functions of organs and the behavior of molecules. In terms of axiomatic design theory, the mapping of biological systems from the functional domain to the physical domain can be represented through decomposition as shown in figure 9.2. The domain on the left is the functional domain, which is characterized by functions (FRs) of the biological system. The domain on the right shows the physical domain, which consists of DPs (i.e., physical entities) at several levels, i.e., organs, tissues, cells, organelles, and molecules.

Decomposition can be achieved only by zigzagging between the two domains, as indicated in figure 9.2. The lower-level FRs and DPs are the children of the FRs and DPs of the level above. The lower-level FRs are the functions of the parent DP and must be consistent with the parent FR. Children DPs are determined by mapping the children-level FRs into the physical domain. Then the children-level DPs as an aggregate constitute the parent DP. DPs may not be separate physical entities but subsets of a physical entity. According to the Information Axiom, the integration of DPs in a physical entity would be a good thing to do if it reduces the information content. This is illustrated below using the functions and DPs of the lung.

Decomposition of the lung
The highest-level DP could be an organ (e.g., the lung),[5] which is there to satisfy a functional requirement:

FR = Supply O_2 to blood and remove CO_2 from blood.

Normally in engineering design, we state the FR first and then create a solution in terms of DPs. However, in biological systems we have to go from the DP (the lung) in the physical domain to the FR (exchange O_2 with CO_2) in the functional domain because the

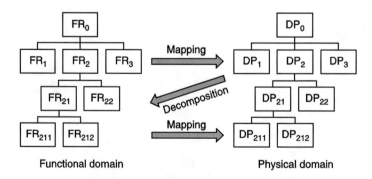

Functional domain   Physical domain

Figure 9.2 The process of decomposition: zigzagging between the functional domain and the physical domain. The lower-level FRs and DPs are the children of the FRs and DPs at the next-higher level.

Table 9.1 Design matrix at the highest level of the lung

	DP_1	DP_2	DP_3	DP_4
FR_1	X	0	0	0
FR_2	X	X	0	0
FR_3	0	0	X	0
FR_4	0	0	0	X

design of the system has already been done by nature. There may be other FRs of the lung such as "remove any undesirable liquids from the lung." In general, it is not an easy task to go from the physical domain to the functional domain because mapping is always a one-to-many relationship. To simplify this illustration, we will consider only one FR.

FR and DP must be decomposed since DP does not have sufficient detail to fulfill FR. The children FR_1 through FR_4, given below, are the tissue-level FRs that describe the functional requirements of DP (the lung). These children-level FRs, which, in aggregate, perform the function FR, may be stated as:

FR_1 = Supply O_2 to blood as required.
FR_2 = Remove CO_2 from blood.
FR_3 = Filter particulates in air.
FR_4 = Remove particulates and infectious agents from lung.

The DPs that can satisfy the children-level FRs are:

DP_1 = O_2 supply mechanism.
DP_2 = CO_2 removal mechanism.
DP_3 = Mucus in large airways.
DP_4 = Coughing/cilia/macrophages.

To understand the system behavior, we must understand the relationship between these FRs and DPs by constructing the design matrix, which is shown in table 9.1. Symbol X indicates a strong relationship between the FR and the DP; symbol x indicates a weak relationship; and symbol 0 indicates no relationship. For example, X indicates that the answer is "yes" to the question: "Does DP_i affect function FR_j?" In the above matrix, DP_1 is assumed to affect FR_2, because if fresh air does not come into the lung, there will not be any way to expel CO_2.

FR_1 and DP_1 may be decomposed as:

FR_{11} = Bring in fresh air to the lung cavity to create O_2 gradient.
FR_{12} = Allow diffusion across the alveolar membrane.

DP_{11} = Expansion of the lung cavity.
DP_{12} = O_2 diffusion passage in alveolar membrane.

The design matrix for the children-level FR_{1x} and DP_{1x} is a diagonal matrix.

FR_2 and DP_2 may be decomposed as:

FR_{21} = Expel CO_2 from the lung cavity to create CO_2 gradient.
FR_{22} = Allow diffusion across the alveolar membrane.

DP_{21} = Compression of the lung cavity.
DP_{22} = CO_2 diffusion in alveolar membrane.

The design matrix for the above set of FR_{2x} and DP_{2x} is also a diagonal matrix.
FR_4 and DP_4 may be decomposed as:

FR_{41} = Separate mucus from membrane.
FR_{42} = Transport gases and liquid droplets from lung.
FR_{43} = Absorb residual particulates.

DP_{41} = Ciliary action.
DP_{42} = Rate of contraction of lung.
DP_{43} = Macrophage cells.

The design matrix for the above set of FR_{4x} and DP_{4x} is also a diagonal matrix. Some of these children FR/DP pairs should be decomposed further, since they are not at the leaf level.

The above decomposition is based on the idea that as long as fresh air is brought into the lung when the chest cavity expands, diffusion occurs because the oxygen partial pressure in the lung, on average, is greater than that of the concentration of oxygen in the blood. A similar argument was used for removal of CO_2.

Significance of a master design matrix
Once the mapping is complete, we then have to determine the relationship between the FRs and DPs using a master design matrix, which relates the highest-level FRs to the molecular-level DPs—the goal of systems biology. For the design done to this level, the master matrix is very simple. To achieve the goal of relating the FRs of the biological system to molecular-level interactions, the decomposition should be continued further to the leaf level.

9.2.4 Integration of molecular entities to create a biological system: the right leg of the V-model

Once we obtain all the molecular-level DPs through the decomposition process (coming down the left leg of the V-model), we need to assemble these molecules and other moieties to form a biological system, such as a cell or the lung, by going up the right leg of the V-model. In other words, we have to put the "physical" pieces together to form a biologically functioning system by figuring out the best way these physical moieties can be made to fit with each other and work as an integrated system. In this process, certain DPs are placed in geometric proximity, because they interact or satisfy the geometric or steric conditions in a most optimum way.

This process of integration may be aided if we create a (DP vs. DP) matrix[6] to determine physical or biological interactions among DPs. The matrix will show which DP pairs interact more than others. This information on physical interactions will enable us to determine the geometric arrangement of these DPs in a biological system. Since, in biology, we already know the morphology of cells, and so on, the purpose of this integration of DPs is to shed greater insight on, for example, why the nucleus must be located where it is in a cell and why centrosomes must be located where they are in order to facilitate cell division during mitosis.

9.3 Cells as a Biological System

In the introduction to this chapter, it was stated that the systems issue—understanding how the higher-level functions of biological systems are related to molecular-level interactions—is one of the three main issues of systems biology. The other two issues are related to complexity. In this section, we will consider the systems issue in detail by identifying the FRs of a cell and relating them to their DPs, which are the physical entities that control the FRs.

A biological system satisfies or fulfills many functions at the same time. A function of a biological system is influenced by many physical parameters. Therefore, when a set of functions must be satisfied simultaneously, the system can behave chaotically if any of the physical parameters simultaneously affects all the functions. This leads to coupling of functions. Because it is sensitive to random variation of physical parameters, such a system cannot be robust and achieve its functions when the external conditions suddenly change. Conversely, not all physical parameters that affect the system behavior are equally important. A particular set of functions that a biological system must satisfy determines the relative importance of the physical parameters.

The task of associating higher-level functions to molecular-level functional behavior is difficult for biological systems since, unlike man-made systems, the functions and physical parameters have evolved over billions of years, responding to the changing external environmental conditions. It is somewhat similar to taking the software code developed by an unknown person and trying to figure out the intended function and the logic behind the functions of the software system. It is also similar to taking a circuit diagram of electrical systems and trying to figure out the functions of the electrical systems. The study of biology poses problems such as these on a far more complicated level.

Let us first consider the FRs of a cell.[7] The highest-level FRs and DPs may be stated as:

FR = Sustain cell life.
DP = Cell mechanisms.

These highest FR/DP may be decomposed as:

FR_1 = Isolate the cell and its components from its environment.
FR_2 = Obtain fuel.
FR_3 = Convert the fuel to energy.
FR_4 = Communicate with its surrounding.
FR_5 = Reproduce itself.
FR_6 = Control cell functions.

The DPs of a cell that satisfy the above FRs may be stated[8] as:

DP_1 = Plasma membrane (phospholipid bilayer).
DP_2 = Diffusion of ions and transport of proteins.
DP_3 = Synthesis of ATP in mitochondria.
DP_4 = Receptors and signal transduction protein.
DP_5 = Reproduction mechanism.
DP_6 = Functional periodicity.

Table 9.2 Design matrix for a cell

		DP_1	DP_2	DP_3	DP_4	DP_5	DP_6		
							DP_{61}	DP_{62}	DP_{63}
FR_1		X	0	0	0	0	0	0	0
FR_2		X	X	0	0	0	0	0	0
FR_3		0	X	X	0	0	0	0	0
FR_4		X	x	0	X	0	0	0	0
FR_5		x	X	X	x	X	0	0	X
FR_6	FR_{61}	x	0	0	0	0	X	0	0
	FR_{62}	0	0	0	0	0	0	X	0
	FR_{63}	X	X	X	X	0	0	X	X

FR_6 (Control cell functions) and DP_6 (Functional periodicity) may be decomposed as:

FR_{61} = Regulate growth.
FR_{62} = Prevent mutation.
FR_{63} = Control cell functions.

DP_{61} = Growth factor.
DP_{62} = DNA repair and apoptosis.
DP_{63} = cdk protein.

The design matrix for the above set is as shown in table 9.2.[9] This design matrix is a triangular matrix. Therefore, to satisfy all the FRs, DPs must be changed in the given sequence, that is, change DP_1 first, DP_2 second, and so on.

The diagonal terms of the design matrix have the most important effect on FRs. The off-diagonal terms are coupling terms that represent a secondary effect of a chosen DP on other FRs. If the particular set of FRs of a biological system changes, the off-diagonal coupling terms may not be important. Although one can study the causality of any pair of FRs and DPs, the coupling terms may not have the most significant effect on the outcome. Therefore, when biologists choose to investigate a particular FR/DP pair, it is important to identify them first from the viewpoint of systems biology.

After the decomposition is completed to the lowest level (leaf level), the X's and x's must be replaced by mathematical expressions through modeling. Then the entire system behavior in terms of the FRs can be analyzed and simulated based on the master design matrix. Although attempts are being made to model the relationship among the molecular-level entities, this will not reveal the functions of the biological system without zigzagging between the domains.

The above matrix states that the function of obtaining the fuel (FR_2) is primarily controlled by the diffusion of ions and transport of proteins (DP_2) and is also secondarily affected by the plasma membrane (DP_1). Therefore, given a plasma membrane, the transport of the fuel across the membrane must be adjusted to meet the need for fuel. The matrix states that the plasma membrane (phospholipid bilayer) should be formed first as a new cell is formed, and that the plasma membrane affects the "obtain the fuel" function (FR_2). The plasma membrane also affects many other functions, such as FR_4 (communicate with its surrounding), FR_5 (reproduce itself), FR_{63} (control cell functions), and to a lesser degree, FR_{61} (regulate growth). FR_6 and DP_6 are decomposed into FR_{61}, FR_{62}, FR_{63}, and DP_{61}, DP_{62}, DP_{63}. DP_1 (plasma membrane) affects FR_{61} (regulate growth) and FR_{63} (reproduce cell). DP_2 (diffusion of ions and transport protein) affects FR_3 (convert the fuel to energy), FR_4 (communicate with its surrounding), FR_5 (reproduce itself), and FR_{63} (control cell functions).

Many of these FRs and DPs can be decomposed further. For example, FR_5 (reproduce itself) and DP_5 (reproduction mechanism) may be decomposed as:

FR_{51} = Initiate the replication process.
FR_{52} = Start chromosome replication.
FR_{53} = Replicate proteins.
FR_{54} = Replicate DNA structures.
FR_{55} = Create cytoplasm skeleton.
FR_{56} = Form membranes of organelles.
FR_{57} = Induce mitosis.

DP_{51} = Start kinase/"restriction point" for G_1 phase.
DP_{52} = Chromosome replication kinase.
DP_{53} = Cyclin-dependent protein kinase (cdk protein).
DP_{54} = DNA polymerase, ligase/mRNA.
DP_{55} = Polymerization mechanisms.
DP_{56} = "Membrane" kinase.
DP_{57} = "Tensile force."

The choice of DP_{57} is based on the assumption that tensile force is required to separate the centrosomes aided by mitotic cyclin, which can be decomposed into lower-level FRs and DPs.

The design matrix for the above set of $\{FR_{5x}\}$ and $\{DP_{5x}\}$ may be represented in table 9.3. This design matrix indicates that the function of reproduction at this level of decomposition is a decoupled design.

Table 9.3   Design matrix for FR_{5x} and DP_{5x}

	DP_{51}	DP_{52}	DP_{53}	DP_{54}	DP_{55}	DP_{56}	DP_{57}
FR_{51}	X	0	0	0	0	0	0
FR_{52}	X	X	0	0	0	0	0
FR_{53}	X	0	X	0	0	0	0
FR_{54}	X	0	0	X	0	0	0
FR_{55}	X	0	0	0	X	0	0
FR_{56}	X	0	0	0	0	X	0
FR_{57}	0	0	0	0	X	X	X

Table 9.4   Design matrix for FR$_{57}$ (induce mitosis) and DP$_{57}$ (tensile force)

	DP$_{571}$	DP$_{572}$	DP$_{573}$	DP$_{574}$	DP$_{575}$	DP$_{576}$
FR$_{571}$	X	0	0	0	0	0
FR$_{572}$	x	X	0	0	0	0
FR$_{573}$	x	X	X	0	0	0
FR$_{574}$	0	X	X	X	0	0
FR$_{575}$	0	X	0	0	X	0
FR$_{576}$	0	X	0	0	0	X

In addition to DP$_{51}$ (Start kinase/"restriction point" for G$_1$ phase), there are many checkpoints in the cell cycle. For example, completion of duplication of DNA is checked and if it is found that something is wrong, the cells are eventually discarded by apoptosis, cell death with purpose, if the attempt to repair the DNA fails. In most cells of an organ in adults, the cells are maintained in a dormant state G$_0$. Only when there is need to replace some cells (there is constant replacement of a certain portion of cells in an organ) or if a large chunk of liver is cut out, the liver cells will get into the cell cycle and grow until the original mass is attained (Chae, 2002).

FR$_{57}$ (Induce mitosis) and DP$_{57}$ ("Tensile force") may be decomposed as:

FR$_{571}$ = Adhere to a reference surface.
FR$_{572}$ = Activate WASp family protein.
FR$_{573}$ = Nucleate cellular actin.
FR$_{574}$ = Polymerize G-actin to make F-actin filament.
FR$_{575}$ = Terminate the polymerization reaction.
FR$_{576}$ = Depolymerize the F-actin filament.

DP$_{571}$ = Active sites on the surface (protein fibronectin).
DP$_{572}$ = Integrin receptor on cell membrane that binds to fibronectin and
            initiates signal transduction.
DP$_{573}$ = Arp2/3 complex.[10]
DP$_{574}$ = Free actin/profilin.
DP$_{575}$ = Capping protein.
DP$_{576}$ = ATP hydrolysis within actin filaments and dissociation of
            phosphate/ADF (actin depolymerization factor).

The design matrix for this third level FR/DP relationship is given in table 9.4. The design indicated by the above design matrix is a decoupled design. If there are more DPs than FRs, they should also be listed and subjected to the same kind of analysis. This problem is treated in Suh (2001).

Systems biology representation
The master matrix, shown in table 9.5, relates the highest-level FRs of a cell to some of the lowest-level proteins, and so on, of the cell (up to the extent of the decomposition done). It is not a complete master matrix since the decomposition of FRs/DPs has not

Table 9.5 Master design matrix of a cell (created using Acclaro). This matrix relates the highest-level functions of a cell to the lowest-level molecules.

	DP1	DP2	DP3	DP4	DP51	DP52	DP53	DP54	DP55	DP56	DP571	DP572	DP573	DP574	DP575	DP576	DP61	DP62	DP63
FR1	X	O	O	O	O	O	O	O	O	O	O	O	O	O	O	O	O	O	O
FR2	X	X	O	O	O	O	O	O	O	O	O	O	O	O	O	O	O	O	O
FR3	O	X	X	O	O	O	O	O	O	O	O	O	O	O	O	O	O	O	O
FR4	X	X	O	X	O	O	O	O	O	O	O	O	O	O	O	O	O	O	O
FR51	O	O	O	O	X	O	O	O	O	O	O	O	O	O	O	O	O	O	O
FR52	O	O	O	O	X	X	O	O	O	O	O	O	O	O	O	O	O	O	O
FR53	O	O	O	O	X	O	X	O	O	O	O	O	O	O	O	O	O	O	O
FR54	O	O	O	O	X	O	O	X	O	O	O	O	O	O	O	O	O	O	O
FR55	O	O	O	O	X	O	O	O	X	O	O	O	O	O	O	O	O	O	O
FR56	O	O	O	O	X	O	O	O	O	X	O	O	O	O	O	O	O	O	O
FR571	O	O	O	O	O	O	O	O	O	O	X	O	O	O	O	O	O	O	O
FR572	O	O	O	O	O	O	O	O	O	O	X	X	O	O	O	O	O	O	O
FR573	O	O	O	O	O	O	O	O	O	O	X	X	X	O	O	O	O	O	O
FR574	O	O	O	O	O	O	O	O	O	O	O	X	X	X	O	O	O	O	O
FR575	O	O	O	O	O	O	O	O	O	O	O	X	O	O	X	O	O	O	O
FR576	O	O	O	O	O	O	O	O	O	O	O	X	O	O	O	X	O	O	O
FR61	X	O	O	O	O	O	O	O	O	O	O	O	O	O	O	O	X	O	O
FR62	O	O	O	O	O	O	O	O	O	O	O	O	O	O	O	O	O	X	O
FR63	X	X	X	X	O	O	O	O	O	O	O	O	O	O	O	O	O	X	X

been completed. Once a full master design matrix is created, we can deal with systems biology, which is to understand the functions of a biological system in relation to its molecular behavior. If this can be done for all organs and tissues, we will begin to understand how systems biology can be formalized.

System modeling
Eventually, all the X's in the design matrix must be replaced with equations and numbers through modeling the relationship between the specific FR and DP. Once all of these are replaced by equations, we will eventually have a model for any biological system, including the entire physiology of the human body. Some of these relationships will be dynamic and some will be static.

Causality
These models can be used to study how various internal and external factors affect the ultimate outcome of biological systems. This will aid the development of medicine. By knowing how different agents will affect the master matrix, we can predict the outcome.

Furthermore, we should be able to study how certain random changes will affect the health of human beings. These are ambitious goals, but that is the ultimate goal of systems biology.

Flow diagram for cells
To characterize the behavior of the biological system in term of the FRs of the cell, we must understand the effect of all the design elements in the design matrix on each of the FRs. Also when certain changes are made in any one of the DPs or elements of a design matrix, we must be able to determine the impact of the change. For this purpose, we develop a flow diagram for the biological system, which is equivalent to an electric circuit diagram.

This can be represented best if we use the concept of *module*. Module includes the effect of all the physical parameters given by the row of the design matrix for a given FR. Module is defined as the physical entity that generates the output (the FR) given an input (a DP) to the module. The module, *Mi*, is defined as (Suh, 2001):

$$Mi = \sum_{j=1}^{j=i} \frac{\partial FR_i}{\partial DP_j} \frac{DP_j}{DP_i} \tag{9.1}$$

A module consists of the reusable diagonal part and the unreusable off-diagonal part, because the diagonal part represents the strongest relationship between the FR and the DP.

A graphical representation of the relationship can be made in terms of a flow diagram that consists of modules. The flow diagram for the above set of FRs and DPs can be represented in figure 9.3.

9.4   Complexity of Biological Systems

How do you measure the complexity of biological systems?
Complexity is defined as the measure of uncertainty in satisfying the FRs. This can happen when the system ranges of FRs are not inside their design ranges, which is defined as time-independent real complexity. Time-independent imaginary complexity is a result of not knowing the exact relationship between the FRs and the DPs even when the design range and the system range completely overlap. When the system range changes as a function of time, we have time-dependent combinatorial or periodic complexity. The physiological processes determine the system range, whereas the laws of nature that govern the survival of biological systems establish the design range.

There are two ways of measuring the complexity of biological systems, depending on the issue involved:

1. *Complexity associated with the behavior of biological systems.* When the complexity associated with how well a biological system performs its functions is to be measured, we have to establish the FRs of the system that must be satisfied and then measure how well these FRs are satisfied by the system. If these FRs are satisfied 100% of the time within the design range, the person should not get sick and should have an immortal life.

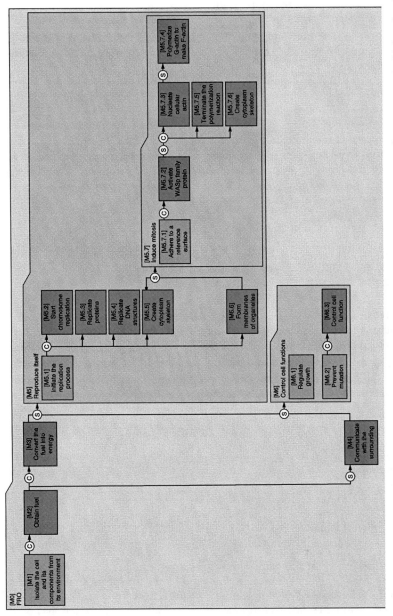

Figure 9.3 Flow diagram of the FRs of the cell in terms of modules defined by equation (9.1). This diagram represents how the highest-level FRs are related to the leaf-level molecular interaction. A decoupled design is indicated by C in a circle, whereas an uncoupled design is indicated by S in a circle. The uncoupled design can also be identified because the module boxes are parallel to each other. Decoupled design is indicated by the serial nature of the modules. There are boxes inside larger boxes. The outer box represents the parent and the inner boxes represent children. (Created using Acclaro, a software system.)

There are two aspects to this complexity. One is time-independent real complexity, the other is time-dependent complexity. It was shown in chapter 3 that a system with time-dependent combinatorial complexity cannot survive and that only the system with time-dependent periodic complexity can be robust and survive external random variation.

2. *Complexity associated with our ability to predict.* When we cannot predict the consequence of certain inputs to the biological system, the system appears to be complex. To predict the system behavior of biological systems, the design matrix must be established.

An example of the first kind—complexity associated with the behavior of biological systems—may be illustrated as follows. Humans must have the ability to convert sugar into energy. How well this function is performed by a person can be measured in terms of complexity. The complexity associated with this FR of a person is zero if the person converts the sugar within the physiologically acceptable design range at all times. A person who is unable to convert the sugar within the design range has nonzero time-independent real complexity. Some of these FRs of biological systems undergo changes as a function of time. Therefore, even a person who did not have problems converting sugar while he was young may develop diabetic problems with aging. This may be a result of the system becoming one of time-dependent combinatorial complexity. If this time-dependent combinatorial complexity cannot be converted into time-dependent periodic complexity by being able to reinitialize the system periodically, the biological system (person) may not survive.

An example of the second kind—complexity associated with our ability to predict—may be illustrated as follows. Suppose that the FR is to predict the period of mitosis within a design range (i.e., within the desired accuracy) under a given set of conditions. In this case, the complexity is measured by the logarithm of the probability of predicting the FR within the desired accuracy based on the molecular behavior of proteins, DNA, and small molecules. For example, to improve our ability to predict the period of mitosis precisely, we must be able to relate the system functions to the molecular mechanisms that control the mitosis of cells within the design range. This aspect of systems biology is yet to be achieved.

In the following section, we will consider the complexity associated only with the behavior of biological systems.

9.5　Functional Periodicity in Cells

Biological systems have periodic complexity!

When the physiological process (DP) that satisfies an FR (e.g., metabolize glucose) is a time-varying function and undergoes a combinatorial process, we can have time-dependent combinatorial complexity. If a certain FR of a biological system has time-dependent combinatorial complexity, it will eventually go into a chaotic state and cease to perform the function. Only the biological system that has time-dependent periodic complexity will survive. This has been partially accomplished through external intervention, such as the administration of medication and surgery.

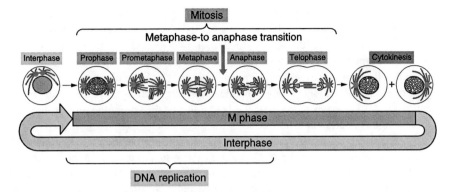

Figure 9.4 The events of cell division as seen under a microscope. The M phase consists of the processes of nucleus division (mitosis) and cell fission (cytokinesis), which take only a small fraction of the cell cycle time. The interphase, the rest of the cell cycle, takes most of the cycle time. During M phase, an abrupt change in the biochemical state of the cell occurs at the transition from metaphase to anaphase. (From Alberts et al., 1994.)

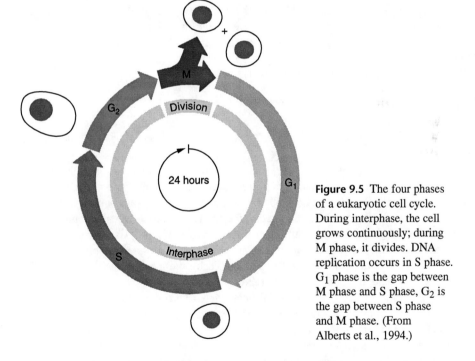

Figure 9.5 The four phases of a eukaryotic cell cycle. During interphase, the cell grows continuously; during M phase, it divides. DNA replication occurs in S phase. G_1 phase is the gap between M phase and S phase, G_2 is the gap between S phase and M phase. (From Alberts et al., 1994.)

Fortunately, functional periodicity appears to control the functioning of biological systems. For example, most eukaryotic cells divide periodically, as shown in figures 9.4 and 9.5. There are several distinct phases in a cell cycle. G_1, S, G_2, and M are the subdivisions of the standard cell cycle. Most of the cell cycles conform to this scheme. During mitosis (the process of cell division), the envelope of the nucleus breaks down and the contents of the nucleus condense into visible chromosomes. The cell's microtubules reorganize to form the mitotic spindle, which will eventually separate the chromosomes. As mitosis proceeds, the cell goes into metaphase, in which the chromosomes are aligned on the mitotic spindle ready for segregation. In anaphase, the separation of the duplicated chromosomes begins, during which the chromosomes move to the poles of the spindles, where they decondense to form two intact nuclei at the poles. The cell separates in two by a process called cytokinesis, the end of the mitotic M phase of the cell cycle (Alberts et al., 1994).

The cell cycle lasts for different durations, depending on the cell. The standard cell cycle is 12 hours or more for fast-growing tissues in a mammal. In most cells, the whole M phase takes only about an hour, a small fraction of the total cycle time. In some cells, the duration of the M phase is only about 8 minutes. Between M phase and the next is known as the interphase. Although not much is visible under the microscope during the interphase, preparations for cell division occur in a closely ordered sequence. During interphase, the DNA in the nucleus is replicated (Alberts et al., 1994).

The replication of DNA in the nucleus takes only a portion of the interphase called the S phase of the cell cycle. The interval between the completion of mitosis and the beginning of DNA synthesis is called the G_1 phase, and the interval between the end of DNA synthesis and the beginning of mitosis called the G_2 phase. It appears that the cell does not double its mass before it divides, requiring diffusion of masses into the cell during the G_1 phase. During this phase, the cell replicates DNA when all the conditions are satisfied. The cell remains in the G_2 phase until DNA replication is complete before it plunges into mitosis (Alberts et al., 1994).

What controls the cell cycle?
The start of a new cycle occurs at the end of G_1, when the S phase begins. One of the key questions is "What controls the cell cycle period?" According to Conlon et al. (2001), in Schwann cells the growth of cells and the cycle are not related, whereas in the yeast cell it was shown that there is a correlation between the cell growth and cell division (Mitchison, 1971). Conlon et al. proposed that the cell cycle is controlled by mitogen and the cell growth is controlled by growth factors. Another possibility is that in G_1 the increase in cell mass may determine the transition between G_1 and S if the prerequisite for cell division is sufficient diffusion of the appropriate ions and molecules to polymerize the required amount of an unreplicated complement of DNA. If this were the case, the functional period is determined by the mass diffusion rate into the daughter cell through the plasma membrane.

Biologists have established that for the cell cycle to proceed, it must go through the series of checkpoints, which depends on the feedback from downstream processes and signals from the environment. This is shown in figure 9.6. The beginning of a new cycle is the transition from G_1 phase to S phase. The mechanism that controls the cell cycle depends on two key families of proteins, cyclin-dependent protein kinases (Cdk) and cyclins. Cdk induces downstream processes by phosphorylating cyclins. Cyclins bind to Cdk molecules and control their ability to phosphorylate target proteins (i.e., G_1 cyclin, mitotic cyclin, etc.). Apparently, the completion of one phase does not immediately trigger the next phase. This has to wait for all the conditions for transition to the next phase to be satisfied at each checkpoint (Alberts et al., 1994). Therefore, the

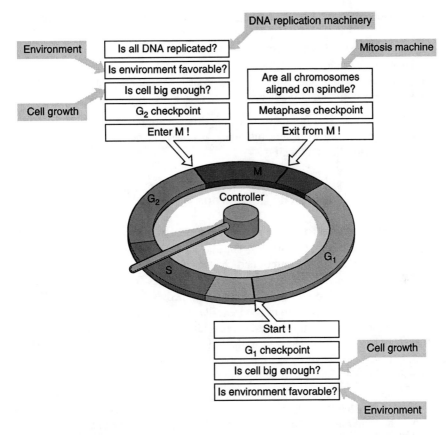

Figure 9.6 Checkpoints and inputs of regulatory information to the cell cycle control systems. (From Alberts et al., 1994.)

functional periodicity for the cell cycle is the completion of each one of the phases, which may take varying temporal durations.

In chapter 6, we have shown that to maximize the productivity of a manufacturing system, we have to introduce appropriate delays (e.g., buffers) to coordinate the robot transport in a most efficient way, in contrast to transporting the part to the next machine as soon as it is finished with one phase of the manufacturing operation. In this limited sense, it appears that there are similarities between the cell cycle and the manufacturing cycle. In the case of the manufacturing cycle, it was done to introduce periodicity and maximize productivity.

Functional periodicity and the cell cycle
The foregoing review of cell division indicates the following:

1. There is a set of functions that repeats on a periodic basis.
2. The period is not temporal. It is controlled by the readiness of each phase to move into the next phase. Its cycle is controlled by replication of DNA and other molecules, which may depend on the diffusion of ions and molecules as well as on activation by external agents.

3. Cell division is the fundamental means by which living beings made of cells survive.
4. The functional periodicity prevents healthy cells from going into time-dependent combinatorial complexity.
5. When the cell cycle breaks down and thus loses functional periodicity, cells will die or transform into an unpredictable form, creating chaotic changes.

The cell division described in this section supports the argument that for a biological system to be robust, it must have a functional periodicity and the system must be of time-dependent periodic complexity.

9.6   Functional Periodicity in Neurons

Neurons are also cells, but they do not undergo mitosis. Neurons also have plasma membranes, which are about 10 nm thick. The plasma membrane that surrounds the nerve cell has proteins that act as ion pumps and channels that conduct an electric pulse from one end of the nerve cell to the other in a fraction of a second. They periodically discharge ions through the axon, which is a long stem tube attached to the nerve cell body. It seems that neurons send electric charges periodically when certain conditions are met. Therefore, we may speculate that functional periodicity exists in nerve cells as well.

There may be other DPs that control the functional period. Neurons in crayfish fire electrical charges modulated by the mechanical tension in a stretch receptor cell. At a fixed tension, its electrical cycle is regular but is transiently upset by arrival of an impulse from a synapsing inhibitory neuron (Perkel et al., 1964; Schulman, 1969).

The above hypotheses may have some merit. According to Chae (2002), in the brain, there are neuronal cells that control circadian rhythm.[11] clusters of neurons that fire neurotransmitters continuously for a certain length of time, and so on. Also certain brain cells that secrete a hormone show certain oscillations in the period of secretion, for example, a one-hour period. Even in the cells that respond to hormones, when the receptor is overloaded, we see a phenomenon called receptor downregulation; that is, no signal is transmitted and even the amount of receptor is decreased.

9.7   Other Observations

Based on complexity theory, there are several observations that one can make about biological systems:

1. To fully characterize the system behavior of a biological system, we must determine the FRs and DPs together with the design matrix of the system, which was illustrated using a cell as an example. Large complex organs may involve many FRs and DPs.
2. To relate the system-level FRs to molecular interactions, we must decompose by zigzagging between the functional and the physical domains. However, because of the difficulties involved, many biological studies are often conducted at the level of one-to-one molecular interactions in the physical domain, although biochemists and biologists have made important contributions to understand systems biology.

3. To fully determine the behavior of an FR, we must know which module takes into account the effect of all DPs that affect it. However, since biological investigation in molecular biology is often done to determine the relationship between DPs, it is difficult to determine the real complexity.

4. When a biological system behaves erratically because of the disruption of FRs (e.g., the system range is not inside the design range), it may cause disease. In this case, the DP level must be adjusted to change the system range or a new DP must be introduced. If the FR cannot be maintained within the allowable tolerance, it may lead to a catastrophic failure such as cancer.

5. If biological systems are decoupled systems, we must determine a means of eliminating time-independent imaginary complexity problems. This should reduce unnecessary effort in research and drug discovery.

6. It appears that for biological systems to survive, they must have time-dependent periodic complexity. If they have time-dependent combinatorial complexity, they will not be robust enough to survive when external conditions suddenly change. Even when cells remain as dormant cells (like brain cells) in terms of the cell cycle for the lifetime of a person, cancer can occur if mutation makes the cell become a system with time-dependent combinatorial complexity.

7. The evolutionary process has produced biological systems with time-dependent periodic complexity. For example, cells, including neurons, have a cell cycle, which has a functional period, not a temporal period. Another example may be that all animals have a circadian cycle. When this cycle is violated, such as when human beings are subjected to a long period of sleep deprivation, the biological system does not function well. Sleeping may be a reinitialization process.

9.8   Challenges to Understanding Systems Biology

This systems biology based on axiomatic design theory and complexity theory is equivalent to the application of thermodynamics to energy-conversion devices at a macro-level. If the systems view is correct, it should be able to provide an aggregated view of how an ensemble of biomolecules behaves as a group.

Some of the research issues are:

1. *Can we convert the signal pathway diagram into a design matrix?*   Signal pathways for biological systems have been established, which are equivalent to the flow diagram for engineering systems that can be constructed based on the master design matrix that includes the design matrices at all levels of the decomposed system. If we can convert the signal pathway diagrams into a design matrix through "reverse transformation," it may enable us to determine the FRs and the design of biological systems in terms of coupling. This then in turn will enable us to understand which DPs are affecting the various FRs of a biological system.

2. *Do we know if the DPs are not controlled in a correct sequence?*   Once the design matrix is established, we should know if a biological system is uncoupled, decoupled, or coupled. If the biological system is decoupled with a triangular matrix, we can eliminate time-independent imaginary complexity and identify the sequence of

setting DPs and FRs, which should expedite the control of biological cells and the development of suitable drugs (i.e., DPs).

3. *Is illness caused by the magnitude of an off-diagonal element becoming larger than a diagonal element?*   When the magnitude of an off-diagonal element of the design matrix becomes larger than that of the diagonal element during the lifetime of a biological system, the physical parameter (DP) that normally controls the FR can no longer control the specific function of the biological system. Since biological systems are highly robust, normally the off-diagonal elements are expected to be smaller than the diagonal elements. However, certain illnesses may be caused when one or more of the off-diagonal elements becomes larger than the diagonal elements, or conversely, when the magnitude of the diagonal element decreases to a small value.

4. *Is cancer caused by a loss of functional periodicity?*   Cancer is often attributed to mutation of cells that affect the cell cycle. The prevailing view is that mutation is the result of failure of the error-correcting functions. Is the mutation a result of the cell cycle breaking down by somehow converting time-dependent periodic complexity into time-dependent combinatorial complexity? This transformation would happen if the cell cycle were reinitiated before the system is ready, which may be a error-correcting function. In contrast to the case discussed above, in some cases the cell may lose the ability to reinitialize the cycle, letting the biological processes propagate randomly without being able to stop the process. This would bring about uncontrollable cell growth.

5. *Can we apply axiomatic design theory and complexity theory in drug development?* The identification of modules is a key problem in biology. Armed with an increasingly comprehensive knowledge of system components, biologists can now create complex interaction models. For example, figure 9.7 depicts the set of protein interactions that transduce the signal when the EGF (epidermal growth factor) receptor is stimulated. The EGF receptor plays a key role in both normal development and cancer. This thorough depiction of circuitry can predict kinetic properties of the system but does little to suggest how this pathway could be pharmaceutically targeted in a way that would specifically inhibit its role in tumor progression without disrupting EGF receptor functions critical for normal cell activities (J.D. Thomas, private communication, 2002). The application of axiomatic design concepts to models such as this will decompose these complex interaction models into physiologically relevant functional modules that can be targeted using pharmaceutical agents.

The big question
There are many alarmists who worry that experimental approaches in biotechnology and biology may create a major disaster for humanity because of the potential of creating genes and mutations that will wipe out a large fraction of the animal kingdom. Certainly, it sometimes appears that these fields have the characteristics of a system with time-dependent combinatorial complexity. To stop this combinatorial process, we must reinitiate the system, which will not be easy. The combinatorial process might stop either when we have a major disaster (like a stock market crash) or if we develop basic scientific principles that can guide the future development of these fields. Wouldn't it be great if we could develop a basic principle of biology that is equivalent to the second law of thermodynamics so that people will not undertake foolish projects?

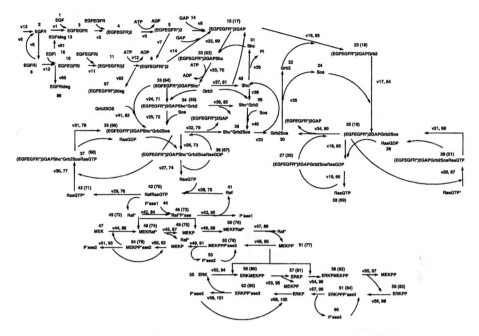

Figure 9.7 Scheme of the EGF receptor-induced MAP kinase cascade. The MAP kinase cascade can be initiated by Shc-dependent and Shc-independent pathways. Each component is identified by a specific number. Numbers in parentheses specify the components after internalization. The arrows represent the reactions and are characterized by reaction rates v1 and v125. The second number identifies reaction rates after internalization. (From Schoeberl et al., 2002.)

9.9 Summary

1. This chapter presented an outline of systems biology. One of the goals of systems biology is to relate the behavior of biological systems to the behavior of billions of molecules that make up the system. Based on axiomatic design theory and complexity theory, a V-model was presented to relate the functions of biological systems to their molecular behavior.

2. The master design matrix relates the highest-level FRs of a biological system to the lowest-level molecules in the system. When the master design matrix is developed for all organs and the exact relationship represented by the elements of the design matrix can be replaced by accurate mathematical models, it should be possible to predict the behavior of biological systems in terms of the behavior of the molecular interactions.

3. Biological systems are robust and can adjust to the changing environment, probably because they are decoupled designs with functional periodicity. Cells undergo periodic changes based on their functional periodicity. One may speculate that the cell cycle and other cyclic behaviors of biological systems have enabled them to survive through the long evolutionary process.

4. To understand the behavior of a biological system, it is necessary to consider all the relevant FRs at a given level and create a design matrix after relevant DPs are

identified. Investigating the relationship between one FR and one DP, in isolation from other FRs and DPs of the system, may not provide an understanding of the systems biology.

5. In view of the existence of time-independent imaginary complexity in all fields, we must be sure that we are not working on imaginary problems of biological systems.

6. The flow diagram is a useful means of assessing the system function and the consequences of changes introduced to a biological system. It represents the workings of a biological system, which is equivalent to an electric circuit diagram for an electrical system.

Notes

1. The fact that the diet may affect the cancer rate is given by the following statistics: According to *SEER Cancer Statistics Review 1975–2000* of the National Cancer Institute, in the U.S., the incidence of stomach cancer is 7.7 per 100,000 persons for White and 17.3 per 100,000 for Asian/Pacific Islanders. The mortality rate from colon and rectum cancer is 20.7 per 100,000 for White and 13.1 for Asian/Pacific Islanders. Apparently the stomach cancer rate in Asia is much greater than that for Asians in the U.S., while the colon cancer rate is Asia is much less than that of Asians in the U.S.

2. Recently, the author was delighted to read the work of Winfree (2001), who has documented the functional periodicity of neurons in his interesting book. See p. 413 for firing of electric charges in neurons.

3. This section is partly based on the paper presented by the author (2002). Readers may also refer to Thomas et al. (2004).

4. Similar V-models have been used in software design (see chapter 5 of Suh, 2001).

5. This decomposition was done in consultation with J.D. Thomas and Taesik Lee.

6. This is similar to design system matrix (DSM) used to determine how physical things should be arranged in a factory (see Ulrich and Eppinger, 1999).

7. This may not be the complete set; it was chosen for the purpose of illustration.

8. Reference materials from Alberts et al. (1994).

9. The construction of these matrices was done with input from Taesik Lee and J.D. Thomas.

10. Arp2/3 complex is a stable complex of seven subunits: two actin-related proteins—Arp2 and Arp3—with five noble proteins, p40 (ARPC1), p35 (ARPC2), p19 (ARPC3), p18 (ARPC4), and p14 (ARPC5).

11. Winfree (2001) shows many examples of circadian cycles of biological systems.

References

Alberts, B., Bray, D., Lewis, J., Roff, M., Roiberts, K., and Watson, J.D. 1994. *Molecular Biology of the Cell*, 3rd ed., Garland Publishing, New York.

Chae, C.-B. 2002. Private communication, Pohang University of Science and Technology, November.

Conlon, I.J., Dunn, G.A., Mudge, A.W., and Raff, M.C. 2001. "Extracellular control of cell size," *Nature Cell Biology*, Vol. 3, pp. 918–921.

Kohn, K.W. 1999. "Molecular interaction map of the mammalian cell cycle control and DNA repair systems," *Molecular Biology of the Cell*, Vol. 10, pp. 2703–2734.

Lipson, H., Pollack, J., and Suh, N.P. 2002. "On the origin of modular variation," *Evolution*, Vol. 56, pp. 1549–1556.

Mitchison, J.M. 1971. *The Biology of the Cell Cycle*, Cambridge University Press, London.

Perkel, D.H., Schulman, J.H., Bullock, T.H., Moore, C.P., and Segundo, J.P. 1964. "Pacemaker neurons: effects of regularly spaced synaptic input," *Science*, Vol. 145, pp. 61–63.

Schoeberl, B., Eichler-Johnson, C., Eiles, E.D., and Mueller, G. 2002. "Computational modeling of the dynamics of the MAP kinase cascade activities by surface and internalized EGF receptors," *Nature Biotechnology*, Vol. 20, pp. 370–375.

Schulman, J. 1969. "Information transfer across an inhibition to pacemaker synapse at the crayfish stretch receptor," Ph.D. dissertation, University of California at Los Angeles, Department of Zoology.

Suh, N.P. 1998. "Axiomatic design theory for systems," *Research in Engineering Design*, Vol. 10, pp. 189–209.

Suh, N.P. 1999. "A theory of complexity, periodicity, and design axioms," *Research in Engineering Design*, Vol. 11, pp. 116–131.

Suh, N.P. 2001. *Axiomatic Design: Advances and Applications*, Oxford University Press, New York, chap 9.

Suh, N.P. 2002. "System design perspective in understanding cellular complexity," *Towards Computational Models of a Mammalian Cell: The Neuron*, Conference, December. 6, New York Academy of Sciences.

Suh, N.P. and Lee, T. 2002. "Reduction of complexity of manufacturing systems through the creation of time-dependent periodic complexity from time-dependent combinatorial complexity," *Proceedings of the CIRP International Seminar on Manufacturing Systems*, Seoul, Korea, May (Keynote lecture).

Thomas, J.D., Lee, T., and Suh, N.P. 2004. "A function-based framework for understanding biological systems," *Annual Review of Biophysics and Biomolecular Structure*, Vol. 33, pp. 75–93

Ulrich, C.T. and Eppinger, S.D. 1999. *Product Design and Development*, McGraw-Hill, New York.

Winfree, A.T. 2001. *The Geometry of Biological Time*, 2nd ed., Springer-Verlag, New York.

Exercises

9.1. State the highest-level functional requirements (FRs) of the human heart and identify the corresponding design parameters (DPs). Write the design equation for the highest-level FRs and DPs. Is the heart a coupled system, a decoupled system, or an uncoupled system?

9.2. Explain the blood circulatory system of human beings in terms of a biological functional periodicity.

9.3. Identify the functional periodicity of cell division.

10
Complexity of Sociopolitical–Economic Issues

10.1 Introduction

Complexity in action ...

WASHINGTON, Nov. 22—A fiercely polarized House approved legislation on Saturday that would add prescription drug benefits to Medicare, after an all-night session and an extraordinary bout of Republican arm-twisting to muster a majority. The Senate opened its debate under threat of a filibuster.

Within hours of the House vote, the Senate began considering the Medicare legislation, aiming toward a final vote on Monday. Republican strategists had expected passage to be easier in the Senate.

But Senator Edward M. Kennedy of Massachusetts, decrying the House vote, said he would attempt to filibuster the Medicare bill. Mr. Kennedy conceded, though, that it would be an uphill fight given that the Democratic leadership does not support blocking a vote and other Democrats have indicated their intention to vote for the overall measure.

Mr. Bush, who lobbied hard for the Medicare bill and called several wavering House members from Air Force One, hailed the vote in the House and urged the Senate to follow its example.

New York Times, November 23, 2003

In this chapter, we will examine the complexity of socioeconomic–political issues confronted by a nation, a government agency, and universities. The cost of mismanaging socioeconomic–political issues is extremely large—a dysfunctional nation, wasteful government agencies, and generations of uninspired university graduates. The complexity begins when concerned parties cannot agree on the functional requirements (FRs)—the goals of the institution. Furthermore, some of the poorly managed institutions tend to spend enormous resources and effort on imaginary complexity problems and suffer from time-dependent combinatorial complexity. The importance of functional periodicity in assuring the stability and viability of our institutions can hardly be overemphasized.

Most people would acknowledge that sociopolitical–economic issues are quite complicated and complex. Why are they complex? Should they be complex?
Sociopolitical–economic issues may be grouped under three categories:

1. Protection from "fear" factors—healthcare, defense, natural disasters, man-made disasters, welfare, social security.
2. Responding to "greed" factors—economy, stock market, productivity, wealth accumulation, investment.
3. Fulfilling of human aspirations—scientific discoveries, invention and innovation, space exploration, freedom, liberty, human rights.

Governments are organized to manage these three basic issues of human society for everyone's ultimate benefit. Governments have adopted various means of dealing with these three basic issues. To deal with the "fear" factors, society supports hospitals, police, army, research on biology and medicine, earthquake mitigation, safety rules and procedures, welfare systems, and religion. To deal with the "greed" factors, society uses such instruments as reward for investment, protection of financial and capital markets, the right to own real estate, promotion of international trade, protection of intellectual property, and support for the value of the currency. Finally, society deals with human aspirations by supporting noble causes such as space exploration, scientific and technological explorations, and schools, and by protecting the right to be free. Some of these instruments affect all three issues.

These sociopolitical–economic issues are complex for three reasons: (1) society's inability to define the common goals, (2) the tendency not to design the policies to assure the desired future outcome, and (3) the uncertainty of the future outcome. Although we may be able to design more rational economic and political solutions to some societal problems, we have not been able to do so because of our inability to define the FRs that all constituents can accept, lack of explicit effort to systematically design policies and solutions that satisfy all FRs, and the uncoordinated operations of government systems. This results in inefficient resource allocation and utilization. The uncertainties associated with these issues increase with an increase in the number of organizations or individuals that can affect the outcome, making them more complex.

The complexity of sociopolitical–economic issues also increases with the increasing uncertainty of future events and circumstances. During the past fifty years alone, society has changed in totally unexpected ways because of many developments: new technologies, new drugs, tyrants, religious movements, scientific discoveries, migration of people, free trade, free flow of information across all boundaries, revolutions, natural disasters, wars, and terrorism. Thus people and information can no longer be managed and manipulated by only a few.

Management of complexity by government
The role of government is to manage and reduce the complexity in improving the quality of life of its citizens and in resolving the sociopolitical–economic issues that are related to fear, greed, and aspirations. An ideal government's role is to provide freedom, welfare, health, and prosperity to its people, as well as protection from external and internal adversaries. To achieve these goals, many types of government have been formed. In the United States, citizens delegate and entrust a few elected officials with power to govern through an electoral process. There are three branches of government—legislative, executive, and judicial—to provide checks and balances in running the country. Congress designs and enacts laws, which are implemented by the executive branch, which also sets general policies. The courts make certain that laws and policies are consistent with the overall architecture of the nation, the Constitution.

The effectiveness of government varies a great deal among different countries, depending on the political system, the shared culture of its people, the resources available, the general level of education, and in some cases, religion. The lack of certainty in achieving various policy goals of government results in unexpected consequences and problems. Uncertainty generates diverse opinions on societal goals and solutions. The greed factor further complicates the management of society. As a result of all these elements that contribute to uncertainties, there are many conflicting arguments on policies at all levels of government—federal, state, and local government—on economy, education, healthcare, taxation, welfare, and so on. Governing a nation or a group of people is a complex business.

The causes that are responsible for the complexity of governing a nation are many. It is difficult to define a set of FRs that a significant majority of people would accept and support. It is equally difficult to develop design parameters (DPs) that will not couple the chosen set of FRs. Even when policies are created and implemented for a set of clearly defined FRs and DPs, their original goals may be circumvented or subject to subterfuge through loopholes. Therefore, many government organizations and policies constitute highly coupled systems. Governments employ qualitative reasoning, empiricism, and optimization techniques based on trial and error to develop best approaches to dealing with their tasks. They often accept compromise solutions.

Why can't we solve these sociopolitical–economic problems more effectively?
Although there are always many ideas and various courses of action proposed by many groups of interested people, some problems seem to be impervious to any policies that emanate from these deliberations. For instance, take the case of healthcare. Many nations are trying to provide good healthcare and medical services for their people, but they are beset by the escalating cost of healthcare and the deteriorating quality of medical services. The increasing number of elderly people is compounding this healthcare problem. In many advanced countries, there are enough doctors who want to practice and there are patients who need their services. Yet it is not working. Why? What are the root causes of the healthcare problem and what are the solutions? Probably the poor design of the healthcare system creates both time-independent and time-dependent complexities. The healthcare system in many countries may have combinatorial complexity. Can we make it effective and manageable by introducing functional periodicity?

Take the issue of jobs. Throughout the world, many people—both well educated and those with less education—cannot get jobs, although many are eager to work. In some ways, the whole job situation appears to be irrational. For example, if unemployed people are hired to produce goods and services, they will in turn purchase goods and services made by others, which should in turn increase the overall economic activity. To a noneconomist, it seems that all we have to do is to initiate the process and create a chain reaction by producing goods and services that people want to purchase.

How do we initiate the economic activity so as to facilitate this beneficial chain reaction? The problem may be traced to the complexity of the economic system. The first question we need to answer is: "What are the FRs of an economic system?" Uncertainty in the expected outcome of an economic policy (i.e., the FRs) would make the economic system complex. The fact that when an economy begins to deteriorate, it continues to get much worse before it gets better, may indicate that it has the characteristics of time-dependent combinatorial complexity.[1] If this is the case, we need to intentionally introduce a functional periodicity to the economic system to reduce time-dependent complexity, in addition to eliminating time-independent real and imaginary complexities, rather than depending on a random event to create the functional periodicity.

There is also the matter of basic needs to sustain lives. There are many children on Earth not developing to their full potential because of malnutrition and starvation. The problem is not caused by inability to produce agricultural goods. It is caused by the complexity of the global economic system. The people who need the food must give something of value to the people who are producing the agricultural products and to those who transport the foodstuffs to them, but the process cannot be initiated because the poor people have nothing to give. Therefore, an economic cycle cannot exist unless the cycle is initiated by making an investment in these starved regions to create jobs. The time-dependent combinatorial complexity of this economic situation is sustained by poor people getting poorer for a variety of reasons—many children, no education, political instability, and so on. To solve this problem, we need to introduce functional periodicity!

In addition to the sociopolitical–economic issues discussed thus far, potentially one of the most destabilizing long-term societal issues may be wealth distribution. Some economists say that society will increasingly be stratified into two classes, depending on wealth and education: those who have and those who do not. Those who can create new technologies will be able amass a fortune by being able to reach the worldwide market.

Those who cannot participate in this technology-based society will be left behind, not benefiting from the new wealth generated by a technology-driven economy. The result will be a greater gap between the rich and the poor at a national and international level, which invites political instability. Unless this problem is managed well, it may introduce additional complexity in solving sociopolitical–economic problems and destabilize the political system. This may eventually lead to an abrupt and costly reinitialization.

Some sociopolitical–economic issues involve risk. What is risk? Do complex issues have a greater risk? How do we reduce risk and complexity?
When there is uncertainty in the outcome, the task appears to be complex and there is a *risk* associated with undertaking the task since there is a finite probability that the goal will not be achieved. Risk may be defined as the monetary or nonmonetary cost of not achieving the FRs, whereas complexity is a measure of uncertainty in achieving the FRs. For example, when someone buys a stock of a company hoping to get a large return on the investment, the FR is "make money through investment," but there is a finite uncertainty that the investor will incur loss rather than gain when the stock does not appreciate as hoped. The possible loss of some of the investment represents the risk (Shiller, 2003). The uncertainty of not achieving the original goal of making money is measured by complexity, since complexity is a measure of uncertainty in achieving the FR.

The common method adopted in the financial world to reduce risk is to spread the risk over many people. This takes the form of insurance policies, hedge funds, mutual funds, and venture capital. However, complexity, unlike risk, cannot be reduced by spreading it over many people. Even risk cannot be completely removed by spreading it; for instance, mutual funds lost many investors' money during the downturn of the stock market and also some mutual funds short-changed their investors through corrupt practices. Furthermore, although individual risk may be reduced, the risk to society at large may not be reduced. In fact, it may increase if society does not carefully regulate the limit of risk that each institution can take relative to its assets. Regulation is a means of controlling the time-dependent combinatorial complexity of a large number of people and organizations.

Does the risk-mitigation argument hold true in other sociopolitical–economic arenas?
Uncertainty, risk, and complexity in the financial arena are reasonably explicit and thus easy to understand. How about in other fields—politics, economy, education, and health-care? Can we reduce the risk, and thus the complexity, of these nonfinancial fields by spreading the risk?

As we think of means of decreasing the complexity of sociopolitical–economic issues, there are many questions that come to mind. One of the major difficulties in dealing with societal issues is in defining FRs that everyone shares. For example, "What are the FRs of an educational system?" There are also questions related to complexity and risk that we need to answer, such as: "Is the conventional way of reducing complexity and risk by spreading the risk the best way?" "What is the time-dependent real and imaginary complexity in the sociopolitical–economic context?" "What is the time-dependent complexity in these fields?" "How can we reduce the complexities of a healthcare system?" However, many of these questions cannot be answered until we can define the FRs of a given task, which is often the most difficult part of attempting to reduce the complexity of these nontechnical fields.

In chapter 3, the concept of functional periodicity that applies to socioeconomic political systems was introduced. The functional periodicity that can reduce the complexity of socioeconomic–political fields was stated as:

1. Organizational functional periodicity.
2. Economic functional periodicity.
3. Political functional periodicity.
4. Academic functional periodicity.

In this chapter, the experience of the author in dealing with these nontechnical issues will be described.

10.2 Theorems Related to Sociopolitical–Economic Systems

A sociopolitical–economic system may be characterized as a system with many more DPs than FRs—equivalent to a system with many interconnecting physical parts in the physical domain. Often the most difficult aspect of the nontechnical system is agreeing on and accepting the societal goals, that is, the FRs. The difficulty is that many people can agree on one FR, for example, "Minimize the public expenditure by government." However, government must deal with many FRs at the same time. The difficulty is in agreeing on a set of FRs from a long list of desirable FRs. Even after one of the FRs is selected, there can be too many different solutions (DPs) that can affect the outcome for each FR. Every agent who controls one of these DPs can affect the successful outcome of the FR. For instance, even when the President of the United States issues an executive order to the executive branch of the government, there is no assurance that the executive order can achieve its FRs. Some of the bureaucrats can take actions that increase the uncertainty in achieving the original goals, that is, the FRs.

In chapters 2 and 3, we presented a number of theorems that deal with systems with many interconnecting parts. Social systems epitomize systems in which many constituents (i.e., the interconnecting parts) may have means of affecting the outcome. In example 4.1, a simple example of a social system (using an academic department for the purpose of illustration) was presented to indicate how a social system may be managed. Additional theorems are stated here that are particularly pertinent to sociopolitical–economic systems:

THEOREM C8 (Complexity of Sociopolitical–Economic Systems)

The complexity of sociopolitical–economic systems increases with the number of entities (i.e., organizations or individuals) that can affect the ultimate outcome.

THEOREM C9 (Reduction of Complexity of Sociopolitical–Economic Systems)

If all the constituents of a social system can agree on the common set of FRs and if the FRs can be satisfied independently, the complexity of the decision-making process can be reduced when the final decision is made by a single entity after understanding and taking into account the uncertainties introduced by other constituents of the system.

> **THEOREM C10 (Reduction of Complexity of a Sociopolitical–Economic System through Reinitialization or Redesign)**
>
> When a sociopolitical–economic system is moving into a chaotic state because of time-dependent combinatorial complexity, the system should be reinitialized or redesigned to reduce complexity.

In the subsequent sections of this chapter, the application of complexity theory to sociopolitical–economic issues will be presented for the purpose of reducing the complexity and risk that is associated with not satisfying the FRs. The examples presented in this chapter are based on the author's personal experience and therefore may be somewhat subjective and biased.

10.3 Complexity of Economic Systems

10.3.1 Design and complexity of an economic system

At the beginning of the twenty-first century, the economy of the entire world seemed to have lost its growth engine, which made many people pessimistic about their future. This was a major change of events from the roaring 1990s, in just one decade. These economic events are cyclic, so in another ten years these people may again become optimists. These complex problems have always existed but they are more glaringly apparent now than ever.

The economic cycle is a well-known phenomenon. Often new economic opportunities created by a new technology (railroads, automobiles, computers, etc.) or some "craze" such as tulips in the Netherlands may initiate a boom in economic activity. When a new technology is created, initially real wealth is created when new jobs and new products spur economic activities and promote productivity and growth. However, when these things turn into a speculative spiral, productivity cannot be sustained, since speculation is not a real wealth-creating activity. The investment made with wealth tied to speculative assets that feed the "greed phenomenon" will eventually collapse, bringing the entire economic activity to a halt. This occurs when the investment in a technology exceeds the potential wealth that can be generated by that technology. When these economic excesses are moderate, the economic collapse is limited both in duration and magnitude. When the speculative bubble collapses, a "new" economic cycle is initiated by "reinitializing" the system, which is typically accompanied by unemployment and recession.

The economic cycle described above is a good example of time-dependent combinatorial complexity at work. The outstanding issue is whether or not a functional periodicity can be introduced to modulate economic activity so as to prevent the "boom to bust" cycle. As of now, the economic system has no automatic means of introducing a functional periodicity before the system becomes unstable. The questions are: "What kinds of functional periodicity can we introduce to reinitialize the system on a periodic basis?" and "When should they be introduced?"

The FRs of an economic system of a nation may be stated as:

FR_1 = Engage the people in productive activities.
FR_2 = Produce sufficient goods and services to satisfy demand.

FR_3 = Invest in the infrastructure to improve the quality of life.
FR_4 = Protect the economic system.
FR_5 = Increase productivity so as to create wealth.
FR_6 = Invest in education and intellectual activities.
FR_7 = Maintain financial instruments.
FR_8 = Regulate the circulation of currency.
FR_9 = Form public capital to pay for collective economic activities.

The DPs that can satisfy the above FRs may be stated as:

DP_1 = Job creation.
DP_2 = Production and service capacity.
DP_3 = Civil construction program.
DP_4 = Defense budget.
DP_5 = Technology.
DP_6 = Educational and R&D systems.
DP_7 = Investment/banking/trade systems.
DP_8 = Monetary policy.
DP_9 = Taxation system.

The design matrix may be given as shown in table 10.1. This table shows that the economic system can be a decoupled design. The taxation system has to be set first before other things are controlled. It also shows that monetary policy can be set at the same time as educational and R&D systems. DP_7 (Investment/banking/trade systems) has to be controlled after the monetary policy is set. The rest of the DPs (i.e., DP_6 through DP_1) must be controlled in the sequence given (DP_6 first, followed by DP_5, and so on) after DP_9 (Taxation system), DP_8 (Monetary policy), and DP_7 (Investment/banking/trade systems) are fixed. The decoupled design matrix of the economic system shows that when taxation or monetary policy is changed, other DPs must be changed to satisfy all FRs.

What can ruin the economic system is either over- or underinvestment that creates time-dependent combinatorial complexity. The consequence of underinvestment is

Table 10.1 Design matrix for an economic system

	DP_1	DP_2	DP_3	DP_4	DP_5	DP_6	DP_7	DP_8	DP_9
FR_1	X	X	X	0	X	x	X	X	X
FR_2	0	X	X	x	X	0	X	X	X
FR_3	0	0	X	0	x	0	0	X	X
FR_4	0	0	0	X	X	X	0	0	X
FR_5	0	0	0	0	X	X	0	0	X
FR_6	0	0	0	0	0	X	0	0	X
FR_7	0	0	0	0	0	0	X	X	X
FR_8	0	0	0	0	0	0	0	X	X
FR_9	0	0	0	0	0	0	0	0	X

easy to understand. If investment is less than what it takes to keep up with population increase or if it cannot generate the growth needed to meet the aspirations of the people, the economic system will deteriorate. On the other hand, if investment is greater than the system can absorb through productivity growth and real wealth generation, it will lead to inflation. Control of the rate of investment is done by means of taxation and monetary policy. Monetary policy controls the short-term circulation of currency and the allocation of investment funds between equities and bonds. One of the difficulties of monetary policy is that the timing of interest rate changes is not an exact science.

Risks and complexity associated with economic policy
The economic system designed above appears to be reasonable since it is a decoupled design and therefore we can satisfy the FRs if we change the DPs in the sequence given. The only other condition we have to satisfy is that the system ranges of FRs remain inside their design ranges. There is a finite uncertainty of not satisfying the FRs associated with the economic system when the system ranges of the FRs are not completely inside the design ranges of FRs, that is, the economic system may not achieve the desired FRs. Therefore, because the system range is not always inside the design range, which may occur for a variety of reasons, the economic system has real complexity.

Until the design equation for an economic system has been determined, the complexity can be great since a random variation of different DPs will not yield the desired result within the time available for adjustment of economic activities. Since the FRs and the design matrix for a nation's economy are never stated explicitly by policy-makers and since economic policy is often based on past experience (empirically, not systematically or rationally), economic policies have sometimes generated unexpected negative results. For every economist who advocates tax cuts, we may find another economist who advocates just the opposite. Only when the economic system is completely uncoupled (i.e., diagonal design matrix), a very unlikely situation, can government change any one of the FRs without affecting other FRs. Otherwise, every decision may have unintended consequences, which require additional remedies. The formulation of economic policies based on experience alone increases the complexity of and the risk to the economic system.

In the following section, an application of the complexity theory presented in this book to economic policy is described. In this case study, the use of economic functional periodicity and resetting of FRs to improve the Korean economy in 1980 are described. By any measure, it was a successful application of the complexity theory, although it was done without the explicit use of the theory presented in this book. The axiomatic design theory, which is the basis of the complexity theory, might have influenced the design of the economic policy.

10.3.2 Five-Year Economic Development Plan of the Republic of Korea, 1980–1985

Introduction
In this section, two things will be illustrated: how the economy of a nation can become a time-dependent combinatorial complexity system (Theorem C8) and how it can be reset by reinitializing the economic system (Theorem C10).[2]

The case involves the Republic of Korea, which has become an industrialized nation in a span of four decades. It goes back to 1980 when Korea's economy was stalled and investment in new industries was not progressing as well as planned. The Five-Year Economic Development Plan of 1980–1985 played a significant role in restructuring (reinitializing) industry and in developing the economy of Korea. Today Korea has become a major industrialized nation.

Brief review of the modern history of the Republic of Korea
Korea was under Japanese occupation for 36 years until 1945 when the Allied forces defeated Japan. Unfortunately, Joseph Stalin, the leader of the Soviet Union, convinced U.S. President Franklin D. Roosevelt to divide Korea into two units with a border along the 38th parallel (of latitude) as a reward for the Soviet Union's attack on the Japanese military in Manchuria. This one decision cost a few million Korean lives and a large number of casualties in the U.S. Armed Forces six years later when the Korean War was started by the communists in North Korea.

From 1945 until 1948, when the Republic of Korea was established, South Korea was under U.S. military rule. Then in 1950, North Korea invaded South Korea and within three days occupied Seoul, the capital city of the Republic. Many South Koreans suffered under the communist occupation, which led to many of them becoming strong anti-communists. The Korean War ended in 1953 in an armistice, South and North still being divided along a Demilitarized Zone, nearly parallel to the original border between the North and the South. In 1950, both South and North Korea were economically poor, and they continued to become poorer after the war.[3]

In 1961, South Korea changed dramatically from a fledgling democracy to a military dictatorship when Major General Junghi Park led a military coup. His timing for the coup was good. The demonstrations by students toppled the government of the first president of Korea, Syngman Rhee, and the new government that was installed through a free election was quite ineffective. After a few years of military rule, General Park was elected to the presidency of Korea.

Today, most Koreans would attribute the birth of modern industrialized Korea to President Park. He was a dictator, but also one of the few uncorrupt presidents of Korea. As president, he did several important but controversial things. First, he engineered capital formation to finance industrialization by concentrating the capital of the nation in a few businessmen through low-cost loans, indirect subsidies, and other privileges.[4] At that time, Korea as a nation was too poor to obtain external financing for his ambitious projects. Second, he built the public infrastructure such as highways, power plants, and telecommunications systems. Third, he built basic industries such as steel making with public money. This was done at the expense of human rights, which has tarnished his accomplishments.

At the beginning of the industrialization, Korea invested primarily in labor-intensive manufacturing industries, such as textile mills, shoe factories, and apparel factories, for Korea did not possess any technologies and lacked trained and experienced industrial leaders and workers at that time. The economic policy was extremely successful. In the 1960s and 1970s, the Korean economy grew at a rate of 10% to 13% a year, one of the fastest-growing economies in the world.

Based on this success in industrialization, the Korean government decided to invest in heavy-machinery industries to transform the Korean industry from labor-intensive businesses to high value-added businesses beginning in the early 1970s. They invested heavily in capital-intensive manufacturing businesses, such as automobiles, earthmoving equipment, diesel engines, shipbuilding, machinery, power plant equipment, and chemicals. In some cases, even with the preferential treatment given by the government, individual industrial firms could not undertake major projects such as steel making. For these large projects, the government created government-owned firms such as the steel-making firm, Pohang Iron and Steel Company. The Korean government also owned and operated telecommunications, which they monopolized until recently. President Park also invested heavily in public infrastructure projects: highways, telephone systems, and banking. Although there were many problems that accompanied the government-inspired and controlled industrialization, forty years later Korea has become a major world manufacturer of automobiles, shipbuilding, steel making, and, following these, semiconductors.

The road to industrialization of Korea in capital-intensive business has had a few bumps, creating time-dependent combinatorial complexity, which had to be transformed into time-dependent periodic complexity. It happened in 1980, when Korea was facing serious economic and political problems as a result of indiscriminate and excessive investment in capital-intensive business in the 1970s. Major changes were introduced, which had a profound impact on the subsequent development of Korea.[5]

The purpose of the story is to illustrate how an economic system with time-dependent combinatorial complexity can be created when it lacks sound business practices and free-market forces. It will also demonstrate how the economic system can be reinitialized to create functional periodicity.

Background on the development of the Five-Year Economic Plan for 1980–1985
In the early 1970s, the Korean government decided to invest in automobile and heavy industries to transform the economy that had depended on labor-intensive business to high value-added business. It was an ambitious program. To achieve this goal, the government designated special industrial sites and created infrastructure. It guaranteed loans secured from overseas by large conglomerates to finance new machines and plants and also provided low-cost loans from government-controlled banks. Anyone who could obtain the government guarantee went out to build massive plants. Politicians, bureaucrats, and bankers were happy to collaborate with these companies. (It is conventional wisdom that these politicians and bureaucrats were beneficiaries of the unwarranted largesse provided by these companies.) Foreign machine manufacturers were happy to fill the orders. Foreign banks were willing to provide loans because the government guaranteed these loans. The borrowers and lenders were all happy at the beginning. It was how an economic system with time-dependent combinatorial complexity was born.

Although there were many problems, this decision to invest in automotive and heavy-machinery industries was the turning point in the industrialization of Korea. That decision has transformed Korea into an advanced industrial nation. However, the road to industrialization was not easy; it had a few major problems along the way—Korea had to overcome the problems created by time-dependent combinatorial complexity. In retrospect, it appears that if Korea had introduced economic functional periodicity when it launched the industrialization project, some of these problems could have been avoided or minimized.

The excesses of the 1970s economic policy became apparent when, in 1979, President Park was assassinated by his intelligence chief, and a new military junta under the leadership of General Doo-Hwan Chun took control of the government. When the generals took over power (by brute force), they found that Korean industry was in deep recession and faced dire consequences. One of the major causes was excessive investment made by various conglomerates, which had borrowed money from overseas under the guarantees provided by the Korean government. In addition to commercial banks, the World Bank also invested in Korean development projects. With the borrowed money from these banks, big conglomerates were building huge shipbuilding facilities for supertankers, plants for gas turbines, plants for large diesel engines for ships and power plants, chemical plants, automobile plants, and machine tools. Unfortunately, it took much longer than anticipated to generate profits from these new investments.

Since they were not generating any profit from these new investments, many conglomerates had to create additional new businesses and projects to maintain positive cash flow by securing new loans for the newly launched projects. Huge conglomerates were born with the borrowed money. Unfortunately, as these conglomerates invested more and more in new projects, their debt grew faster than their assets. Some of them survived and grew by maintaining a positive cash flow, not by making a profit. This economic and political process—the spiral of rampant investment made by many conglomerates, the economic system that allowed it, and the political corruption that siphoned away resources from the projects—created an economic system with time-dependent combinatorial complexity. This system worked as long as they could borrow money or secure new businesses or projects. When the economy of the world hit a downturn in 1979–1980, the Korean economy rapidly deteriorated and many of the conglomerates became insolvent and had to be bailed out by the government.

Mission for the World Bank
In 1980, at the request of the World Bank, the author visited Korea to evaluate the status of the industrialization.[6] Many of these new plants—some of which were built on reclaimed land—were operating at perhaps 20% capacity and 20% efficiency. These plants had brand-new machines, some of which were not yet completely installed. Most plants were idle. Not many products were being manufactured. At first, it seemed that these plants would never operate at their rated capacity and produce profit. After a month (June 9–July 3, 1980) of intensive study of the issues and gathering data from various companies and government agencies, a report was produced (Suh, 1980), which was adopted as the Five-Year Economic Development Plan of the Republic of Korea for the period 1980–1985.

The 1980–1985 Five-Year Economic Development Plan reinitialized (i.e. creation of a new functional period) the Korean industrialization program by terminating the activities that had caused time-dependent combinatorial complexity. The economy of Korea grew very rapidly from 1980 on, correcting some of the structural problems by realigning the industrial effort. This may be contrasted to the stagnation of the Japanese economy since 1990. Many attribute Japan's problem to its reluctance to reinitialize its industry and financial sector.

A summary of the recommendations made to the Republic of Korea
Based on the key findings from the month-long review of Korean industries, the following recommendations were made to the Korean government:

1. Give the highest priority to the automotive business by restructuring the industry through a merger of the three passenger car manufacturers into one.

2. Reorient the power-plant equipment business by restructuring the investment made in the business.
3. Invest more money in shipbuilding to complete the projects and make them viable.
4. Make machine-tool companies specialize in specific products and adopt modern manufacturing systems.
5. Support small- and medium-sized industrial firms to create an industrial infrastructure.
6. Spend more on education and R&D.

The details of the findings and recommendations are given in the next section.

10.3.3 Summary of the assessment made in the 1980 report

10.3.3.1 General recommendations made in 1980

"The machinery industries[7] in Korea have now reached a critical and difficult juncture in their development. Some hard decisions must be made by the government to resolve some of the critical issues. The quality of these decisions may irreversibly shape the future of the machinery industries. The stakes involved for Korea make it imperative that rational policies be established without being over-constrained by the current realities of the Korean [economic] scene. In this sense, the current economic recession might provide an opportunity to induce necessary changes and adjustments by devising reasonable financial incentives to the industrial firms in the machinery sector" (Suh, 1980).

10.3.3.2 Recommendations for each sector for reinitialization in 1980

Highest priority to automobile industry
It was concluded that among all the industries examined, the automobile industry should be given the highest priority for the following reasons:

1. Korea has the domestic market size to support the automobile industry because of its large population (over 40 million). Once the domestic market can support the industry, it can provide the basis for creating the export market, since only about a dozen countries will be producing cars for the entire world.
2. Furthermore, an automobile is a consumer product and therefore Korea can compete in the world market even if it does not possess the best technology.
3. Korea has an industrious, resourceful, and determined labor force that can produce cars.
4. Korea needs automobiles to provide transportation for its people and goods.
5. The automotive industry is not a very energy-intensive industry.
6. There is a large potential market outside of Korea.

Recommended policies for the creation of a competitive automotive industry were:

1. Combine the three automotive companies that are producing passenger cars into one[8] until economy of scale is reached.

(a) To be competitive by having economy of scale, at least 300,000 to 400,000 cars must be produced annually (i.e., the design range is greater than 400,000 cars per year).

(b) In 1980, Korea produced a total of about 125,000 cars a year by three competing automotive companies (i.e., the average system range for each company was less than 42,000 cars per year).[9]

2. One of the remaining two companies should manufacture only buses and the other only trucks.

3. Create policies and an environment to help increase the domestic demand for cars of a given basic design to 400,000 cars per year from the current level of 125,000 cars per year within the next five years.

4. Undertake massive R&D projects to develop fuel-efficient cars that can meet the U.S. emission standard within the next four years.

5. Seek foreign licenses for manufacture of advanced internal-combustion engines.

6. Once the domestic market can support a viable automobile industry, open the domestic market systematically and gradually so as to insure the competitiveness of the Korean firms in the world market.

7. Develop the vendor community (i.e., small- to medium-sized industrial firms) that must supply the parts.

Heavy industries for power generation and heavy equipment in 1980
The overall findings were:

"Korea has made a major investment in the heavy machinery industry since the mid-1970s.... The industry suffers from overcapacity and fragmentation of the domestic market among several firms. The overcapacity has resulted in underutilization of equipment, inefficient job-shop type production operations, and ultimately, higher costs. The fragmentation has resulted in the dispersion of engineering talents with a consequent loss of opportunities to accumulate technological capability. The degree of overcapacity in equipment can be discerned from the fact that even if the entire hardware order for future domestic power plants is given to Changwon Plant of Hyundai/Hyundai International, it can utilize the facility for only 50% of the time. A large-scale export of power plants will be difficult until the Korean firms have accumulated sufficient experience in building domestic power plants. The Changwon plant may have enough business if Korea can get 50% of the 4 million kW/year of power plant business available in 14 developing nations of the world. However, this is a very unlikely prospect" (Suh, 1980).

The recommended policies were:

1. Do not grant any more licenses for power-plant manufacturing facilities.

2. Establish "National Heavy Machinery Plants" by taking over the Changwon facilities of Hyundai/Hyundai International and sell management rights to the highest bidder for a fixed period of time. Management rights may be granted for the use of each individual plant. Since the government owns the plants, technology accumulation will largely remain with the plant even though the management team changes.

3. Award construction and engineering contracts separately from the hardware-manufacturing contract, with the stipulation that parts that can be manufactured by the National Heavy Machinery Plants must be manufactured in Korea.

This restructuring was actually achieved by letting Korea Electric Power Company operate the facility, which has now become a private-sector company. This form of reinitialization was necessary since the investment made could not have been recouped in any other way.

Shipbuilding industry in 1980

The assessment and recommendations for the shipbuilding industry were:

"The shipbuilding industry is unique in that it is the only industry of those examined which is primarily oriented for export business. The industry has reached an internationally competitive position in a relatively short period of time. It has capitalized on the availability of skilled labor at low cost. Today the industry is on the verge of transition from construction of technically less sophisticated hulls and superstructures to the manufacture of items with greater added values such as equipment and power plant for ships.

One of the important characteristics of the industry is that it is a cyclic business. In order to operate a shipyard profitably throughout the boom and bust cycle of the industry, many shipyards throughout the world also construct other metal structures such as plant equipment.

Korea has expanded its shipbuilding capacity a great deal in recent years. When they are fully operational, Korea will have one of the largest shipbuilding capacities in the world. One of the largest facilities is being completed at Okpo[10] by the Daewoo Group.

The financial structure of the Okpo shipbuilding complex, according to the information provided by Daewoo, is such that the shipyard cannot be financially solvent with the facilities to be completed this year. Daewoo needs additional business to make the shipyard financially viable, which will require additional investment. A similar situation currently prevails at Samsung.

The difficulties faced by the Okpo shipyard illustrate the problem created when a government tries to save an investment made in a bankrupt firm without financially restructuring the firm. Obviously there are many reasons why the government prefers this mode of operation that are not apparent to outsiders. However, this mode of operation distorts the relationship between the government and major business groups, creating abnormal constraints on both.

One possible solution for the Okpo shipyard and the Changwon facilities would be to combine the two operations under one corporate umbrella. For this to be possible, one of the involved firms (Hyundai and Daewoo) should sell their automotive operation to the other and acquire the combined facilities at Okpo and Changwon.

As an overall policy of the government, it is recommended that the shipbuilding industry be left alone to the forces of the marketplace. This will strengthen the competitive position of the firms in the business on a long-term basis" (Suh, 1980).

Since then, Korea has become the world's largest shipbuilding country.

Machine tool industry in 1980

"The machine tool industry is a highly specialized and competitive business that has to cater its products to professional users, who demand reliability, accuracy, and after-sale service. It requires a strong technological base, especially when new equipment is designed and manufactured. The total market for machine tools is relatively small and the business is cyclic. ... These characteristics of the machine tool industry suggest that the Korean machine tool builders may be competitive in the world market by selling only relatively simple, general-purpose machine tools for the next five years.

The comparative advantage of Korea in the machine tool business lies in the availability of young skilled workers who will be experienced tool-and-die makers in five to ten years. Their talents coupled with relatively low labor costs can be a formidable force in the world where an increasingly larger number of young people shy away from that type of industrial employment.

There are weaknesses in the Korean machine tool industry which must be addressed. It needs more engineers with practical know-how and also engineers with a deep theoretical understanding in all sub-disciplines of machine tool technology.

The world of machine tool business may undergo a drastic change during the coming decade due to the use of computers and new sensing devices. For the Korean machine tool industry to be competitive in this changing world, it must undertake significant R&D. It must digest the available technology, assimilate new knowledge and technology which is being advanced within and outside of the machine tool industry for incorporation in their new generation of machine tools, and innovate new processes and techniques" (Suh, 1980).

The recommended policy objectives were:

1. Specialization and strengthening of the machine tool manufacturers through mergers.
2. Establishment of a university-based (or research institute-based) R&D team in the field of machine tools with significant government support.
3. Continuing educational programs for practicing engineers.
4. Strengthening of government/university/industry/research institute relationship for coordinated R&D, better-focused investment plan, and improved manpower training.

Korean machine tool manufacturers have specialized in a few areas, and today Korea is a net exporter of machine tools. However, Korean machine tool manufacturers still cannot compete with those in Japan, Germany, and the United States. It is a relatively small industry.

Small business in 1980
The importance of small business is summarized below:

"All industrialized countries depend heavily on small business as a source of innovation and employment. In discrete-parts manufacturing, small businesses can be particularly innovative, since their survival hinges very much on their adaptability and the unique products they can offer. Many firms in discrete-parts manufacturing are often not innovative since they rely on capital-intensive mass-production items to sustain their business. Furthermore, small- and medium-sized firms collectively employ more people than do large business firms. Therefore, it is important to nurture small business in industrialized countries.

Small business is important to large firms. Small- and medium-sized firms are often vendors to large industrial companies. Consequently, the quality of products manufactured by major industries such as the automotive industry depends on the technology possessed by small business firms.

Small- and medium-sized companies in Korea do not possess three essential elements for survival: financial resources, technological assistance, and the power to influence the governmental decision-making process (i.e., clout)..." (Suh, 1980).

It was concluded that the government should consider the establishment of the following policies:

1. Make long-term financing available at low interest rates.
2. Establish government-sponsored regional "Engineering Experimental Stations" in major industrial regions of the country with the specific goal of assisting these firms with technological problems.
3. Require the prime contractors of all major projects awarded by the government to subcontract a fraction of the work to other firms.
4. Simplify the government approval procedure.

10.3.3.3 Concluding remarks on the Five-Year Economic Development Plan of Korea

This case study illustrates how a national economy can become a victim of time-dependent combinatorial complexity. Unfortunately, in many nations, the combinatorial complexity cannot be terminated without a major national crisis. The proposed economic plan, based on a short-term study of the status of Korean industry in 1980, had two components: how to stop the combinatorial process and how to reinitialize the system. The system was slightly readjusted, but the original plan of building up capital-intensive business as a national goal was not altered. In that sense, the system was simply "reinitiated" by the adoption of the Five-Year Economic Plan.

Beginning in the late 1980s, Korea had allowed another combinatorial process—financial and real estate speculation—to grow unstably through ineffective policies. In 1998, the time-dependent combinatorial complexity ultimately bankrupted the country — the so-called IMF crisis—when Korea could not service its foreign debt with the available foreign exchange. The crisis resulted in a major collapse of financial institutions and industrial firms. Since then, the government of the Republic of Korea has partially reinitiated the economic system. The government, and thus the people, took on the debt created by the insolvent industrial firms and banks. The system is still in transition and may require another reinitialization, since the labor cost increase relative to the productivity increase is spiraling out of control.

It appears that many countries should review their economic policies in light of the complexity theory presented in this book. It is clear that many countries, including both industrialized nations and developing nations, are suffering from time-dependent combinatorial complexity. For example, since the 1990s, Japan's economy has not grown because of a process that appears to have the characteristics of time-dependent combinatorial complexity. Japan could cope with this lack of growth because its population has not grown. Japan will grow again if it reinitializes its economy.

10.4 Complexity in Government's Role in Science and Technology

10.4.1 Introduction

This section analyzes the complexity of a small segment of the U.S. Government, namely, the Engineering Directorate of the National Science Foundation. It shows how the organization was redesigned and reinitialized to fulfill the goals of an organization that had been besieged by time-dependent combinatorial complexity problems (see Theorem C10). This case study provides an example of the use of organizational functional periodicity and redesign to improve the performance of a government agency.

National Science Foundation Act of 1950
In 1941, President Franklin D. Roosevelt established the Office of Scientific Research and Development (OSRD) and placed it under the leadership of Dr. Vannevar Bush, who coordinated the research and development activities of the federal government for the war effort. OSRD mobilized the talents at universities to conduct research and development so as to support the war effort. It achieved spectacular results. Bush created a research structure that enabled researchers to conduct research "under conditions as closely akin to normal, peacetime conditions as possible." Only when the senior management of OSRD determined that a project had been completed would they turn the new-found technology over to the military.

On May 10, 1950, the National Science Foundation (NSF) was established through the National Science Foundation Act of 1950, which was signed into law by President Harry S. Truman. The creation of NSF was the culmination of the document produced by Vannevar Bush, entitled "Science—the Endless Frontier." This document was requested by President Franklin D. Roosevelt on November 17, 1944, and transmitted to President Truman on July 5, 1945. In this document, Bush advocated active government support of research at universities to continue the major role universities had played for the war effort during the Second World War. Today much credit is given to Vannevar Bush for creating the infrastructure for support of the modern research university in the United States.

In "Science—the Endless Frontier," Bush outlined his vision for the post-Second World War infrastructure for university-based research. It was a treatise composed during the final months of the war. Bush proposed that the U.S. Government provide financial support to universities for their self-directed, individualized peacetime pursuits. His vision was to have universities and government collaborate in their research efforts. He envisioned the National Science Foundation becoming the federal agency with control over all federally supported research, military as well as civilian. However, because of the Korean War of 1950 and the long delay in creating the National Science Foundation, the scope and funding of NSF was reduced when it was finally established in 1950.

Functional requirements of NSF
The National Science Foundation Act of 1950, as amended, states the purpose of creating the Foundation as follows: "To promote the progress of science and engineering; to advance the health, prosperity, and welfare; to secure the national defense; and for other purposes." This act, which created NSF, provides the legal *raison d'être* for the Foundation. All lower-level FRs of the Foundation must be consistent with and derivable from this law.

Complexity of NSF
NSF has become the principal government agency that supports research in physical sciences and engineering.[11] NSF is the lead government agency in formulating many of the research- and education-oriented policies in science and engineering.[12] According to the law, the director of the National Science Foundation reports to the President of the United States, although few NSF directors have had direct access to the Oval Office. Even the President's Science Advisor, who is also the director of the Office of Science and Technology Policy in the White House, has had very limited access to the President. They normally derive their clout through their access to the senior staff members of the White House, who are close to the Oval Office.

The National Science Foundation, like many government agencies, is a relatively complex organization to administer. First of all, NSF must satisfy many different constituents—Congress, OMB, universities, professional societies, researchers—with diverse viewpoints and needs. It is a system with many interconnected parts. Second, the budget of NSF is much less than is needed to support all worthy research projects and do its job adequatelyly.[13] Third, the annual budgeting process is long and uncertain in terms of the ultimate outcome, and therefore it is difficult to plan programs on a long-term basis. Fourth, it is difficult to identify important research opportunities—no crystal balls. Finally, the merit review process used by NSF tends to favor well-established areas and existing research paradigms. It is basically a conservative organization with a great deal of inertia, with the usual pros and cons.

Role of mission-oriented agencies with strong extramural activities
In the United States, research in physical sciences, computer science, and engineering is also supported by other agencies. Most of the life science-oriented research is supported by the National Institutes of Health. The existence of this other R&D support must be understood to comprehend the specific role of the National Science Foundation.

During the Cold War between the democratic countries of the West and the communist countries led by the Soviet Union, the need for national defense led to the establishment of the Defense Advanced Research Projects Agency (DARPA) as well as other mission-oriented research organizations. Over the last fifty years, these mission-oriented agencies have played a major role in shaping research universities. DARPA's support of computer science and materials science has greatly contributed to the development of the academic infrastructure of American universities in these two fields.

DARPA played a significant role in strengthening the research infrastructure of the United States because the Department of Defense was the end-user of research with a clear mission and the purchaser of the products that incorporate the research results. DARPA's mission was to support exploratory research for three years, then later to let other agencies take over the support of promising ideas. One factor that has made DARPA a successful agency might have been the fact that it has not had its own intramural research establishment, but instead has been free to support the most promising ideas wherever they were generated.

Today, the agency that is having the most important impact on American universities is the National Institutes of Health (NIH). NIH conducts intramural research activities and also supports extramural research. NIH has grown to be the largest research-funding agency in the United States. With the immense promise of the biological sciences, the societal importance of healthcare, and the economic potential of the biotechnology industry, Congress has been a strong supporter of NIH, which in turn has made biology-related areas, including bioengineering, the most rapidly growing part of research universities.

The government's role in shaping the R&D agenda cannot be overstated. The development of the Internet, modern airplanes, computer science, and materials science owes its beginning to the investment made by the federal government. Recently the federal government of the United States has been increasing its investment in nanotechnology—exceeding $700 million in financial year 2003—which will have a significant impact on research universities. The increasing resources and talents going into this and other new areas are likely to generate exciting new science and technologies.

Budgetary process at NSF and the federal government
To fulfill its mission and to support the science and engineering community, NSF must secure a sufficient budget from the Office of Management and Budget (OMB) of the White House and Congress. Securing the budget is one of the most important functions of the leadership of NSF. The budgetary process is quite complex because it is uncertain and complicated. The budgetary process is affected by many factors, some of which are completely outside of NSF control—war, recession, budget deficit, trade deficit, and the like. NSF must continue to seek the support of powerful people in Congress as well as in the White House.

The fiscal year of the federal government runs from October 1 through September 30. Congress approves the budget for one year at a time. It is a long process. It begins first within NSF in the early spring of each year for the fiscal year that is about eighteen months away. The budgetary process begins soon after the White House submits the federal budget at the end of January for the next fiscal year that begins on October 1 of that year. Each directorate of NSF submits its budget to the director of NSF. After a great deal of deliberation and argument, the NSF prepares a budget for submission to OMB in the fall. OMB examiners review the budget and after many informal discussions OMB decides on the NSF budget, which is often less than the NSF request.

The budget approved by OMB is then returned to the agency for final reconciliation—the budget of each directorate is readjusted to conform to the OMB number. OMB then assembles the President's budget for all agencies and departments in one document, which consists of many volumes of bound books. As soon as OMB summits the President's budget to Congress, the political process of developing the final budget of the federal government begins.[14] It is not uncommon to see newspapers print such headlines as "Dead on Arrival." Various committees of Congress hold hearings on the budget of the agencies and departments that come under their jurisdiction. The Budget Committee allocates an overall budget to different committees. The chairperson of the committee wields absolute power in deciding the budget and its line items. What is even more amazing is the power held by the unelected government staff, such as the chief of staff of the committee, because the politicians often rely on their specialized knowledge.

Any new program or project must be proposed in the budget as a line item if it is to be funded. Therefore, the initiation of a new program takes at least two years to secure funding.[15] Because of this long budgetary cycle, new presidential appointees to high-level government positions must operate with the budget that his/her predecessor has secured. By the time their budget is approved, it is about time to leave Washington, since the average tenure of many Assistant Secretaries is two years.

How well has NSF dealt with its budget complexity?
In 1985, the NSF budget was about 60% of that of NIH, but in 2003 the NIH budget is about 500% larger. This rapid growth of the NIH budget is a testimony to both the importance of biological sciences and medicine in the twenty-first century and the ability of the biological community to appeal to the "fear" factor (i.e., serious illnesses such as cancer and cardiovascular disease) and to articulate economic opportunities offered by advances in biological science—the "greed" and "aspiration" factors. NIH has mastered the complexity of the budgetary process well.

Whatever the reason, the current NSF budget is minuscule relative to the NIH budget, which indicates that NSF has not done well in dealing with the complexity of the federal budget process. It has either failed to articulate the vision for physical sciences and engineering or has not provided a convincing argument on how technology can become the engine for new economic growth. Branscomb (2002) showed that the "nucleation" of new technologies needs the investment of the federal government in basic research in physical sciences and engineering.

10.4.2 Reinitialization of engineering programs at NSF

The engineering programs of the National Science Foundation were reinitialized in January 1985 when the NSF Engineering Directorate was reorganized. The reorganization followed three months of intensive study to define the goals of the NSF Engineering Directorate that are consistent with the NSF Act.

Before implementing the reorganization, the revised budget and the reorganization plan were submitted to the House committee that had jurisdiction over the NSF and to OMB for their approval. After hearing the rationale for the proposed changes, the plan was approved in a relatively short time.

Status of the NSF Engineering Directorate prior to 1985
The old engineering programs of the NSF prior to reorganization in 1985 may be characterized as follows:

1. The NSF program structure made it difficult to foster research at the interface of disciplines. There were sharp demarcation lines among engineering programs, each of which narrowly focused on then existing disciplines of engineering.

2. The NSF engineering programs were designed to support research in well-established fields. The idea that traditional disciplines must embrace new ideas from other fields to deal with future engineering challenges could not find a home at the NSF Engineering Directorate.

3. The structure of the NSF Engineering Directorate was nearly identical to that of typical engineering colleges. There were four divisions—mechanical, chemical, civil, and electrical engineering—with programs in mechanics, and so on (figure 10.1). This structure was not suited for NSF, because NSF, as a government agency, had responsibility for promoting the progress of engineering, securing national defense, and providing health, welfare, and prosperity to people.

4. The budget was allocated more or less in proportion to the number of proposals received in each division. This budget allocation scheme strengthened the areas in which universities were already producing too many Ph.D.s and did not support new emerging areas or areas in which universities were weak.

5. The NSF program structure precluded support for research in new emerging areas such as biotechnology, microelectromechanical systems (MEMS),

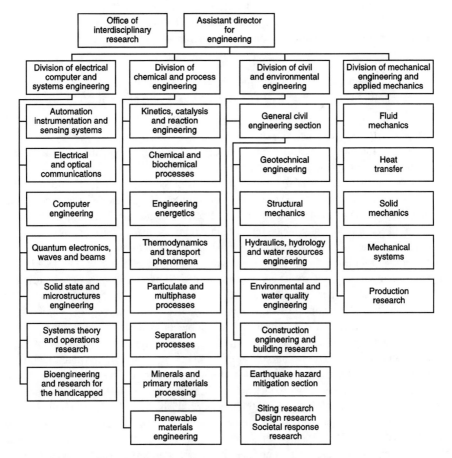

Figure 10.1 Pre-1985 organizational structure of the NSF Engineering Directorate.

optoelectronics, and so on. Therefore, universities, in turn, did not produce graduates in these new areas, since they did not (or could not) hire faculty members in new emerging fields.

6. There were no programs that supported research in areas that still lacked an engineering science base, such as design, computer-aided engineering, and so on.
7. Nearly 100% of the research support was given in small chunks in the form of small single-investigator research projects.
8. NSF did not support large or multidisciplinary research programs that dealt with engineering systems.
9. The long-term needs of the nation were not specifically addressed, because the programs were designed and executed to sustain the long-standing research activities of the well-established people.
10. Young researchers at universities had difficulty in getting NSF grants. Furthermore, the NSF peer-review process tended to force young professors to continue their doctoral research all over again, since creative ideas that have not been accepted by others could not get good review under the existing system. Proposals had to have many references with a strong "engineering science" flavor to get good reviews.
11. The long review process based on detailed description of research ideas often precluded the support of really creative ideas that should be verified quickly. The success rate of "synthesis"-oriented research proposals was low.

In summary, the NSF engineering programs were not closely aligned with the three main goals stated in the NSF Act—to promote progress of science and engineering, to promote welfare, health, prosperity to people, and to secure national defense.

Second-class status of engineering at NSF
To understand the basic difficulties the Engineering Directorate had at NSF, it is necessary to understand the history of engineering in Washington.

In short, engineering was treated as a second-class citizen at NSF and at other Washington establishments.[16] The science directorates (e.g., the Mathematical and Physical Sciences Directorate) dominated NSF from the beginning. This might have been a remnant of the fact that physicists, rather than engineers, made major engineering contributions in developing new weapons during the Second World War. These physicists have dominated the policies of governmental and industrial organizations for many decades.

Engineering tried to justify its existence within NSF by claiming that engineering is an "applied science" where the reductionism of science is equally applicable. This view was strengthened during the post-Sputnik era when engineering schools emphasized "engineering science" and deemphasized such subjects as design, manufacturing, and power engineering. This was a major mistake, since engineering consists of both synthesis and analysis. The primary goal of engineering is design and manufacture— creating "artifacts" that have not existed in the past—which require both synthesis and analysis. Reductionism deals with only half of these engineering tasks. Synthesis and the systems aspect of engineering constitute the other half, but could not get much support.

The idea that engineering is an "applied science" had affected many programs adversely. Although it was a way of justifying the support of engineering research at NSF, it sent the wrong message to engineering schools and reinforced the idea that the reductionism model of engineering research is what engineering research was all about. It downgraded technology innovation, design, manufacturing, and other related fields, instead of encouraging the creation of a new science base for these empiricism-dominated fields.

This philosophy of engineering research at NSF was reflecting the post-Sputnik era engineering culture of engineering schools, which was dominated by analysis and reductionism of science in the 1960s and 1970s. Even the advisory groups that provided the community input to NSF were dominated by leading engineering scientists, who were mostly male and academic types. Physicists headed even the Engineering Directorate.[17] The fact that there was a clear need to strengthen engineering science was not the issue, but the fact that it was done at the expense of engineering had to be corrected.

This inferiority complex at the NSF Engineering Directorate had to be replaced with the idea that engineering is as important as science, that design is the essence of engineering, that NSF has a role in strengthening the *engineering* part of the academic discipline, that NSF has to support the creative aspects of engineering that will be important to the nation, and that there has to be a better balance between systems research and individual research projects.

Time-dependent combinatorial complexity at NSF

As a result of the prevailing situation and culture at NSF and in Washington, the NSF engineering programs were not closely tied to the long-term national interest as envisioned at the time the NSF Act was enacted. NSF programs evolved rather than being designed to satisfy specific needs of the field and the nation. Therefore, the political establishment was not pleased with the situation at NSF. Many people knew that something was wrong with the NSF engineering programs, but could not do much about it. This is an example of time-dependent combinatorial complexity, where the current programs were simply the continuation, extension, and amplification of past practice, philosophy, and outlook. The result is a refinement of what is already well known and a diminished output of the research conducted with the taxpayers' money. Clearly, without making an impact on the fundamental knowledge base and technology base, this combinatorial process was not serving either the engineering community or the taxpayers' interests well.

Reinitialization of NSF Engineering Directorate

Beginning in April 1984,[18] active planning was initiated to determine the future direction of the NSF Engineering Directorate by consulting with representatives of engineering societies, the Dean's Council of the American Society of Engineering Education (ASEE), prominent engineering professors, and the political establishment in Washington. Then, beginning on October 16, 1984,[19] several committees were formed within the NSF Engineering Directorate to review the proposed redirection of the engineering programs. Most of the program directors of the Engineering Directorate participated in this review and policy formulation process. Some program directors had difficulty dealing with these changes.

Based on the input of these committees, a new organizational structure was proposed, as shown in figure 10.2. After obtaining the support of the cognizant authorities, the new structure was implemented three months later in January 1985. The new organization was to achieve the following objectives:

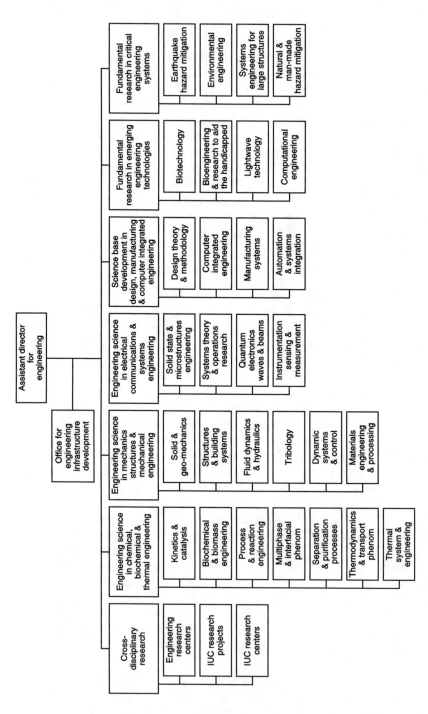

Figure 10.2 New 1985 organizational structure of the NSF Engineering Directorate.

1. To strengthen the engineering fields in which the science base is well established.
2. To establish a science base in engineering fields that did not have an engineering science base.
3. To encourage research in newly emerging technology fields.
4. To support research in critical technology area.
5. To promote multidisciplinary group research.
6. To support engineering education.

The new organization and programs were implemented in January 1985.[20] It was a drastically different organization designed to achieve a new set of goals for the engineering group at NSF. Many new programs were created to achieve the stated goals and many outstanding scholars and engineers were recruited to run the programs.

It has been almost twenty years since the reorganization of NSF was implemented. Some of the results of the "reinitialization" of the NSF Engineering Directorate done in 1985 are:

1. The Engineering Research Centers (ERC) Program is continuing to thrive and has made a major impact at many universities in fostering a group effort in education and research and in strengthening ties to industry.
2. Many young design professors who are now teaching in many engineering schools might not be around today without the Design Theory and Methodology Program.
3. Research in emerging fields has yielded dividends. For example, MEMS, which was initiated with a modest $1 million investment, has now become one of the major research areas at many universities.
4. Research in a number of areas has changed in direction and emphasis, with more research at the interface of traditional disciplines.
5. Research support for young investigators has increased.
6. The idea that engineering deals with the creation of new artifacts that have not existed in the past through both synthesis and analysis is much better accepted.

The budget and the disgruntled
One of the most important roles of a leader is to secure the budget for one's organization. It is a difficult task in the Washington scene where so many fight for their budget. In some ways, it is a zero-sum game. However, it was relatively easy for the NSF Engineering Directorate to obtain a modest budget increase because of its many new and exciting initiatives. Securing a fair budget for engineering programs will continue to be a challenge at NSF, which is still an organization dominated by sciences.

It should be noted that right after the reorganization of the NSF Directorate in January 1985, the only budget the Engineering Directorate had was the reprogrammed fund without any net increase in the total budget.[21] Therefore, the budget of some of the existing programs had to be reduced by 10% or so to support the new program areas. This reduction of the budget hit raw nerves in some of the affected communities. They were most concerned about the funding for their own areas, which is the way it should be. With further increases in the budget of the Directorate, the engineering community had a substantial increase in funding in comparison to other fields, and all of the engineering fields ultimately benefited.

When it comes to budget and salaries, it is not the absolute amount that bothers people; it is the relative amount that gets their attention. Some people refused to believe that their budget was cut not because their field was not important but because other fields, which has been denied for decades, needed funding. Some of these people still cannot accept the fact that decisions at NSF were made to serve the long-term interests of the nation. Some may still be bitter about the changes made at NSF two decades ago. Public service, which is a noble cause, has its price and cost.

It is interesting to note that the basic framework and philosophy of the NSF Engineering Directorate has not changed much during the last two decades, although there were some modifications to the program structure that was introduced in 1985.

Time-dependent complexity, reinitialization, and redesign

The basis for the reorganization and redesign of the NSF Engineering Directorate—a small branch of the federal government—was to extract the agency from a system that was mired in time-dependent combinatorial complexity. By reinitializing the existing programs, it was assumed that the organization would better serve the intent of the people as represented by the NSF Act of 1950. Also, many new programs and divisions were created to satisfy a new set of FRs created for the NSF Engineering Directorate.

Has the reinitialization and redesign achieved the goals of the NSF Act? It may depend on one's perspective. The answer should be left to the historians of the National Science Foundation.

10.5   Complexity in Higher Education Systems

Higher educational systems have all four types of complexity: time-independent real complexity, time-independent imaginary complexity, time-dependent combinatorial complexity, and time-dependent periodic complexity. In administering an academic institution, complexity must be reduced to achieve its FRs. The case study presented in this section shows why educational goals and programs should be examined from time to time. They must be redesigned and reinitialized on a periodic basis to improve the quality of education and research. The redirection of the Department of Mechanical Engineering at MIT will be used as a case study of complexity of educational institutions, the role of academic functional periodicity, and the value of reestablishing FRs.

Time-independent real complexity of a university is a result of the system range of the FRs not overlapping the design range, because universities are complex entities with many DPs—individual professors, administrators, funding agencies, students, and alumni—who can intervene and introduce additional uncertainties in achieving a university's FRs. Therefore, the default mode of operation of some universities is to use a fuzzy set of goals, rather than well-defined FRs, to satisfy diverse constituents.

Universities have their share of issues that arise because of time-independent imaginary complexity. The redundant and repetitive work created by imaginary complexity creates unnecessary demand on professors' and administrators' time and effort, because the system is not carefully designed and operated. Many decisions made at universities are ad hoc—not systematic decisions—to deal with crisis and to please its most vocal constituents. Often they work on the same problem—disguised as a different problem—repeatedly. Longevity of administrators is often determined by the ability to avoid controversial decisions rather than make important achievements for the long-term advancement of the university.

The complexity that bedevils a university the most in developing a university is the time-dependent combinatorial complexity. Universities are particularly susceptible to time-dependent combinatorial complexity because of the repetitive nature of academic life and the immense effort required to change curriculum and course content, the temptation to use well-honed teaching materials, and to dwell on one's narrow expertise. Furthermore, most universities have the same set of faculty members for a few decades, who often prefer a status quo.

History shows that many educational institutions evolve rather than change through purposeful design or metamorphosis. However, outstanding educational institutions—leading in cultivating young minds, contemplating the future through research, and advancing the knowledge and technology base—cannot be created through evolution alone. Apathy and complacency will creep in and dominate the evolution. Academic institutions and programs must be designed and nurtured with a clear set of FRs from time to time. Theorems C8 through C10 presented at the beginning of this chapter hold true for universities and academic departments.

Before the case study on the MIT Department of Mechanical Engineering is presented, the characteristics of higher education in two different countries, the United States and Korea, will be briefly examined as a representative case.

10.5.1 Introduction to higher education systems

The quality of human resources determines the future of a nation
Countries need a large number of competent, well-educated people to advance and administer all segments of their society. The educational system of a nation is the primary means of generating the educated human resource that fulfills the needs of the nation. It is also an indispensable means of satisfying the aspiration of their people for a better and more meaningful life. In this regard, the higher education system, along with primary and secondary education, must be strong, effective, and efficient.

In most countries, a higher education system consists of colleges, universities, and professional schools. In some countries, these institutions produce many of their leaders. A strong higher education system must be designed correctly and operated effectively. Universities cannot be designed to be producers of Newtons, Churchills, Gandhis, Mozarts, and Einsteins, but can be designed to produce a large number of able and competent people. Great universities can and have produced great minds, but it is a result not only of the formal curriculum but also of the exposure to ideas, the people they come in contact with, and the cultural values of the institution.[22]

At an individual level, educational background determines the future of a young person, perhaps more than anything else. Even a gifted person may not have the opportunity to fully develop his/her potential without a formal education, although there are always exceptions. Therefore, in most countries, there is great competition to get into good universities. Great universities, in turn, must attract and choose the best students to maintain their excellent quality. Indeed many of the better universities enjoy their stellar reputation because of their ability to attract good students.

Managing a higher education system can be a complex task since there are many people—individual professors, administrators, students, support staff, and government programs and policies—who can affect the outcome of the educational process, thus increasing the uncertainty of achieving the FRs of the educational system. Furthermore, the efficacy of the educational process is anything but certain since it depends on many intangible factors, such as the culture of the school, students' ability and attitude, and societal conditions.

The risk that a poor educational system creates for a country cannot be overstated, since the educational system affects the economy, technology, healthcare, and all other sectors of its society. Unlike certain financial risks, which can be spread equally over all segments of the society through an insurance system, educational risks cannot be reduced by spreading them. For instance, we cannot reduce the risk created by second-rate universities by increasing the number of universities; the quality must be there. In the financial world, a dollar is a dollar regardless of who possesses it, but in education the quality of mind is specific to an individual and cannot be redistributed to many as a means of reducing the risk and improving the odds. Mass education is required and beneficial for the whole society, but human resources cannot be simply aggregated to make up for poor quality with quantity. A few of the best minds with wisdom may contribute more to societal well-being than groups of many well-meaning people with a less than stellar knowledge base and limited wisdom.

There have been at least two different approaches used by various countries to enhance the overall quality and quantity of higher education. In countries like the Republic of Korea and Hong Kong,[23] a few new high-quality science and engineering universities were set up to provide an excellent education and at the same time become a catalyst in changing the existing universities. In the absence of competition, some of the existing institutions might not change at all. In some countries, a large number of four-year programs have been created to produce more college-educated young people, some at the expense of better-known traditional universities. In some countries, three- or four-year colleges have been upgraded to universities.

To have high quality, universities must be free to set their own FRs and DPs without the intervention of government. The role of government is to provide support, to provide level playing fields, and to promote the right environment for higher education, but it must not stifle progress by applying a rigid set of FRs and DPs that all schools must adopt. It is surprising that in many countries, a small number of bureaucrats in the ministry of education control all educational programs of their country from kindergarten to universities. In many countries, governments are making the mistake of being too paternalistic by stifling competition through regulations. Under such conditions, only the status quo can be maintained—better universities do not need to get even better and poor universities will continue to survive. Without a meaningful competition among universities, it is difficult to improve the quality of universities.

10.5.2 Role of the U.S. federal government in higher education

The higher education system in the United States has been a model for many nations. Many countries have studied and emulated the U.S. system, including the role of the U.S. federal government. The effectiveness of the U.S. higher educational systems can be assessed by the fact that universities in the United States generate credible college graduates, although the quality of high school graduates generally lags behind those in Asia and Europe. Also, there are more Nobel laureates—a measure of accomplishments among many others—coming out of the American universities. However, even the U.S. system can be improved. One of the biggest concerns about universities in general stems from time-dependent combinatorial complexity. In this section, higher education in science and engineering will be discussed.

One of the most distinguishing aspects of higher education in the United States is that the state[24] and federal governments play no *direct* role in running the higher education system, although state universities are funded by the state government and their boards of trustees are often made up of friends of politicians. The federal government indirectly affects universities through research funding, student loan programs, patent policies on government-funded research, and the like. Agencies such as NSF and NIH can have a profound effect on higher education through their research funding as discussed in the preceding section. In addition, research universities are also affected by a number of laws enacted by Congress.

There have been a number of congressional acts created during the last four decades to strengthen technological innovations and technology transfer from universities to industry. Of these, perhaps the most important legislation may be the Bayh–Dole Act of 1980. This act has enabled universities to keep the patents generated from government-funded research projects. Universities have the right to license these patented discoveries to industrial firms. This legislation was enacted to overcome the shortcomings of the old system. When the government owned all patent rights, industrial firms were not willing to take the financial risk of converting research results into commercially viable products.

Strong research universities have strong links to mission-oriented agencies and industry—the end-users of new technologies
An important characteristic of strong research universities in the United States is their link to mission-oriented agencies and industry. For example, Stanford University has played a large part in the creation of Silicon Valley, which in turn has had a major role in strengthening Stanford University. Many of MIT's outstanding programs can be traced back to the initiatives of government agencies and also the strong backing it has received from industry. MIT people, in turn, have contributed to the basic knowledge base, transferred technology to industry, and started many venture firms, some of which have contributed to the wealth of the nation and to the Boston area.[25]

Gathering of "brains"
One of the big advantages of American universities is their ability to attract the best brains of the world to their campuses. Prior to the Second World War, it was the Jewish scientists and engineers who fled Europe and came to the United States. These refugees became a part of the cadre of eminent scientists and engineers in a number of important fields, especially physics. The United States has been a haven for a large number of foreign-born scientists and engineers who have come there as immigrants, students, and researchers. They have supplied the United States with an incomparable pool of highly educated and capable human resources. They come to the United States because of the opportunities to learn and work in an environment that is built on the premise of democracy, equality, human rights, freedom, and meritocracy.

Strong link between research and education at a research university
A characteristic of a strong research university is its commitment to provide a quality education for its students, in addition to conducting outstanding research. Great

research universities conduct research not only as a separate agenda for knowledge and technology generation, but also to enhance their educational process. Both graduate and undergraduate students learn through their participation in research. At these universities, research results are incorporated in educational materials, first in graduate courses, and ultimately in undergraduate courses. At many of the leading research universities, the best researchers are also the most dedicated educators.

10.5.3 Role of the Korean government in higher education

In many ways, the Korean and the U.S. higher education systems are very much alike. Furthermore, a large number of the faculty members teaching in Korean universities were educated in the United States, in part because of the Korean War. Both countries have four-year colleges and universities. Also the quality of Korean universities has improved just as the industrial sector has. Notwithstanding all these similarities between the two systems, the complexity—the uncertainty of achieving their institutional FRs discussed later in this section—of the two systems is substantially different. The difference, in part, is due to the strong control of universities by the government in Korea.

The Republic of Korea, like many countries in Asia and Europe, has a Ministry of Education, which controls the educational system of the entire nation. The Ministry sets the educational policies for the entire system—from kindergarten to universities—that educates several million students. The Ministry has about 47 or so managers who make these decisions, which then must be followed by the schools. The Ministry dictates many things to universities, both public and private. For instance, universities need the approval of the Ministry to establish new academic departments. They also control the compensation of the faculty at nationally supported universities, the number of undergraduate and graduate students, the minimum number of professors, the tuition at public universities, and many other things. A consequence of the near-dictatorial power of the Ministry over the universities has increased the complexity of the higher education system in Korea by increasing uncertainty and by creating a coupled educational system that does not encourage innovation.

In addition to the Ministry of Education, there is also the Ministry of Science and Technology (MOST). MOST has jurisdiction over the Korea Advanced Institute of Science and Technology (KAIST), which has been rated as the best university of its kind in Asia. KAIST is a unique university, which was established in the early 1970s to generate highly educated scientists and engineers. Students are given living allowances in addition to free tuition to attend this university. Unlike their U.S. counterparts, the KAIST professors are not under great pressure to seek outside funding for their research because they do not have to support their graduate students, although many professors partially supplement the stipend of graduate students from their research projects.

Many universities in Korea are fairly new and some four-year colleges have been elevated to the status of university. For many years, universities were guaranteed student enrollment because a larger fraction of high school students pursue higher education in Korea than in the United States. Admission to universities is determined by scores in the national entrance examination; to get into the most competitive departments at the most selective university requires high examination scores. This system of admitting students based on a one-dimensional measure of students' ability has created a thriving business for private tutors.[26]

The total investment in education as a fraction of Korea's GDP must be one of the highest in the world. Koreans take education, especially the university pedigree, more seriously than their counterparts in the West because the reputation of their alma mater plays a big role in advancing their careers, in addition to traditional values accorded scholarly pursuits. As soon as Koreans meet, they quickly determine whether they are alumni of the same school. They tend to trust their fellow alumni.

The quality of universities varies a great deal in Korea, notwithstanding the control of universities by the national government. Some Korean universities—for example, Seoul National University (SNU), KAIST, Yonsei University, and Pohang University of Science and Technology—are competitive with some of the best universities in the world, but at the other extreme, some universities do not have an adequate number of faculty members and have only minimal physical facilities and thus cannot provide hands-on education in science and engineering. The disparity in quality between universities may be greater in Korea than in the the United States.

The overall quality of the leading research universities in Korea, especially in science and technology, is not as good as the best research universities in the United States, despite the fact that the quality of the faculty may be equally good.[27] The exact cause is not known, but it may be due to the lack of competition among universities and among faculty, nearly guaranteed job security once they become a member of the faculty, and lack of the culture to be the best in the world. To improve the quality of these research universities, the Ministry of Education and the Ministry of Science and Technology have initiated a number of programs such as "Brain Korea 21," "Frontier Projects," and "Engineering Research Centers," in addition to the regular research funding programs administered by government-funded foundations and agencies. The availability of research funds to leading Korean universities seems to be comparable to that of U.S. universities.

10.5.4 Complexity of the U.S. and the Korean higher education systems

What is the cause of complexity in higher education? How do we reduce the complexity?

Both the United States and Korea have outstanding educational systems in the world and give strong support to education. They have improved universities more than many other countries. Universities in the United States have a longer history in comparison to Korean universities, some of which are still in their development stages. Notwithstanding the advanced nature of these two systems, there are things that can be improved for future development.

Complexity, uncertainty, and risk cannot be evaluated without first defining the FRs of a higher education system. The FRs of educational institutions differ widely, depending on the country and the region within the country. Therefore, it is difficult to generalize the complexity of higher education systems for all nations or for all regions within a country. Even if we can define the FRs for these universities, it is very difficult to change universities because of the tremendous inertia that is associated with higher education systems. This is evidenced by the list of the top twenty best universities in the United States, which has changed only a little during the past fifty years.

Two different systems will be contrasted here: the U.S. educational system and the Korean educational system. For this analysis of complexity of the two systems, we will assume a set of common FRs for higher education, which is clearly subjective. This analysis is intended to promote discussion rather than to be judgmental about the merits of each, for there are many historical reasons for the present status of each system, some of which transcend purely educational reasons.

What are the FRs of a higher education system?
A university is a system that must satisfy many FRs at the same time through a proper design of the system. The FRs of a higher education system are not explicitly stated in most countries or in most universities. It is assumed that everyone knows what an excellent university is—primarily based on its reputation, like a beauty contest. Also most people know what kinds of resources a great university should have, such as good facilities, strong faculty, and outstanding students. However, to create and operate an outstanding university, it is difficult to chart the right course of action without establishing the FRs first. Yet most administrators of universities are primarily concerned with "things" in the physical domain—faculty, buildings, curriculum, and so on—before they define the customer needs and the FRs of the university.

The FRs of a higher education system of a nation must be stated in order to judge the quality of the system. The FRs may vary depending on the specific needs of a country. However, it is reasonable to assume that the ultimate outcome desired of their higher education system is similar in both countries. Therefore we will apply the same FRs to both countries.

The FRs of a higher education system of a nation may be stated as:

FR_1 = Provide opportunities to learn for all those who want to learn.
FR_2 = Create future leaders for all sectors of a nation.
FR_3 = Advance fundamental knowledge and technology.
FR_4 = Create professionals in all fields.

How do we achieve these goals? How well can we achieve these FRs? What is the uncertainty in achieving these FRs? What is the complexity of a higher education system in terms of these FRs?
These FRs seem to be reasonable goals for universities in any nation. However, many countries are having difficulty in satisfying these FRs for a variety of different reasons, some external factors and many internal factors. But the consequence of not satisfying these FRs is the same: lack of competent professionals, slow economic development, unfulfilled aspirations of young people, crippling of opportunities for future generations, and the emigration of bright young minds to other countries.

To satisfy the four FRs, many countries have created physical entities (DPs) for the higher education system. For instance, typical DPs they have chosen to satisfy the above four FRs may be stated as:

DP_1 = General four-year colleges.
DP_2 = Highly competitive undergraduate programs at leading universities.
DP_3 = Graduate and research programs at leading universities.
DP_4 = Professional schools.

The complexity of a higher education system is determined by the uncertainty associated with knowing how well the FRs are satisfied by the chosen DPs. In some countries, separate institutions are created to deal with each one of these FRs and DPs, but in many countries, DP_2 (Highly competitive undergraduate programs at leading universities), DP_3 (Graduate and research programs at leading universities), and DP_4 (Professional schools) are administered under the same umbrella called "university."

Given the above set of FRs and DPs, what is the complexity of the U.S. higher education system and the Korean higher education system?
To be able to answer these questions, each one of these FRs and DPs must be decomposed. For example, when FR_1 and DP_1 are decomposed, the lower-level FRs of FR_1 may be different for different colleges and for different fields. Therefore, the lower-level DPs of DP_1 may also be different for each university. The other FRs and DPs should similarly be decomposed and the lower-level FRs and DPs will be specific to a university and the field of specialization. That is, as we decompose these high-level FRs and DPs for a nation, the lower-level FRs and DPs can be developed for the specific needs of each university. Therefore, each university must design its system to satisfy a specific set of goals and objectives.

According to one of the theorems of axiomatic design, when a system has more FRs than DPs, it is a coupled system. The real complexity of a coupled system is large since it cannot deal with random variation. It is more complex than an uncoupled or decoupled system. If we apply this theorem to the Korean higher education system, the Korean university system is a highly coupled system because of the strict government regulation in developing DPs. Therefore, the complexity of Korean universities is greater than that of the U.S. system, where each university has the freedom to chart its own direction, policies, and actions. In the United States, even state universities are allowed to operate with a greater degree of latitude than are Korean universities, although they do depend on the state government for basic support of their teaching activities.

Because the Korean higher education system is more complex than the U.S. system, it is more difficult for Korean universities to achieve their educational objectives because too much effort is spent maneuvering through the complicated array of bureaucratic hurdles. Such a system tends to advance slowly, and changes to such systems can rarely be introduced. Under such a system, universities do not have any competition; universities with better reputation do not have rivals, while less well-known schools cannot shed their disadvantages and aspire to be leading universities.

As a result, many faculty members do not need to excel because their jobs are secure and stable. It is in their best interest to protect the system because the system now guarantees their survival irrespective of their productivity and contributions. A similar situation also exists in many universities in the United States, but the difference is that universities in the United States must compete with their peer universities, making it more difficult to maintain a status quo, to secure funding, and to attract students. Life is more "insecure" for faculty members of the U.S. universities!

How can Korean universities be more competitive?
The solution to the Korean higher education system, and to many others with similar complexity in their higher education systems, may be fairly simple—reduce the complexity of the system.

First, the government should reduce its dictatorial power and give universities freedom to administer to the best of their ability and provide a competitive environment for public funding. Japanese national universities are now given more freedom to chart their own future, but it will take many years to establish roots. It may take at least two generations of new professors to establish a more competitive system in Japan.

Second, let Korean universities compete with their peer groups in other countries, including universities in the United States and Asia, which will require financial support for global activities. Initially, it may be useful to commission a comparative evaluation by an impartial entity.

If it is so simple, why is it so difficult to change universities?
Fear of risk and lack of trust may be one of the main reasons. Most people do not like change, because change can be disruptive. Furthermore, there are beneficiaries of any existing system, who may lose their comfortable positions as a result of the change. These people can come up with hundreds of reasons why change is not acceptable. Some may argue against the idea that government grant their universities more freedom for fear that it will be misused. For example, they may argue against abolishing the entrance examination admissions policy; they say it will create a corrupt system that will allow some people to "buy" their admission. But the risk has to be measured against the benefits. Certainly such a free system has not corrupted leading universities in the United States. Furthermore, under the current educational policy of Korea, more money is spent by parents to gain admission to good schools through private tutoring, which favors those who can afford to pay for the special tutoring. Korean parents send their children to special cram schools after their regular schooling.

In Korea as well as in many other countries, many things are decided by examination in addition to entrance to universities—selection of judges, bureaucrats, and so on. The rationale for this one-day examination system is that it is a fair and objective system.[28] The shortcoming of the system is that it is a one-dimensional measure of a student's capability, whereas people and the society in which they serve are multidimensional. A student's performance in the one-day examination is not a good measure of that student's future success.

A nation needs an educational system that produces both one-dimensional and multidimensional people. Multidimensional thinking is needed to solve many systems problems, create new solutions that have not existed hitherto, overcome difficulties of all sorts, have intelligence and vision that can see beyond the next hill, and provide intellectual flexibility in generating ideas and solutions. These people are needed to provide the necessary leadership that requires multidimensional thinking. Society also needs many one-dimensional people to serve in well-defined roles and perform more repetitive and well-defined jobs in society. Society needs more of the latter kind, since there are more jobs that require high performance in well-defined functions than there are jobs that require a more creative role.

Complexity in the U.S. system
The U.S. higher education system thrives because it is so diverse. Some schools cater to mass education of the public, some cater to niche markets, and some are primarily research universities. There is strong competition among all schools; for example, the best research universities compete with each other for better students and faculty. However, a few well-known institutions have advantages over other universities because of their long historical reputation, financial resources, research infrastructure,

and the quality of students and professors. These advantages continue to perpetuate the system.

The complexity of the U.S. system is less than that of the Korean system because each university can pursue its own FRs without being centrally required to adopt a system that does not quite fit its institutional goals. In fact, the probability of graduates of less well-known universities becoming leaders of American society is as good as that of graduates of leading research universities.[29] The fact that the U.S. university system is robust can be seen by the fact that many countries are trying to emulate the American system[30] and that the United States attracts the largest number of foreign students to its diverse colleges and universities. The U.S. higher education system is a market-driven system, dictated by supply and demand as well as by competition.

Every system has shortcomings. One of the major weaknesses of the U.S. higher education system is that because it is market-driven,[31] social sciences and liberal arts are not given as much emphasis as science, engineering, and business fields in terms of research support, faculty salary, and recognition. However, these nontechnical, non-business fields are equally important, since advances in these fields are needed to deal with societal issues that require better understanding of humanity. The rapid advances in scientific and technological fields must be augmented by active intellectual research and inquiry into the long-term development of enlightened human interactions. It appears that these fields are very complex. The market economy is a great invention, but it may not be able to deal with many irrational things people do as a group. For example, historically humanity has spent more on wars than on means of securing peace!

10.5.5 Complexity of an academic department

So far we have discussed the complexity of a national higher education system in general. How about the complexity of a research university? Is it complex? How can we deal with complexity at a departmental level? In this section, the Department of Mechanical Engineering at MIT will be used as an example to discuss the complexity of an academic department.

Introduction to the Department of Mechanical Engineering at MIT
The Department of Mechanical Engineering is one of the original departments that was established when the Massachusetts Institute of Technology (MIT) was incorporated in 1861 in Boston. The department is designated as Course II within MIT. It has approximately 60 faculty members, 350 undergraduate students, 400 graduate students, and 50 technical and support staff members. Of its undergraduate students, about 35% are women, which is less than the average enrollment of women (about 50%) at MIT.

The department has been rated as the number one mechanical engineering department in the country ever since the game of rating academic departments started. It has produced some of the pioneering textbooks and monographs in a number of fields, including solid and fluid mechanics, thermodynamics, design, manufacturing, tribology, control, internal-combustion engines, engineering analysis, system modeling, robotics, gas turbines, and others. It also has innovated, perhaps more than any other academic department of its kind, many new technologies that are being used in industry. The quality of the students is excellent. Its graduate student body is made up equally of American students and foreign students.[32]

Just like the rest of MIT, the department has a strong faculty. Of the 60 or so faculty members, about 75% hold tenured appointments, which is low in comparison to other universities, which have mostly tenured faculty members. This tenure ratio is not set by the department or by the university. It is the product of the tenure process. The tenure review process is based on their performance during the seven years[33] of their service at MIT. It is so stringent that the success ratio ranges from 30% to 40%, which keeps the tenured faculty ratio at about the 75% level. A few highly qualified faculty members[34] were not given tenure, indicating that the review process is subjective and imperfect. The criteria for tenure are not clearly stated, but the key idea is to tenure those who have become intellectual leaders in their field as well as being outstanding teachers. Experience shows that people who joined the faculty with some experience elsewhere, either before or after their Ph.D., have advantages over those who join the faculty right after finishing their Ph.D. because of their experience and exposure to other disciplines.

Combinatorial complexity and the need for reinitialization
The department has gone through a few transformations during the last fifty years. The modern mechanical engineering department that is deeply rooted in research, especially sponsored research, may be traced to the transitional period right after the Second World War, when many outstanding faculty members came to the department from various institutions and departments. Milton C. Shaw, who had a Ph.D. in physical chemistry from the University of Cincinnati and worked at NASA in the field of tribology, came to MIT and started metal-cutting research. Warren M. Rohsenow, who received his doctorate from Yale, came after serving two years in the U.S. Navy and started research in boiling heat transfer. Stephen H. Crandall, who joined the department after getting his Ph.D. in mathematics at MIT and teaching mathematics at MIT, produced fundamental textbooks in solid mechanics and engineering analysis. J.P. den Hartog, who came to MIT after working at Westinghouse during the Second World War, produced seminal books in mechanical vibration. Egon Orowan, the metal physicist who is one of the originators of dislocation theory, came to MIT after years of research at Cambridge University in England.

These and other distinguished people joined the department, which already had such eminent scholars as Joseph Keenan, a leading scholar in thermodynamics, Carl Soderberg, a visionary in engineering, Ascher H. Shapiro, the author of the seminal book on compressible fluid mechanics, and Fay Taylor, a leading engineer in the field of internal combustion. These people, who constituted the department, established its reputation through their textbooks and research. The department started a major effort in controls and system modeling during this era. The department started a major doctoral program for the first time during the 1950s, since many engineering professors at that time did not have doctoral degrees. In the 1950s, many Ph.D. candidates were appointed as assistant professors even before they finished their doctoral degrees.[35] Many of these young faculty members eventually left MIT to teach at other universities.

The culture and the mode of operation of the department that were established during the post-Second World War period remained intact for about fifteen years. The emphasis during this era was on improving education by strengthening engineering science and by emphasizing research-based graduate education. Then the department experienced a major turbulent period in the late 1950s and the early 1960s, due to several factors, including the launching of Sputnik by the Soviet Union, the decreasing enrollment in the department, and a strong dean by the name of Gordon Brown, an electrical engineer who led the Servomechanisms Laboratory, which developed numerically controlled machines, at MIT. Brown used Ford Foundation funding to produce more Ph.D.-level engineering professors and advocated strengthening engineering science subjects to the exclusion of other fields. During this period, those who advocated that the department should concentrate on engineering science subjects[36] and give up traditional areas such as design and manufacturing (which did not have a strong science base) dominated the department. Some of those who strongly disagreed with this philosophy left MIT, including Milton C. Shaw, who went to Carnegie-Mellon University (CMU) in 1961 as the head of its mechanical engineering department, and others who went to Case Institute of Technology. During this turbulent period, the department had four new department heads.

This change in the early 1960s "reinitialized" the department, moving it into plasma physics and other fields related to space and "engineering science" areas. In retrospect, it was a mistake to deemphasize the core engineering subjects that dealt with the main mission of engineering, that is, design and manufacturing. The department should have attempted to create a science base for these fields that did not have one rather than deemphasize them. This was in part caused by the fact that the faculty members in these fields themselves believed that these are not subjects that can have a science base. This philosophy and the culture that reinitialized the department under the leadership of Professor Ascher H. Shapiro lasted for about fifteen years. During this period, the department established major efforts in bioengineering and strengthened fluid mechanics, solid mechanics, control, and thermodynamics.

Beginning in the 1970s, it was realized that MIT must do more for the fields of design and manufacturing because the United States had its first trade deficit and American companies were no longer dominating the manufacturing industry worldwide. Beginning in 1975, the department, under the leadership of Professor Herbert Richardson, decided to create a major activity in the field of manufacturing. The department created the Laboratory for Manufacturing and Productivity (LMP) in 1976,[37] which became an interdepartmental laboratory of MIT two years later. LMP concentrated on the two ends of the research spectrum: creation of a science base in design and manufacturing, and innovation of new technologies in polymer and metal processing. During this period, the department established strong links to industry by creating the first industrial consortium in the field of polymer processing and later in internal-combustion engines. This period lasted for about fifteen years under the leadership of Professors Herbert H. Richardson and David N. Wormley as the department heads.

"Decade of reinitialization—from 1991 to the millennium"
In the early 1990s, there was a general consensus that the department had to reinitialize its research and educational activities. There was a feeling that it had been on a path of time-dependent combinatorial complexity for too long. During the ten-year period beginning from 1991, the department reinitiated its goals and introduced a new functional periodicity transforming the system with time-dependent combinatorial complexity into a system with periodic complexity.

Beginning in 1991, a major effort was undertaken to redefine the discipline of mechanical engineering and make a significant impact on the knowledge base and technology innovation. At the beginning, it was not an easy task to embark on a new path, although the author took the department head job because the search committee convinced him that the department was ready for a change.[38] One group of the faculty could not comprehend why the best mechanical engineering department in the country had to change. Their view was: "We must be doing something right to be rated number one all these years. Why change?"

Changes were made to achieve the following three highest FRs of the department by reinitializing the department:

FR_1 = Transform the discipline of mechanical engineering from one that is
based on physics to one that is based on physics, information, and biology.
FR_2 = Make impact through research on the knowledge base and technology
innovation—the two ends of the research spectrum—rather than
being in the middle of the research spectrum.
FR_3 = Provide the best teaching to students.

The DPs to satisfy these three FRs were chosen as:

DP_1 = New research groups/efforts in information science and technology
and in the "new" field of bioengineering.
DP_2 = Shift in research emphasis.
DP_3 = New undergraduate curriculum.

The process variables (PVs) that can satisfy the DPs were:

PV_1 = Faculty members who can bring new disciplinary background into
mechanical engineering.
PV_2 = Reward structure based on impact (rather than the number of
papers or the amount of research funding).
PV_3 = Gathering of financial resources to support the new curriculum.

To achieve these goals (FRs), the following things were done (not necessarily in chronological order):

1. Creation of the Pappalardo Laboratories[39] to house new undergraduate laboratories for design, instrumentation, and student projects by converting and renovating 20,000 square feet of dilapidated laboratory space on the ground floor into a modern laboratory.
2. Creation of a new research laboratory in information science and technology— the d'Arbeloff Laboratory[40] for Information Systems—by renovating old space to house new research groups in mechatronics, automation of healthcare, and automatic identification of products.
3. Establishment of the Der Torrosian Computational Laboratory to provide computational facilities for students taking mechanics and design courses.
4. The emphasis of bioengineering was changed from prosthesis to bioinstrumentation to create the "third leg" in the tripartite arrangement of medicine, biology, and engineering.[41]
5. A new energy-related laboratory entitled the "Laboratory for 21st Century Energy" was created to develop the intellectual basis and technologies for the era when demand for petroleum is greater than supply.

6. The Hatsopoulos Microfluid Dynamics Laboratory[42] was created by converting the traditional fluid mechanics laboratory.

7. The AMP Teaching Laboratory[43] in Mechanical Behavior of Materials was created by renovating the old laboratory.

8. A completely new undergraduate curriculum was adopted that offers integrated undergraduate subjects rather than traditional subjects to provide a better context for learning.

9. The Ralph E. and Eloise F. Cross CAD/CAM Laboratory[44] was created for undergraduate teaching in design and manufacturing.

10. The Cross Student Lounges were created for undergraduate students.

11. Since 1991, nearly 50% of the faculty members were replaced. Nearly 50% of these new faculty members came into the department from other disciplines, such as physics, mathematics, optics, computer science, physiology, materials, electrical engineering, and chemistry. Many of these new faculty members are now tenured, which should give continuity and permanence to the transformation of the department.

12. New chairs were created to recognize those who made special contributions.

13. Modern lectures halls, the B.J. and Chunghi Park Lecture Halls,[45] were created to accommodate new teaching methods to support the new curriculum and provide a better environment for learning and teaching.

14. A faculty prize for teaching innovation, called the Keenan Award[46] for Teaching Innovation, was created that carries a reasonable stipend.

15. The Pappalardo endowment fund for book writing was created to support the new textbook writing activities. Oxford University Press agreed to establish the MIT/Pappalardo Series for Mechanical Engineering and to publish all the books written with the Pappalardo fund. A number of books have been published and more will be published in perpetuity.

16. Many new research projects were created that are outside the traditional fields of mechanical engineering, such as two-photon microscopy for detection of cancer cells without incision, quantum mechanical computers, the use of the Internet, RF sensors to identify products, and others.[47]

17. Two years in a row, two untenured professors of the department received the prestigious Edgerton Award, which is given to the most promising young faculty members at MIT, one of whom went to the University of Cambridge as a chaired full professor in the Department of Applied Mathematics and Physics.

18. Two new faculty members, one with a traditional mechanical engineering doctorate degree and the other with a computer science doctorate degree, and a research staff member with a doctorate in mechanical engineering, started a new center on automatic identification (Auto ID Center) using radiofrequency (RF) receptors. Their idea of creating a unique identification for each product or component has laid the foundation for modern commerce worldwide supported by 130 industrial firms.

The department was successfully redesigned and reinitialized during this period. It is clearly the best department of its kind in the world, according to the surveys published annually by a magazine.

Complexity in academic departments

There are always "victims" during a period of major transition. Life can be uncomfortable for some, because change requires learning new things, abandoning ideas and books that one has developed, confronting a new culture that is foreign to traditional mechanical engineers, and accepting new paradigms that may be of questionable value. These concerns about change can play a useful role in making sure that we do not undertake foolish ventures.

Each faculty member who is in the department may increase the uncertainty of achieving the new FRs and thus raise the complexity level (Theorem C10). Each one of the faculty members in the department may be considered to be a DP, each of whom can add uncertainty in achieving the FRs. That is, FRs are functions of DPs, which may be expressed as

$$
\begin{aligned}
FR_1 &= f(DP_1^a, DP_1^b, \ldots, DP_1^x) \\
FR_2 &= g(DP_2^a, DP_2^b, \ldots, DP_2^y) \\
FR_3 &= h(DP_3^a, DP_3^b, \ldots, DP_3^z)
\end{aligned}
\tag{10.1}
$$

To achieve these FRs effectively, we need to choose the right set of DPs so that the design matrix that relates the FR vector and the DP vector will be either diagonal or triangular; that is, to reduce the time-independent real complexity of the system by avoiding the creation of a coupled system, the right DPs must be chosen to satisfy the FRs.

To be sure that the system range is inside the design range of the FRs and thus reduce real complexity, we need to eliminate uncertainty by compensating for the possible random changes introduced by other DPs. After fully understanding the concerns and ideas of other faculty members and after developing a consensus, the uncertainty introduced by these DPs can be compensated by making a right decision. This way of making decisions may be expressed as:

$$
\delta FR_1 = \frac{\partial FR_1}{\partial DP_1^s}\delta DP_1^s + \sum_{\substack{for\ all\ i\ except\ s}}^{x} \frac{\partial FR_1}{\partial DP_1^i}\delta DP_1^i
$$

$$
\delta FR_2 = \frac{\partial FR_2}{\partial DP_2^s}\delta DP_2^s + \sum_{\substack{for\ all\ i\ except\ s}}^{y} \frac{\partial FR_2}{\partial DP_2^i}\delta DP_2^i
\tag{10.2}
$$

$$
\delta FR_3 = \frac{\partial FR_3}{\partial DP_3^s}\delta DP_2^s + \sum_{\substack{for\ all\ i\ except\ s}}^{z} \frac{\partial FR_3}{\partial DP_3^i}\delta DP_3^i
$$

where the lower-case delta (δ) indicates random variation.

The above equation states that DP_x^s may used to compensate for the random variation introduced by other DPs, which is represented by the second term of the RHS of equation (10.2), to reduce real complexity. When the department head can have the final decision-making authority, the probability of achieving the FRs or the likelihood of reducing the complexity of the system increases if the first term is larger than the second term of the RHS of the equation. This is the MIT system. In many other universities, decisions are made by a simple majority vote, which makes it difficult to achieve the stated FRs.

Analogy—"rowing a boat"

Equation (10.2) may be applied to the case of eleven people on a rowboat. Ten of these people are oarsmen with one oar each who row the boat. Although they are all doing their best and are supposed to be rowing the boat in unison with the same power, sometimes the boat may not go straight or turn as intended. It is a lot easier if there exists an eleventh person who controls the rudder to point the boat in the right direction. Academic departments are equivalent to rowboats with lots of oars in the water. Even though all of the oarsmen are smart and extremely capable (but probably not of equal weight and size), the boat may wander around without a set destination, dissipating lots of energy, unless there is somebody at the rudder.

10.6 Summary

1. The complexity of sociopolitical–economic systems can be explained in terms of the complexity theory presented in this book. What makes the sociopolitical–economic issues much more complex than some of the scientific and engineering issues is the difficulty in defining FRs, the coupling of FRs by the chosen DPs, wasting efforts on problems created by imaginary complexity, the increasing uncertainty of the future outcome as a function of time, the tendency to cling to a system with time-dependent combinatorial complexity, and the difficulty in transforming a system with combinatorial complexity to a system with functional periodicity.
2. To reduce the complexity, the sociopolitical–economic system should be designed to be either uncoupled or decoupled.
3. Time-dependent real complexity of an organization can be reduced by compensating for the uncertainties introduced by other members of an organization.
4. Many socioeconomic–political systems have unnecessary inefficiencies introduced by imaginary complexity.
5. Many organizations and sociopolitical–economic systems may benefit by timely reinitialization of the system and by introducing functional periodicity.
6. It is advisable for all systems to have a built-in functional periodicity to prevent combinatorial complexity from eventually weakening the system.

Notes

1. It is ironic that the Federal Reserve Bank always changes interest rates after the economy has been ailing for a while because of time-dependent combinatorial complexity rather than when the FR begins to go into the combinatorial mode.

2. This section is based on the report by the author (Suh, 1980) to the Economic Planning Board of the Republic of Korea, entitled "An Assessment of Critical Issues Confronting the Korean Machinery Industries," July 1980, which became the de facto "Five-Year Economic Plan for the Period 1980–1985" of the Republic of Korea. This work was done during the author's visit to Korea from June 9 to July 3, 1980, as a consultant to the World Bank. When the author arrived in Australia to attend a conference around August 20, 1980, the front page

of the financial section of an Australian paper had an article with a big headline on the new Five-Year Economic Plan of the Republic of Korea, which was in essence a summary of what was in this report.

3. In 1950, Egypt's per capita income was greater than that of South Korea.

4. These select businessmen were given unsecured loans at low interest rates by government-controlled banks. They could also obtain government guarantees on loans they secured from overseas. Sometimes they could borrow money from the international market at lower interest rates than those prevailing in domestic financial markets, which was used to make money through land transactions.

5. In 1980, the Korean economy was fairly simple and small (perhaps equal to the revenue of the three largest U.S. corporations at the time), which made it possible for government to develop a central economic plan. Now the Korean economy is so large and diverse, with a large private sector dominating the economy, that it would be more difficult to implement a centralized economic plan.

6. The World Bank asked the author to go to Korea and assess the status of the industrial development projects, which were partially financed by the World Bank. This request might have been made because he was a Chief Technical Advisor of the United Nations Industrial Development Organization (UNIDO) when he worked with the Korea Institute of Science and Technology, and later with Egyptian and Indian organizations. As part of this assignment, he began to work with a number of outstanding Korean leaders, who were helping the new military junta establish an economic plan. The suggestions he made in his report were adopted by the new government in the form of the Five-Year Economic Development Plan, 1980–1985. Many able economists and others implemented the plan effectively.

7. The term "machinery" includes automobiles, shipbuilding, power plant, and machine tools, which is a result of the translation of Korean into English.

8. This was a radical recommendation, since the three companies were owned by three different major conglomerates, one with a joint venture with General Motors of the United States.

9. In 2002, the Korean automobile industry produced about 4.5 million passenger cars.

10. Okpo shipyard is a modern facility that was started by a company on the southern coast of Korea. After the company went bankrupt, the government asked Daewoo to take over the facility. In 1980, the shipyard was only partially finished and had a major debt load.

11. Research in biological science and medicine is supported by the National Institutes of Health, which, unlike NSF, has both intramural as well as extramural research activities.

12. The National Science Board, appointed by the President, is a statutory body that oversees the work of the National Science Foundation.

13. Many university professors gave up the idea of doing research or becoming better teachers because of their inability to obtain financial support.

14. At NSF, we used to get a memorandum from President Reagan instructing us not to lobby Congress for a larger budget than the one OMB submitted.

15. Congress and OMB can allow changes in the line item such as when the program and organization of the agency change.

16. One of the most important persons in the Washington establishment said in 1984 that NSF needs engineering because physicists need engineers to construct their experiments!

17. This may be a highly biased description of the NSF Engineering Directorate of 1984, since the author was the first presidential appointee to head engineering at NSF. He was appointed as Assistant Director for Engineering by President Ronald Reagan and confirmed by the U.S. Senate. He assumed his office on October 16, 1984. Two months earlier, President Reagan had also appointed Erich Bloch as Director of NSF, which surprised the community since he was not an academic but a distinguished engineer/manager from IBM.

18. The author accepted the job offer from the White House (Science Advisor George "Jay" Keyworth) in April 1984 and waited to clear the nomination and confirmation process, which was completed on October 15, 1984.

19. When the author was confirmed, he was returning from a trip to Europe. He was surprised to read in the front page of the *International Herald Tribune* an article about the appointment. Apparently, a similar article had been carried earlier by the *Washington Post*. The article noted the fact that a "foreign born" American was appointed to such a position in engineering, indicating that engineering is increasingly dominated by foreign-born engineers. However, the irony is that the article did not say anything about the fact that Erich Bloch, who was appointed as Director of NSF, was also a foreign-born American. Apparently the Asian name caught the attention of the newspapers.

20. This reorganization was implemented quickly since the author accepted the position on condition that he would serve in the position for only one year. He ended up staying at NSF for 39 months.

21. Any reprogramming involving over $250,000 required the approval of the congressional committee that oversees the NSF.

22. It is similar to the seedlings that spread over an area. They may be genetically all alike, but where they land (away from the big trees), their ability to withstand adversity, and the quality of the soil make some of these seedlings gigantic timbers. Other seedlings merely survive. A formal curriculum is designed to fit the majority of the student body, which depends on the overall quality of students.

23. Korea Advanced Institute of Science and Technology (KAIST) and Hong Kong University of Science and Technology (HKUST) are two examples. KAIST, which was established in 1972, has become one of the leading universities of its kind, not only in Korea but in Asia as well. HKUST has also established itself as a good university.

24. One of the weaknesses of some state universities is the real complexity associated with the unpredictability of the state-supported budget for universities. This uncertainty in state support adds to the complexity of state universities.

25. Universities such as MIT have a strict set of "conflict of interest" rules that faculty members cannot violate in starting these firms.

26. Young high school students spend more than twelve hours a day in their regular school and in private cram schools, which they attend after a full day at the regular school. By the time they become college students, many of them are burnt out.

27. This is based on the author's limited number of sample cases. The author's former students who are teaching in Korea and the United States were equally capable students at MIT.

28. The current President of South Korea, Roh Moo Hyun, became a lawyer without going to college by passing the examination. This is an advantage of an examination-based system, where pedigree is very important. The negative side of the system is that many college students prepare to pass this examination rather than acquire a broad comprehensive education.

29. It appears that the future CEOs of Fortune 500 companies are likely to be graduates of less well-known schools, because graduates of well-known schools shun the idea of working for large corporations, except those who go to business schools.

30. The government of Singapore gave MIT $100 million over five years to strengthen two national universities in Singapore. The government of the United Kingdom is funding a similar program to promote collaboration between MIT and the University of Cambridge.

31. The limited support given to social sciences, arts, and humanities is a direct result of the market-driven, highly competitive U.S. higher education system. In areas such as business, faculty salaries tend to be highest because universities have to compete with Wall Street in attracting the best minds. Similarly, engineering schools must compete with industrial firms. In Korea and other countries where faculty salaries are fixed by the government, the salary in the liberal arts is on a par with that in engineering.

32. The department accepts about 30% graduate students from other countries by design, but many American students do not purse a Ph.D. degree after they complete the Master of Science degree. Therefore, the steady-state ratio of foreign students is about 50%.

33. MIT has an eight-year tenure cycle, but the decision must be made at the latest by the end of the seventh year.

34. Judged by their achievements in their professions after they leave MIT. It should also be noted that many who applied for faculty positions at MIT but did not get offers also have done outstanding work at other institutions. We have erred at both ends of the decision-making. The quality of the decision is a function of the quality of the people involved in evaluation.

35. This might explain the then low salaries of assistant professors, which is no longer the case.

36. Engineering science typically refers to the study of scientific issues that apply to engineering core subjects.

37. The author was asked to be the founding director of LMP.

38. The author decided to remain at MIT and run the department rather than taking up a position as the president of a Midwestern university, because the NSF experience convinced him that educational change can best be achieved at a departmental level. However, after four months into the job, he was told by the chairman of the Visiting Committee that at their biennial meeting, about half of the senior faculty members indicated that they were not in favor of the changes being made. The chairman advised that the author gracefully resign, but others advised him against the idea. He decided to remain and carry out the task of transforming the department. Resigning in the face of opposition would make it impossible for his successors to implement any changes in the future and would create a bad precedent. Without the contemplated change, he reasoned that the time-dependent combinatorial complexity would ultimately lead the department to a second-rate status among academic departments of its kind. The reinitialization and redesign effort took ten years—more than twice as long as intended.

39. Named to honor A. Neil and Jane Pappalardo, who have supported many other projects of the department and MIT.

40. To honor Alex and Brit d'Arbeloff, who made many other contributions to MIT.

41. When the author visited several universities and industrial firms, he found that they were also conducting research on issues related to prosthesis, which was similar to that conducted at MIT. He realized that it was time for the MIT Mechanical Engineering Department to reinitiate and redesign the FR of the department in this area. Able faculty members in bioengineering decided that bioinstrumentation should be the new focus.

42. In honor of George N. and Daphne Hatsopoulos. Dr. Hatsopoulos was a former professor in the department and a well-known thermodynamicist.

43. Fund provided by AMP, Inc., which later merged with Tyco, Inc.

44. In honor of Ralph E. and Eloise F. Cross, who created a chair in addition to establishing this laboratory and the student lounges.

45. In honor of Dr. B.J. Park and Mrs. Chunghi Park, who also supported the activities of the Manufacturing Institute.

46. In honor of the late Professor Joseph Keenan, who was the department head in the late 1950s.

47. It is interesting to note that all the new faculty members who came into the department from other disciplines received their tenure when their mandatory year came up.

References

Branscomb, L. 2002. "Between Invention and Innovation: NIST GCR 02-841," National Institute of Standards and Technology Advanced Technology Program (ATP), U.S. Department of Commerce, November.

National Science Foundation, "Strategic Plan and Program Description," Directorate for Engineering, National Science Foundation, March 9, 1987.

Shiller, R.J. 2003. *The New Financial Order: Risk in the 21st Century*, Princeton University Press, Princeton, NJ.

Suh, N.P. 1980. "An Assessment of Critical Issues Confronting the Korean Machinery Industries," Report to the Economic Planning Board of the Republic of Korea, July.

Exercises

10.1. The current structure of the Department of Defense (DoD) was established when the United States was engaged in the Cold War with the Soviet Union. However, it is no longer appropriate since the nature of conflict has changed, especially after 9/11, as the recent terrorist activities throughout the world attest. In the future, DoD must be able to anticipate and respond to unforeseen events. It must also have the ability to reconfigure itself rapidly and respond to changing FRs.

Propose a new set of FRs for DoD. How should DoD be organized? Is there a way of introducing the concept of functional periodicity to DoD?

10.2. Economists use various economic models and analytical techniques to predict future economic performance. However, economic forecasting has been too unreliable and complex to establish economic policies based on their forecasts. As a result, the economy of a nation behaves as a system with time-dependent combinatorial complexity. It is reinitialized only when there is a major crisis such as the collapse of the stock market and recession.

One way of improving economic activities would be to design and model an ideal economic system rather than simply analyze the performance of an existing economic system. Although it is difficult to design the economy of large nations top-down, it may provide a useful insight to understanding the performance of an economic system and clarify many questions such as the effect of reinitialization of an economic system, appropriate tax policies, the benefits and risks of a social safety net, and the effect of changing monetary policies. Your job is to show how such an economic system can be designed using axiomatic design theory and complexity theory.

10.3. The United States needs a rational energy policy for two reasons: to minimize the environmental problems created by its enormous energy consumption and to reduce its dependence on imported oil. The current energy system is in a time-dependent combinatorial complexity mode. In enacting a new energy policy, all three factors of human behavior—greed, fear, and aspiration factors—must be considered, in addition to technology and future energy requirements. Your job is to establish a new energy policy. You should establish FRs and constraints, create a functional periodicity, and reinitialize the system.

10.4. Universities have large inertia. There are many reasons for this: democratic governance, long tenure of professors, financial cost, and the general reluctance to change. It is both the strength and the weakness of academia. It enables universities to pursue important educational and scholarly goals on a long-term basis. At the same time, it makes it difficult to respond to external changes, especially in rapidly changing fields. It is highly desirable to introduce a functional periodicity to a university so as to achieve both of these goals, long-term academic pursuit and timely changes of educational goals and programs. Propose how a functional periodicity can be introduced to the university system.

10.5. The global ecological system is a system with time-dependent combinatorial complexity. At the rate it has been deteriorating, it will lead to a serious calamity unless steps are taken now to renew the ecological system by introducing a functional periodicity. What kinds of actions should the United Nations take to alleviate the global ecological problem? Design a policy and show how a functional periodicity can be introduced to minimize the damage and sustain the ecological system.

Glossary of Key Words

Academic functional periodicity. Academic institutions have a set of functional periodicity such as semesters, graduation, examinations, and vacations, which rejuvenates students and teachers alike. Also some universities have sabbatical leaves that serve a similar purpose.

Acclaro. A software system for axiomatic design (www.axiomaticdesign.com).

Axiom. Self-evident truth or fundamental truth for which there are no counterexamples or exceptions. An axiom cannot be derived from other laws or principles of nature.

Axiomatic design. Design theory and methodology that are based on the Independence Axiom and the Information Axiom.

Biological functional periodicity. A repeating set of biological functions such as cell division that determines the functional periodicity in biological systems.

Chemical functional periodicity. An example is the periodic table of the chemical elements, which shows the functional periodicity of chemical elements.

Combinatorial complexity. The combinatorial complexity is defined as the complexity that increases as a function of time due to a continued expansion in the number of possible combinations with time, which may eventually lead to a chaotic state or a system failure. Time-dependent combinatorial complexity arises because the future events occur in unpredictable ways and thus cannot be predicted.

Complexity. Complexity is defined as a measure of uncertainty in achieving the specified FRs.

Constraint. Constraints (Cs) are bounds on acceptable solutions. There are two kinds of constraints: *input* constraints and *system* constraints. Input constraints are imposed as part of the design specifications. System constraints are constraints imposed by the system in which the design solution must function.

Corollary. Inference derived from axioms or from propositions (theorems) that follow from axioms or from other propositions that have been proven.

Coupled design. A design that does not maintain the independence of functional requirements (FRs) is called a coupled design. The design matrix is a full matrix.

Decomposition. The process of developing children-level functional requirements (FRs), design parameters (DPs), and process variables (PVs) through zigzagging among the domains is called decomposition.

Decoupled design. A design that maintains the independence of functional requirements (FRs) if the design parameters (DPs) are changed in a specified sequence as per the design matrix. The design matrix is triangular.

Design matrix. A design matrix provides the relationship between the characteristic vectors of design domains, for example, the functional requirement vector to the design parameter vector.

Design parameters. Design parameters (DPs) are the key physical variables (or other equivalent terms in the case of software design, etc.) in the physical domain that characterize the design that satisfies the specified FRs.

Design range. Design range is the allowable tolerance of a given FR or the desired accuracy of a natural phenomenon to be determined.

Domains. The design world is divided into four domain: customer domain, functional domain, physical domain, and process domain, each of which is characterized by a vector.

Economic functional periodicity. Functional periodicity of an economy is established when there is a set of FRs that repeats itself on a periodic basis.

Functional periodicity. Functional periodicity is the period set by a repeating set of FRs. There are many different types of functional periodicity.

Functional requirement. Functional requirements (FRs) are a minimum set of independent requirements that completely characterize the functional needs of the product (or software, organization, system, etc.) in the functional domain. By definition, each FR is independent of every other FR at the time the FRs are established.

Geometric functional periodicity. Functions that are controlled by geometry repeat on a periodic basis. Examples are wire rope, textiles, and undulated surfaces for low friction.

Ideal design. Ideal design is the one that has the same number of FRs and DPs and satisfies the Independence Axiom with zero information content.

Imaginary complexity. Imaginary complexity is defined as uncertainty that is not real uncertainty, but arises because of the designer's lack of knowledge and understanding of a specific design itself.

Independence Axiom. The Independence Axiom is one of the two axioms that constitute axiomatic design. It states that the independence of FRs must be maintained for robustness, simplicity, and reliability of systems.

Information Axiom. The Information Axiom is one of the two axioms that constitute axiomatic design. It states that the system must be designed to minimize the information content, that is, to maximize the probability of success in achieving the FRs.

Information content. Information content is defined as the logarithm of the probability of achieving the FR.

Informational functional periodicity. Refers to the functional periodicity in the field of information. Natural language that has a repeating structure is an example. Music has an informational functional periodicity as well as a software system that reinitializes on a periodic basis.

Manufacturing process functional periodicity. Manufacturing processes that make identical products using a set of processes have a functional periodicity.

Mapping. The process of relating a set of characteristic vectors in one design domain to another design domain is called mapping.

Material functional periodicity. The crytallinity of metals is an example of materials functional periodicity; the wavy nature of matter as stipulated by quantum mechanics is another example.

Organizational functional periodicity. Functional periodicity of an organization defined by the repeating set of FRs of the organization.

Periodic complexity. The periodic complexity is defined as the complexity that exists only in a finite time period, resulting in a finite and limited number of probable combinations.

Political functional periodicity. Political systems can have a functional periodicity when elections are held on a periodic basis.

Process variables. Process variables (PVs) are the key variables (or other equivalent terms in the case of software design, etc.) in the process domain that characterize the process that can generate the specified DPs.

Real complexity. Real complexity is defined as a measure of uncertainty when the probability of achieving the FR is less than 1.0 because the system range is not identical to the design range.

Redundant design. A design that has more design parameters than the number of functional requirements is called a redundant design.

Reinitialization. The process of setting the values at the beginning of a functional cycle of a system with time-dependent periodic complexity is called reinitialization.

Robust design. A design that satisfies the specified functional requirements (FRs) even though the design parameters have a large variation.

System range. System range is the probability density function of a given functional requirement (FR) of the actual system created.

Systems biology. Systems biology is a subdiscipline of biology that deals with the relationship between higher-level biological functions to the molecular behavior of proteins and DNA.

Theorem. A proposition that is not self-evident but that can be proven from accepted premises or axioms and so is established as a law or principle.

Thermal functional periodicity. Many heat engines have a thermal cycle such as the Carnot cycle. The functional periodicity is established by the cycle the thermodynamic properties must undergo repeatedly to convert the thermal energy to mechanical work.

Time-dependent combinatorial complexity. The combinatorial complexity is defined as the complexity that increases as a function of time due to a continued expansion in the number of possible combinations with time, which may eventually lead to a chaotic state or a system failure. Time-dependent combinatorial complexity arises because the future events occur in unpredictable ways and thus cannot be predicted.

Time-dependent periodic complexity. The periodic complexity is defined as the complexity that exists only in a finite time period, resulting in a finite and limited number of probable combinations.

Time-independent imaginary complexity. Imaginary complexity is defined as uncertainty that is not real uncertainty, but arises because of the designer's lack of knowledge and understanding of a specific design itself.

Time-independent real complexity. Real complexity is defined as a measure of uncertainty when the probability of achieving the FR is less than 1.0 because the system range is not identical to the design range.

Uncertainty. Uncertainty is measured by determining the common range between the design range and the system range. Uncertainty is zero when the common range is the same as the system range.

Uncoupled design. A design that has all of its functional requirements (FRs) independent from other FRs in the set is called an uncoupled design. The design matrix is diagonal.

Index